W9-ADO-138

Library of the Chathams
Chatham, New Jersey

Presented by

The Friends of
The Library

This book purchased
with funds raised at
the annual Book Sale

The Alzheimer Conundrum

The Alzheimer Conundrum

Entanglements of Dementia and Aging

Margaret Lock

0/3 2795 16+1

Copyright © 2013 by Princeton University Press

Published by Princeton University Press, 41 William Street, Princeton, New Jersey 08540
In the United Kingdom: Princeton University Press, 6 Oxford Street,
Woodstock, Oxfordshire OX20 1TW

press.princeton.edu

Jacket art: William Utermohlen (1933–2007), detail of *Self Portrait with Cat* 1955.
Pencil on paper, 44 x 32 cm. Courtesy of Galerie Beckel-Odille-Boïcos, Paris.

All Rights Reserved

Library of Congress Cataloging-in-Publication Data

Lock, Margaret M.
The Alzheimer conundrum : entanglements of dementia
and aging / Margaret Lock.
        pages   cm
Includes bibliographical references and index.
ISBN 978-0-691-14978-3
1. Alzheimer's disease—Age factors. 2. Older people—Mental health.
3. Brain—Aging—Molecular aspects. I. Title.
RC523.L63 2013
362.1968'31—dc23   2013011850

British Library Cataloging-in-Publication Data is available

This book has been composed in Verdigris MVB Pro Text.

Printed on acid-free paper. ∞

Printed in the United States of America

1 3 5 7 9 10 8 6 4 2

Dedicated to the memory of my parents,
Albert A. Foreman and Anne Foreman

PROPERTY OF
LIBRARY OF THE CHATHAMS
CHATHAM NJ

Contents

Acknowledgments

Without the unstinting donation of time by the many experts in the Alzheimer's world to whom I have talked over the past ten years, some of them repeatedly, this book could never have been written. Among those from whom I have learned the most are Howard Chertkow, a neurologist who works at the Jewish General Hospital, part of the McGill University Health Center and Carol Brayne, who is the Professor of Public Health Medicine in the Department of Public Health and Primary Care, Cambridge University. Exchanges often took place while consuming delicious meals prepared in one or other of our respective homes. In addition, many other extraordinarily busy scientists, clinicians, and genetic counselors have given generously of their time to teach me about the world to which the majority have devoted their life's work. Exchanges with these experts were, inevitably, exceptionally rewarding.

Numerous patients and families readily agreed to meet with my research assistants or myself, often spending up to an hour or more talking with us. These conversations frequently touched on painful subjects, but, even so, virtually everyone insisted that they were only too happy to assist with the research project. Robert Green, Professor of Genetics, Neurology and Epidemiology at Harvard University medical school, made it possible for my research team to meet with research subjects who had volunteered to be part of the REVEAL randomized controlled trial, for which I am very grateful. Involved researchers flew in from several cities in the eastern United States in order to meet with us for a whole day in Boston in preparation for the interviews we would carry out over the following months. Howard Chertkow made it possible for us to meet with patients who were attending a memory clinic in Montréal, and Tom Dening, an old-age psychiatrist who resides in the United Kingdom, was kind enough to take me along with him to meet with families in his care living in the vicinity of Cambridge.

Over the years a number of research assistants have worked with me, especially in connection with interviewing families where Alzheimer disease is present and then compiling and analyzing the findings. I have lasting memories of hurtling to Boston and back in a rented car, more than once, in very lively company! Many thanks are due to Janalyn Prest, Stephanie Lloyd, Julie Freeman, Gillian Chilibeck, Briony Beveridge, and Miriam Padolsky. In addition, Gillian Chilibeck, Wilson Will, Adam Finchler, and Kristin Flemons were indispensible in carrying out library research, entering data, preparing PowerPoint presentations, and battling with the odd computer glitches.

The Social Sciences and Humanities Research Council of Canada (SSHRC), the Canadian Institutes of Health Research (CIHR), the Wenner-Gren

Foundation, and the Trudeau Foundation provided funding for this research. This generous support ensured that the project could be carried out in Canada, the United Kingdom, and the United States, permitting to some extent comparison of the respective situations in these three countries. It also meant that I was able to delve very deeply into an exceptionally complex problem, one that has taken on an aura of urgency in recent years as the aging of the world's population and what this implies not only in terms of health care but in addition with respect to the global economy at large are now apparent to all.

I am indebted to the anonymous reviewers of the manuscript who provided certain invaluable comments. In addition, colleagues near and far perused parts of the manuscript and made important suggestions. Among them are Jesse Ballenger, Kenneth Weiss, Sharon Kaufman, Howard Chertkow, Carol Brayne, and Peter Whitehouse. Many thanks go to Fred Appel, the Executive Editor of Princeton University Press, as well as to Sarah David, Jenny Wolkowicki, Jessica Massabrook, and Elizabeth Blazejewski, whose assiduous attention and care throughout have been of the very highest professional standard.

As ever, Richard Lock has been the bedrock of support, master chef, close companion for many decades, critical commentator, and source of inspired exchanges in all manner of situations.

The Alzheimer Conundrum

Orientations

The Epidemic of the 21st Century

We live with a plethora of "epidemics"—obesity, diabetes, autism, prostate cancer, breast cancer, HIV/AIDS, child abuse, crime, and terrorism, to name a few. Among this multiplication of catastrophes, reports about a proliferating epidemic of Alzheimer disease are increasingly conspicuous in the media.

In his book *The Longevity Revolution* Robert Butler, gerontologist, psychiatrist, and Pulitzer Prize winner, argues that one of the triumphs of the 20th century has been the dramatic increase in the numbers of people who live to old age, but he quickly adds that this has brought about an increase in the number of individuals suffering from dementia. "Unless we find ways to prevent or cure Alzheimer's and other severe dementing diseases," Butler argues, "the world will shortly be confronted with . . . the epidemic of the 21st century."[1] Extrapolating from this undeniable association of aging and dementia, it would appear that the ever-increasing proportion of elderly individuals in the world constitutes a burgeoning pandemic with the potential of bringing the global economy to its knees.[2]

Use of the term "epidemic" has a long history, from the time of Homer or earlier, but its meaning has changed over the years.[3] Not until the late 19th century, following the formation of the modern discipline of epidemiology, was its use restricted to infectious diseases and their spread. However, by the mid-20th century, contagion was no longer the defining characteristic of an epidemic—numbers alone became significant, and the statistical accumulation of cases of heart disease and cancer, notably lung cancer, began to be described as epidemic in both the medical world and the media. A glance at Google suggests that Alzheimer disease may now have taken pride of place: "Alzheimer's Disease: A Global Epidemic"; "Alzheimer's 'Epidemic' Hitting Minorities Hardest"; "Alzheimer's Epidemic Will Follow the Obesity One"; "10 Million Baby Boomers Face Alzheimer's Epidemic." Also, a 2011 Larry King special was titled "Unthinkable: The Alzheimer's Epidemic."[4]

Since the condition was first formally named as a disease in 1908, repeated efforts have been made to delineate with ever more accuracy the clinical and neuropathological features of Alzheimer's, with the ultimate objective of finding a cure. However, despite many billions of dollars poured into research over the past several decades, no cure has been found, and, at present, only four drugs are available by prescription that variably alleviate symptoms for a period of some months, often with side effects, and by no means in all patients.

It is perhaps not surprising, then, given the projected increase in the numbers of people who will become demented in the coming years that a move is under way in the Alzheimer world to implement research designed to bring about the *prevention* of this devastating condition. This new orientation is facilitated by biomedical technologies developed relatively recently expressly designed to detect molecular changes regarded as incipient signs of Alzheimer disease (AD) in the bodies of individuals. On the basis of the results of clinical trials with human populations, these technologies are in the process of being standardized, with the expectation that they will be put to use in clinical care in the coming years. The objective is to predict what the future may have in store for any one of us as we age and, ideally, to develop drugs that nip the very beginnings of AD in the bud, long before the behavioral changes associated with dementia appear.

In 2002, when I began to amass the data that appear in this book, my original objective was to write about advances in molecular genetics and the associated social ramifications of genetic testing for complex diseases, using Alzheimer disease as an illustrative example. When I settled down to read the relevant literature, I was confronted by a complexity virtually unknown at that time beyond the world of research focused on molecular and population genetics. It was abundantly clear that making individual risk assessments for AD, other than the rare, early-onset form, on the basis of genetic testing alone amounted to little more than hazardous guessing. Moreover, professional guidelines were opposed to personalized testing for the common, late-onset form of Alzheimer's, other than in research settings where, customarily, people sign a consent form stating that they relinquish a right to be informed of their test results. Clinician researchers justified this position because carrying the "risky" gene associated with AD determines nothing; nor does knowledge of its presence permit, on its own, meaningful calculations of individual risk estimates. Clearly my research objective needed modification, or perhaps it needed to be abandoned entirely.

As I slowly began to grasp the labyrinthine convolutions associated with genes that contribute to complex conditions such as AD, it became increasingly apparent that a social science contribution was called for, one not confined to the genetics of AD, but that also addressed the phenomenon of AD itself. Not only was it evident that the genetic contribution to AD was by no means straightforward but, among experts, the very category of AD was being subjected to questioning and possible category fragmentation or reshuffling was in the air, making for a plethora of unknowns.

Two years later, while listening to papers given at an international conference on Alzheimer's held in Philadelphia, I began to fully realize the enormity of such an undertaking. However, my timing was fortunate, because technological innovations had recently resulted in research findings that, together with remarkable epidemiological insights, were forcing researchers to confront head-on the ontological question of what exactly is AD.

Acknowledgment that the AD phenomenon is deeply entangled with the very processes of aging could no longer be conveniently set to one side, as had usually been the case to date, raising difficulties about what counts as "normal" and what as "pathological." In the ensuing years, in a concerted effort to defeat AD and its increasingly debilitating effects on global economies, certain members of the involved scientific community mobilized themselves to bring about what is described by insiders as a paradigm shift, but is better characterized as something less grand—a shake-up in "normal" science, as Thomas Kuhn would have put it.

The ethnographic account set out in this book is a case study about the events leading up to the recasting of the phenomenon of Alzheimer disease as a condition to be prevented, and the debates that have erupted in connection with the implementation of this process. This shift has restoked older debates, heated at times, about how best to delineate what is glossed as AD. Uncertainties persist in connection with AD labeling in part due to inconsistent diagnostic practices in settings other than highly specialized units and, second, because, at present, AD can be definitively diagnosed only retroactively, postmortem. However numerous cases never go to autopsy. Furthermore, uncertainties also continue in part because AD etiology has never been explained. Numerous causative associations have been suggested in addition to aging, including education levels, brain trauma, and life style, among many others. The majority of researchers, however, have long assumed that certain specific molecular changes revealed at autopsy are evidence of "the" cause of this condition, regardless of the contribution made by more distal variables, but this hypothesis has never been proven. An increasing number of experts express skepticism about this theory today, although few, if any, dismiss the significance of the molecular changes outright. Even so, an increasing number argue that the situation is more complex than heretofore was believed to be the case, particularly because it is has been demonstrated repeatedly that one can harbor AD pathology in the brain but exhibit no outward signs of dementia. Alzheimer's is an elusive mobile target in which brain and mind are not necessarily in sync.

Even though an *accurate* assessment of "so-called" AD cases is simply not possible at either national or international levels, when we read about Alzheimer's in the media and in publications put out by AD advocacy organizations, the estimated number of cases and predictions about the pending catastrophe are reported with great assurance. This mobilization of numbers without doubt has a political effect, but it muddies the waters when attempting to come to grips with what exactly constitutes Alzheimer's disease and how best to confront it.

These current disputations and their interim settlement (research findings are very unlikely to bring about closure any time soon) will have direct consequences for individuals, families, policy making, and future research directions, as we strive to contain the "epidemic" of the 21st century. Despite the undeniable complexity and associated uncertainties, hope must be sustained if funding is to keep flowing. Overly optimistic estimates about a cure for AD are no longer tenable; even so, the U.S. government announced in 2012 that it was increasing its funding for AD as part of a bid to "prevent and effectively treat Alzheimer's disease by 2025."[5]

The narrative set out in this book focuses on the generation and transformation of expert knowledge and practices in connection with the phenomenon of AD, in an era of increasing uncertainties and recognition of apparently boundless complexity. Emphasis is given to the way in which debates in the Alzheimer world are played out in research and clinical settings, in medical and media publications, at major conferences, and in talks for public consumption, and with what potential effects on the millions of healthy people who will be systematically monitored if and when an era of prevention becomes routinized. Also included are excerpts from extensive interviews with individuals who have undergone testing in research settings, including genetic testing. One objective of disclosure of such test results in an experimental setting is to assess how individuals respond to learning about their genotype.

The concluding chapter of the book moves to a broader dimension and presents emerging knowledge in both epigenetics and epidemiology,[6] strongly suggesting that forms of prevention that take a public health approach, including lifestyle changes, reduced exposure to toxins, reductions in poverty, increased community support, and other variables, is likely to reduce the prevalence of dementia worldwide to a much greater extent than would an approach confined to expensive molecular micro-medical management of segments of those populations deemed to be at risk that happen to be located in wealthier countries. Furthermore, a molecular approach to AD prevention requires healthy individuals, some as young as 18 years of age, to become research subjects on whom repeated tests will be carried out. Volunteer subjects are currently being recruited at over 50 sites in North America, Europe, Asia, and Australia. The extensive medicalization and years of monitoring that these subjects will have to undergo, some of it quite invasive, demand critical attention.

Conceptualizing the Aging Brain: Fundamental Tensions

Three tensions are evident in the Alzheimer world that have profound implications for how experts are confronting the apparent crisis posed by aging societies. These tensions, entangled with one another, are driving debates and bringing about a reconceptualization of Alzheimer disease, not for the first time in its history. Alzheimer's is a phenomenon that one or two well-known neurologists have recently characterized as a myth, but without denying the frightening reality of states of dementia.[7] And numerous experts concede that in the near future the Alzheimer label may well be fragmented into subtaxa on the basis of recent findings, although it is often noted that for political purposes in connection with raising money and when communicating with the public, the overarching label will continue to be used.

The first tension, evident since the "discovery" of Alzheimer disease in the early 20th century, is the age-old problem of the relationship of mind and body that became particularly prominent in neurological circles during the 18th century. On the one hand are adherents of what came to be known as "localization theory," in which neuropathological changes in the brain are assumed to be causal of specific behavioral changes in persons. Thomas Sydenham, the 17th-century English physician, was among the first to argue that diseases are facts of nature waiting to be "discovered" and that they exist as entities, entirely independent of human intervention. Such an understanding contributed greatly to the acceptance of localization theory. In practice, adherents of localization theory do not acknowledge mind or consciousness as having relevance in AD causation, although their very existence is not dismissed outright.

On the other hand are those experts who favor theories about the way in which mind, persons, life events, aging, and environments interact to precipitate neurological and behavioral transformations that are pathological, a position that I characterize as the "entanglement" theory of dementia. Of these two approaches, localization theory has been dominant in the AD world throughout the 20th century, although, as a result of recent findings in epidemiology, neurogenetics, and epigenetics, a partial rethinking is currently taking place toward greater recognition of mind, body, and environment entanglement during the entire life course.

The second tension also has a long history, from well before the time Alzheimer disease was so named. This debate revolves around the question of whether or not behavioral changes commonly seen in older people, and very often glossed as dementia, are an inevitable part of aging or, alternatively, are manifestations of distinct pathological conditions. This tension has been reinvigorated recently because neuroimaging findings have confirmed what was formerly evident only at autopsy—that approximately one-third of "normal" nondemented persons exhibit "Alzheimer neuropathology" in their brains. Perennial difficulties with delineation of normal and pathological states are brought to the fore when confronting this tension. Can the mind be healthy if the brain is diseased? Many experts would answer "yes," on the assumption that such individuals will inevitably become demented in the coming years—if death from other causes does not intrude.

A third tension arises as a result of a paradox made apparent following the mapping of the human genome. It was assumed by many involved researchers that a single map created from the pooled DNA of a few individuals would be sufficient to delineate the human genome and its significant differences from other living entities. However, it was already well understood at the time that once research moved beyond mapping of the structure of the genome to an examination of the function of genes, molecular differences among humans could no longer be set to one side. Knowledge about genetic variation among human populations is indispensible when attempting to account for biological adaptation to environments over time, and also for the prevalence and incidence of many diseases and disorders. On the basis of findings known for many years and substantiated by postgenomic research,[8] it is now widely accepted that genes are part of dynamic epigenetic networks, one function of which is to activate or deactivate their "expression." These networks comprise environments—material and social—that are distributed unevenly over time and space. This reorientation in connection with the activity of DNA has significant implications for coming to grips with how the genes associated with Alzheimer disease function. Moreover, as we move into an era in which prevention of AD moves to center stage, consideration of social and environmental variables implicated in dementia risk takes on a new importance. In short, this tension revolves around a revitalized and reformulated nature/nurture debate.

Techno-phenomena and Reality

Each of the above tensions—brain/mind, aging/pathology, gene/environment—is subject to repeated reconceptualization over time. New ways of thinking about and understanding objects of investigation result from ongoing technological innovation, experimentation, and interventions into human subjects and their animal model substitutes, bringing about revisions in what is regarded as cutting-edge AD knowledge. The result is that these interrelated tensions are inherently unstable and never fully resolved.

Karen Barad, physicist and philosopher, alerts us to a fundamental difficulty with knowledge production in the sciences about what will count as "real" and "factual." She argues that "realness" in science is not an "essence, an entity, or an independently existing object with inherent attributes" of its own.[9] Influenced by the physicist Niels Bohr, she insists that primary ontological units in scientific investigations are what Bohr termed "phenomena."[10] Scientific findings simply cannot be "unadorned facts,"[11] although the material reality of objects is not in question.

Arguments similar to that of Barad have been put forward earlier by social scientists and philosophers, notably Bruno Latour, Ian Hacking, Donna Haraway, and Hans-Jörg Rheinberger, among others.[12] And the position taken today by an increasing number of social scientists researching medicine is that medical conditions are best understood as "material/social artifacts," that is, as entities that result from various kinds of interventions into the human body designed to bring to light, make real, and standardize named bodily conditions as diseases.[13]

Exactly what is drawn on to constitute and verify named diseases changes over time, usually due to technological innovations that permit new ways of "seeing" and thinking about the "object" in question. Evolving knowledge is debated and contested among scientists and clinicians until one argument becomes widely accepted as the dominant paradigm, as Ludwik Fleck's seminal work on syphilis made evident.[14] Similar debates have taken place recently in connection with the neurological condition known as *kuru* and with HIV/AIDS.[15] The newfound ability to map genes has resulted in numerous modifications to disease taxonomies, often resulting in subtaxa; the field of cancer research is a striking example of such a transformation. But, inevitably, in contrast to diseases in which bacteria, toxins, tumor formation, or specific genes are clearly implicated in disease causation, it is more difficult to sustain consistent arguments and reach consensus about a phenomenon like Alzheimer's where, aside from the contribution of aging itself, causation is undeniably complex and remains, for all intents and purposes, unknown.

In the chapters that follow, it will become apparent that Alzheimer's has a disputed history and, furthermore, that arguments generated by recent findings are becoming quite acrimonious in some quarters. The exponential burgeoning of dementia due to aging populations, of which AD is recognized as the most common subtype, is not denied by anyone, but debates revolving around the tensions set out above make it clear that assumptions about the ontology of AD and what exactly constitutes this condition are subject to contestation among experts. Because a good proportion of individuals who harbor amyloid plaque (the substance hypothesized to incite the molecular cascade that results in Alzheimer neuropathology) continue to be cognitively in good health, the assumed "factness" of Alzheimer's as a disease that inevitably causes people to become demented is being questioned by a growing minority of experts. Many of these individuals ask what it is that appears to protect people from becoming demented, even when their brains are "littered" with pathology (as it has been put to me).

Some experts believe that with more research subtaxa of AD will be identified, based on better understanding of the involved molecular pathways. Others, not necessarily opposed to the idea that AD will eventually fragment into a molecularly defined cluster of entities, argue that increased research energies should be moved upstream, and attention should be paid to generalized bodily changes such as chronic inflammation, oxidative stress, obesity, and so on—factors thought to be implicated in the devastating changes brought about by dementia. Yet others delve into the relationship among so-called cognitive reserve, education levels, and the incidence of AD, or the relationship of AD to repeated brain trauma resulting from participation in football, hockey, boxing, and other sports (dementia pugilistica is a general label for cognitive impairment resulting from such sports). To date, relatively few researchers have paid attention to the relationship among AD incidence and poverty, social inequalities, and family histories, but this situation is beginning to change, spurred on by findings in epigenetics. Approaches to prevention range, then, from molecular manipulations to improvement in individual health and changes in lifestyle and social conditions. The relative emphasis given to these different approaches depends in large part upon the medical specialty to which the AD expert belongs—a molecular biologist and an epidemiologist inevitably approach research very differently (see below). Most researchers would not in principle object to systematically implementing several or all of these preventive approaches given that they are not mutually exclusive, but funding is not equally disbursed among them; efforts to bring about prevention by means of molecular manipulations receive the lion's share of research money because this approach is directly linked to the interests of Big Pharma.

From Cure to Prevention

Given numerous reports about the rapidly emerging pandemic of Alzheimer disease, the hope that this condition can be prevented or at least slowed down has a sense of urgency about it. Among those researchers who work at the molecular level, for some, effective prevention lies with the development of a vaccine made from the peptide beta-amyloid, of which amyloid plaques are composed. Others are working to better understand the molecular changes thought to take place immediately prior to the onset of the final, devastating, neuropathological changes associated with AD. Yet more researchers believe that the most promising approach lies in delineating molecular pathways assumed to be "prodromal" presymptomatic changes of AD that can be detected in individuals up to 20 years or more before the irreversible final stages of dementia set in. Technologies developed over the past decade that facilitate knowledge generation about presymptomatic dementia have resulted in a flood of research findings that make the prevention of AD in this manner a realistic possibility in the minds of many. And the hope is that by modulating molecular changes believed to be causal of AD, future dementia can be averted. But, for this approach to be effective, large numbers of healthy middle-aged or younger people must agree to become research subjects; if rigorous research is to be carried out, then these people should ideally be drawn from carefully delineated populations around the world. Advertising for research volunteers is becoming increasingly visible on AD society websites and elsewhere.

Should these research results be positive, the expectation is that diagnoses of presymptomatic dementia will then be made routinely in clinics wherever governments can afford it, or when insurance companies are willing to pay. Before too long, we may all be required to engage in what could well be thought of as 21st-century divinatory practices in an attempt to avert our respective destinies. The key signifiers of the future will no longer be cracked turtle plastrons or heated ox, sheep, or deer shoulder bones—the tools made use of in the *I Ching*—but rather molecular substances known as "biomarkers"—inscriptions (as Bruno Latour would put it)[16] of changes concealed in the body assumed to signify future events.

Based on individual biomarker testing, many thousands of people will be expected to submit themselves to intense medical surveillance to detect signs of progression from presymptomatic dementia to mild cognitive impairment detectable in the clinic, and then to early Alzheimer disease. This move, if it comes to fruition, has the potential to become the biggest screening program with global reach ever established, at untold expense to health care systems. It will quite possibly have problems similar to those associated with the screening programs designed to detect breast and prostate cancer, in which estimations of individual risk and decisions about the appropriate course of action to take based on these estimates have proven highly problematic in all but certain high-risk cases.[17] As the following chapters will show, these divinatory practices are likely to create a highly potent zone of anxiety about what may be in store in the future for individuals and their family members. Having bodily signs of apparent medical significance uncovered and named by experts would, one might assume, bring about greater insight into the future than do fables told by fortune-tellers. However, paradoxically, the more we learn about the world of molecular biology, a characteristic feature of all forms of divination persists—namely, that in seeking to take control of the unknown, new ambiguities, anxieties, and uncertainties inevitably come to the fore, as the following chapters will make clear.[18] These uncertainties equally pervade the worlds of researchers and the everyday lives of affected people.

In order to delineate what will count as presymptomatic AD, specific biomarkers associated with risk for future AD/dementia have been identified, among which three are particularly prominent. The first is a set of changes in proteins located in cerebrospinal fluid obtained by lumbar puncture; the second is the presence of insoluble amyloid plaques detected in the living brain by means of neuroimaging, and the third is the presence of a genetic variant consistently associated with increased risk for late onset Alzheimer's (as distinct from so-called early-onset, familial AD, a rare single gene disorder). These biomarkers, in common with all divinatory signifiers, indicate increased risk for AD, but provide no certain predictions. Longitudinal research to date has shown that among those individuals who exhibit abnormal biomarkers when tested, many do not thereafter "convert" to AD (the exact numbers are not known, but 30% is the often touted figure).

In summary, the move to early detection of AD will push the process of diagnosis into an arena that involves probabilistic calculations of individual risk in which "normal" and "pathological" aging are confounded, raising disturbing ethical questions. These efforts to prevent dementia represent in effect a revitalized version of localization theory, in which the timeline has been extended to an earlier point in the life cycle, long before clinical symptoms appear. This position is in striking contrast to that of the theory of a fundamental entanglement of dementia and aging,[19] in which the basic question posed is whether AD is indeed a phenomenon readily isolable from the wear and tear of aging. Experts who take this latter position pay attention to assumed risk factors external to the body that are at work throughout the life course, and are not *singularly* preoccupied with molecular changes internal to the body. Inevitably, this position highlights the enormous difficulties associated with making predictions about individual risk for AD, among them the very real danger of creating false positives when diagnosing presymptomatic dementia.

It must be emphasized that these two approaches, one based on localization theory and the other on entanglements of aging and dementia, are not necessarily mutually exclusive (although arguments among experts often make it appear as though this is the case); rather, they set out from different ontological positions in which the phenomenon of AD is not conceptualized in the same way. Localization theorists visualize AD as a demonstrable neuropathological entity, whereas entanglement theorists are more inclined to understand AD as an emergent process—the product of contextualized individual biologies and life experiences culminating in the clinical expression of the phenomenon of AD/dementia. Of course there are relatively few "purists" who flatly deny either one of these fundamental orientations, and all experts assume that the behavioral changes associated with dementia are in some way intimately linked to changes taking place in the brain.

But, as noted above, behavioral and neuropathological changes are not well correlated one with another, and amyloid deposits and other changes associated with dementia are present in a good proportion, perhaps 30%, of healthy people in old age. Until relatively recently, the tendency has been to set this lack of concordance to one side, usually on the assumption that a clinical diagnosis of AD may well not be accurate. The "truth" of the matter can be established only at autopsy when plaques, tangles, and cell loss—the gold standard signifiers of AD—are laid bare. However, on the basis of technological interventions, primarily in the form of in vivo neuroimaging and ongoing research into amyloid activity itself, it has become increasingly difficult to ignore anomalies. Even so, amyloid deposition in the brain continues to be heralded as the earliest sign of prodromal AD, and its early detection is the spearhead of the move to prevention of AD.

In announcing their intention to make use of a diagnosis of presymptomatic AD, involved researchers have repeatedly made the claim that a paradigm shift is taking place. Such a claim is not justified if Thomas Kuhn's intention when creating this idiom is to be adhered to. The move toward AD prevention cannot bring about a scientific revolution; rather it is a disruption of what Kuhn termed "normal" science. The hypothesis that amyloid is the causative agent that sets off a cascade of negative events in the brain is left in place, although its detection has been shifted to an earlier point in time. The amyloid cascade hypothesis is a powerful actor in the contemporary Alzheimer story— over the years it has, in effect, become sacrosanct, making it difficult for some to engage with new ways of thinking, particularly because it has been the target for most drug development. But repeated failures to produce effective drugs have precipitated the present "shake-up." The hope is now that drugs can be developed to tackle amyloid at the very first sign of its appearance, before the disease has progressed too far and done irreparable damage—hence amyloid remains as the foundation of this story. Aging societies ensure that should an effective drug be found, it will bring untold wealth to the company that successfully brings such a drug to market.

In what follows, I use interchangeably the words "Alzheimer's," "Alzheimer disease," "AD," and "the Alzheimer phenomenon," despite the obvious difficulties with definition made clear above. My position, in common with perhaps the majority of leading researchers in the Alzheimer world today, is as follows: Clearly, conditions involving behavioral changes related to concomitant changes in the brain commonly glossed as dementia, are undeniably real. Dementia has many causes, some of which are well understood. However, Alzheimer's, the most commonly diagnosed subcategory of dementia, proves to be an elusive phenomenon, particularly in light of recent findings in both epidemiology and neuroimaging. In short, it is a stubborn conundrum.

An Imperialism of Probabilities

Given that estimates of risk based on probability theory are going to be central to Alzheimer prevention, a brief consideration of the risk concept and its mobilization for clinical, public health, and political purposes is in order at this point. The Canadian philosopher Ian Hacking, in his 1990 classic *The Taming of Chance*, sets out a genealogy of the concept of risk. Hacking argues that a "professional lust for measurement" first emerged during the Industrial Revolution.[20] Historians disagree as to whether a direct link exists between this "avalanche of numbers" and earlier forms of mechanical divination, gambling, and lotteries, in which the idea of probability is inherent.[21] Hacking insists that continuities with the past, although no doubt they exist, are of less interest than is the partial erosion of a science of determinism, so dominant during the 16th and 17th centuries. This transformation culminated in what Hacking characterizes as "the philosophical success story of the first half of the twentieth century" in the form of probability mathematics. He argues that in 19th-century Europe, with the founding of biometrical schools of statistical research, "the stage was set for ultimate indeterminism"—for the acknowledgment of a place for chance in the scheme of things.[22]

Hacking goes on to make clear that the "imperialism of probabilities" with which we live today could have come about only in conjunction with a massive expansion of literacy, computation, and bookkeeping, the invention of the census, and the idea that people can be systematically divided into various groups or "populations," ultimately for the purposes of governing society. These technologies of data collection, intimately associated with the growth of a "research mentality" evident in Europe and North America of the late 19th and early 20th centuries, are today diffused unevenly throughout the world.

Probability theory makes extensive use of the concept of risk that began to be systematically applied in the 17th century by newly founded insurance companies, initially in connection with shipping and associated commerce. Today, the idea of risk is embedded in the majority of business endeavors and is pervasive in insurance practices and in virtually all aspects of research and clinical medicine. The German sociologist Ulrich Beck argues, "The category of risk opens up a world within and beyond the clear distinction between knowledge and non-knowing."[23] But, he is quick to add that the idea of risk does not annul all forms of valid knowledge, rather "it amalgamates knowledge with non-knowing within the semantic horizon of probability."[24]

Beck does not dwell on the statistical notion of risk as set out by Hacking, but writes instead about risk as a general phenomenological concept applied ubiquitously today to situations deemed to be of "unprecedented danger," including terror attacks, the effects of climate change, the recent economic crash, and the "'catastrophe' of greying societies,"[25] among yet other calamities. He points out that, paradoxically, such dangers are, in effect, the products or offshoots of modernization: "[W]e live in a world that has to make decisions concerning its future under the conditions of manufactured, self-inflicted insecurity." Increasingly it is becoming apparent that we can no longer control what we have brought upon ourselves,[26] and one such success appears to be the proportional (some might argue disproportional) increase of elderly in populations worldwide. Beck concludes that the idea of "risk" is not synonymous with catastrophe; rather, risk enables the anticipation of catastrophe; it signifies "the controversial reality of the possible,"[27] as the title of the Larry King show "Unthinkable: The Alzheimer's Epidemic," suggests.

Politicizing Alzheimer's Risk

Beck is concerned not only with how certain events come to signal danger, but in addition with their politicization. The graying of society and the risk it poses in the form of far-reaching economic and social repercussions are today made public through an onslaught of media reports, the contents of which are based, sometimes loosely, on professional documents of various kinds. The figures that appear repeatedly in the media produce statements such as the following: An estimated 35.6 million people were living worldwide with dementia in 2010. This number is estimated to nearly double every 20 years to 65.7 million in 2030 and 115.4 million in 2050.[28] It is evident that these figures are designed to incite political action and increase funding for AD research.

Prominent is concern about the enormous economic and social burden associated with this phenomenal number of demented people, especially given the small size of most families these days, because the majority will have to be institutionalized. In 2005, it was reported that the global cost of caring for dementia exceeded $315 billion,[29] and by 2010 the annual cost of care was estimated to have essentially doubled to $600 billion. The medical journal *The Lancet* writes that if all the microeconomic and macroeconomic factors remain unchanged, this cost is predicted to increase by 85% by 2030 and will "become one of the biggest economic strains for health care systems and communities worldwide."[30] *Maclean's* magazine reports that the aging of the baby boomer generation in Canada will create a tenfold increase in the demand for long-term care beds, and the Alzheimer Society of Canada estimates that this increase will cost the comparatively small economy $107 billion.[31]

Media reports about Alzheimer's, inevitably accompanied by an extensive array of comments and blogs, very often draw on figures obtained from the umbrella organization known as Alzheimer's Disease International, and from associated national and regional Alzheimer societies active in very many countries of the world. Information is also obtained from articles published in medical journals such as *The Lancet* and the *New England Journal of Medicine*, and from government reports, but it is of note that the "facts and figures" of Alzheimer disease reported in medical journals and by governments are often taken directly from AD society documents. A 2008 article published in the prestigious journal *Alzheimer's & Dementia* was authored by an employee of the Alzheimer's Association, the nation-wide AD organization of the United States, and presents the following figures:

> More than 5 million Americans are estimated to have Alzheimer disease. Every 71 seconds someone in America develops Alzheimer disease; by 2050 it is expected to occur every 33 seconds. During the coming decades, baby boomers are projected to add 10 million people per year, with a total estimated prevalence of 11 to 16 million persons.[32]

Over the past decade professional and media publications about Alzheimer's have increased exponentially, and some highlight an entrenched problem not evident in simple tallies about the burgeoning numbers of elderly. If the burden that increasing numbers of demented elderly place on society, families, and individual caregivers is to be engaged with constructively, then the ignorance, fear, stigma, shame, discrimination, denial, and indifference commonly associated with dementia must first be exposed and overcome. This has been one of the objectives of Alzheimer advocacy groups since their inception in the 1970s; originally, the primary goal was to have Alzheimer's recognized as a bona fide medical condition that should be treated, and not simply dismissed as senility, and hence regarded as an unavoidable part of aging. In order to achieve this objective, together with activist families, the newly formed AD societies in the United Kingdom and the United States worked to have dementia medicalized, in the expectation that government funding would then be set aside to support research, drug development, and the creation of facilities for care.

Despite its clinical recognition, Alzheimer's has not been thought of as a major health care priority until very recently—no doubt due in large part to the very stigma and denial that the AD societies sought to overcome—and funding has always lagged far behind that allotted by governments and donated by the public to cures for cancer and heart disease. As the neurogeneticist Peter St George-Hyslop puts it, "Alzheimer's is not that sexy." It's not "hot like breast cancer or HIV," and lobbying efforts fall short when "affected people cannot speak for themselves."[33] In the United Kingdom, for example, it is estimated that until recently for every £1 spent on dementia research, £12 went to investigating cancer.[34]

However, in the past few years it appears that the concerted efforts of medical professionals, AD societies, and the media to make Alzheimer's more publically visible as a condition that demands immediate attention for economic, social, and humanitarian reasons are beginning to pay off. It is increasingly hard to ignore the numbers reiterated again and again in the media about the rapid increase in the near future of demented, dependent people worldwide who may well contribute to the partial derailing of whole economies. The first World Alzheimer's Day took place on September 21, 2010, accompanied by the publication of a World Alzheimer's Report containing damning figures, and over the past few years Alzheimer's walks, bike rides, and marathons have been instituted in Europe, North America, and Asia. In addition, Alzheimer's Disease International has recently called for the World Health Organization to make Alzheimer's a global priority. It is through such activities that direct associations are made in the minds of health policy makers and economic advisors alike about links among the devastating effects of the experience of Alzheimer's on individuals and families; a rapidly changing global demography increasingly populated by the elderly, many of whom are eligible for pensions; and the economic impact on families, the workforce, and productivity, particularly in an era of general economic decline.

Medicalization and Destigmatization

By making schizophrenia, psychosis, depression, menopause, autism, ADHD, and many other conditions into medical diseases or disease-like conditions that affect individual, decontextualized bodies, allocation of responsibility for illness causation is neutralized. Once the disease is medicalized, treatment is primarily focused on changes internal to the body that are assumed to have been caused by a "mistake" of nature, a pathogen or toxin, a life cycle transition, or aging.[35] The very process of medicalization means that, in theory, physical conditions shed much of the shame and stigma associated with them, familial and individual responsibility for their causation is disposed of and, by labeling a condition as a disease, it becomes possible to attract political attention in order to mobilize funding. In practice, however, although stigma has indeed diminished in connection with AD, its persistence continues to be all too often apparent.

When Alzheimer's was first named as a disease in the early years of the 20th century, it was removed from the catchall pejorative category of insanity and classified as a neurological condition intimately associated with aging. But, even so, the undeniable symptoms of "senility" continued to be thought of as part of "normal" aging by the majority of the medical profession and the public alike, and remained hidden and managed as a family matter. It was not until the 1970s that Alzheimer's in elderly people was recognized in the medical world as a disease, first in Europe, North America, and Japan, and then later in many other parts of the world. This recognition has resulted in extensive destigmatization of the condition worldwide and is slowly reducing the abandonment and abuse that so many elderly and demented people today continue to endure. But where comprehensive national health care systems do not exist, poor and disadvantaged people rarely come to medical attention when demented, and their families continue to absorb unaided the care of their ailing relatives. A widely shared assumption persists among many families and indeed among certain health care providers around the world that dementia "is just old age," and therefore nothing can be done about it, as Lawrence Cohen's ethnographic research in India showed all too clearly.[36] This attitude is reinforced by a lack of effective medical treatment for AD, and hence there is no reason to spend precious money on an arduous trip to the doctor, whose practice is often in a city miles away.

Of course, as is well known, medicalization rarely takes place without raising problems of its own, among them, excessive prescription of and often abuse of medication. And, in the haste to find cures, less often marked is a second problem, namely a shift of attention away from social, political, and environmental factors, including poverty, inequality, discrimination, and racism—factors deeply implicated in disease causation. These are the variables that thus far have received the least attention in the AD world.

Global Responses to Aging

Japan was perhaps the first country to alert its population to what lay in store with the dramatic aging of its population. *Kōreika shakai* (the "aging society") has received a great deal of publicity for over half a century in that country. Demographic changes that took 85 years to unfold in Sweden, 130 years in France, and 70 years in the United States have taken just 25 years in Japan. This transformation, coupled with a drop in the birthrate to 1.26 children per household in 2005, but now rising slightly again, has been not only rapid, but also more extreme than in either Europe or North America, and the elderly in Japan today make up 29% of the population.

In contrast, in Sweden, for example, with the largest number of people aged 65 and older in Europe, the elderly at present compose nearly 19% of the population. Middle-aged Japanese women who continue to be the primary caregivers of their aged parents feel the burden of this "graying" directly and, because Japanese women have the longest life expectancy in the world (although it declined a little following the latest tsunami with its horrific loss of life), it is not at all unusual to encounter 80-year-olds caring for their centenarian parents or parents-in-law. Over the past decades, extensive government policies have been implemented both nationally and locally to manage the aging society, designed in part to assist overburdened families and to alleviate the undeniable effect on the country's economy, exacerbated by the recent Fukushima catastrophe.[37] To a greater extent than anywhere else in the world, Japan has introduced robots into the workforce at an unprecedented rate, primarily to supplement the diminished number of blue-collar workers, but also to aid in home care, residential homes, and elsewhere.[38] Increasingly the term *chō kōrei shakai* (super-aged society) is made use of in the Japanese media and elsewhere, particularly when people are reminded of the prediction that by 2050 the elderly will constitute over 40% of the nation's population.

A similar transition to that which took place in the "early developed" countries is currently taking place worldwide, and with much more rapidity than demographers had predicted, so that the numbers of the elderly in "developing countries" are beginning to approach those of the youngest age groups, and are expected to overtake them by 2050.[39] This "triumph" of modernity is characterized as a pending catastrophe—a "greying tsunami"[40]—one that is exacerbated due to a reduction of family size in many countries. China is beginning to come to grips with the extent of the problem that their aging society poses. At present, the future for the elderly in China, and for their caregivers, looks exceptionally bleak. China already has a population of elderly of around 178 million (surpassing that of the number of elderly in all the European countries put together). It is estimated that by 2040 people aged 60 and older will make up about 28% of the population—in absolute numbers 397 million people—by far the majority of whom have no old age security.[41] Worries about the impact that the growing ranks of elderly will have on the economy and on individual families have been dubbed in China as the 4-2-1 problem because, in large part as a result of the one-child policy, a single person in China will soon be expected to support two parents and four grandparents. The media are only now beginning to expose the ignorance and stigma long associated with dementia in China, but even with recognition of dementia as an illness, the family will continue to be the primary caregiver. Shanghai, for example, is proposing what it calls the 90-7-3 plan, meaning that 90% of the elderly will be cared for at home, 7% will be permitted to make occasional visits to community centers, and 3% will live in the very limited supply of nursing homes, many of them still under construction.[42] Recently the government has implemented a national pension program for people older than 60; this enables many more people to break with tradition and live separately from the extended family, an arrangement that suits the increasing number of individuals, especially among the burgeoning middle class, who these days value their independence.[43]

Inevitably, increased longevity is distributed unevenly, and poverty, inequality, and discrimination continue to take a great toll on life expectancy in many parts of the world, including so-called economically advantaged countries. Despite enormous leaps in the gross wealth of certain segments of the populations of India, China, and other recently developed countries, inequities are on the increase, and abject poverty remains strikingly evident.[44] The current global economic recession will no doubt negatively affect longevity among the elderly virtually everywhere; even so, the rapid rise in absolute numbers of older people worldwide is undeniable. The proportion of those older than 80 is increasing at an unprecedented rate, and centenarians are today the most rapidly growing age group in certain countries. It is this extension of life expectancy that has taken everyone by surprise, including demographers.[45] However, significantly, the majority of research on this topic suggests that the incidence of dementia does not exhibit an overall increase among those in their 90s and older, strongly indicating that an association between aging and Alzheimer's is not straightforward.[46] Space does not permit further discussion of these demographic changes and their effects in specific locations, but it is important to keep in mind that the attempt to prevent AD by approaching it through the systematized screening and monitoring in medical settings of molecular changes in people's bodies as they age is simply not realistic in most parts of the world.

The Modern Rise of Longevity

In addition to noting the dramatic and widespread rise of aging populations, it is important to reflect on what brought about this "success" story. The historian of medicine Thomas McKeown, writing in the middle of the last century, argued that the exponential rise in human population that commenced in the 18th century in Europe was due primarily to social and political factors, and did not result from improved medical care per se, as is commonly believed to be the case. McKeown's point was that, initially, public health measures, including access to clean water, the spread of hygienic practices, and better availability and more equal distribution of food, were crucial to increased longevity, and only later did medical interventions have an effect. Drawing on public health records and death certificates, McKeown demonstrated that in Northern Europe it was not until the mid-20th century that medical technologies, notably antibiotics, brought about a substantial drop in infant and maternal mortality at birth and during early childhood. At the same time, medical interventions together with public health measures resulted in a longer survival rate among elderly people.[47]

Certain historians and demographers continue to take issue with some aspects of McKeown's thesis; nevertheless, it is undeniable that over the past two centuries, and longer in several of the countries that first modernized, an exponential rise in population has taken place. Moreover, during the latter part of the 20th century the number of individuals 65 and older in these populations has outstripped those who are younger than 25.[48] Further increased longevity in recent years is in part due to a steady decline in early deaths from cardiovascular disease and stroke as a result of effective medical care, and, recently, also due to significant dietary changes among large segments of many populations. Moreover, received wisdom among the middle classes that the elderly should become sedentary in retirement is now widely regarded with distain, no doubt adding to their longevity (the luxury of "retiring" was never possible, of course, for by far the majority of the world's population). We would do well, I believe, following McKeown's example, to consider the approaching pandemic of aging as a matter of public health and local politics and not merely a medical matter, with its individualized approach to the detection of bodily ills. I will return to this discussion in the concluding chapter.

Styles of Thought

The global enterprise designed to combat Alzheimer's today is a multi-billion-dollar endeavor, a massive consortium of electronically linked research collaborators—networks of experts working with clusters of research technologies. Such consortia are research teams in which some members work in the same institution that, in turn, is linked across many countries to corresponding individuals or research teams elsewhere. Networks such as these, in virtual contact most of the time, frequently publish multiauthored articles, sometimes involving up to 50 or even 100 authors, particularly when the topic is genetics. These AD collectivities specialize in proteomics, genetics and genomics, specific molecular pathways, animal models, drug development, comparative epidemiology, population genetics, clinical research, and so on. For over 100 years the AD kaleidoscope has been turning continuously, so that various parts of the Alzheimer puzzle come to be "seen," worked with, and standardized from perspectives that are continually modified by these networks of researchers.[49] The chapters that follow highlight the way in which emerging knowledge is consolidated on the basis of technological innovation, the introduction of new concepts and hypotheses, the amassing of empirical evidence and its analysis, and policy and guideline changes, only to be subjected continually to further modification.

The long-standing goal of pinning down the Alzheimer's phenomenon, usually by subdividing and fragmenting what is subjected to scrutiny to make it manageable, is a moving target, and so far it has been doggedly resistant to all efforts to neatly define and treat it. At present, the shift to prevention is bringing a renewed excitement to the field, although a moment's reflection makes clear, given the decreasing resources allotted to health care in most parts of the world at a time when elderly populations are everywhere increasing, that an AD epidemic is unlikely to be resolved in the near future. Efforts to detect presymptomatic dementia are at present limited to research settings, as noted above; even if, and this is a big if, biomarker investigation proves reliably effective in detecting the first signs of AD, and should it eventually be adopted routinely in clinical care in wealthy countries, it is unlikely to bring about a major change in the global incidence of AD. This cannot take place until such time as a public health approach to dementia is universally adopted. At present, the best one can hope from the shift to detection of biomarkers is for some important insights into the Alzheimer phenomena itself.

As long ago as the 1930s, Ludwik Fleck argued that the production of scientific knowledge is a social process. He insisted that researchers who belong to a "thought collective" participate in a body of shared knowledge resulting in a common "thought style." Such thought styles are contingent and continually subject to change and modification over time (in this sense, Fleck's argument differs fundamentally from that made by Kuhn when he argued for recognition of scientific paradigms). For Fleck, both stability and change among thought collectives are "controlled by the communication of thought within and between collectives."[50] He posited, again in contrast to Kuhn, that collective communication of thought constitutes an important source of novelty, and that the incommensurability that is inevitably present among collectives due to different epistemological positions can produce a fruitful tension from which innovations may arise.[51]

In tackling a condition as complex as Alzheimer's, groups of specialists bring their own shared assumptions and technical knowhow to the task of creating a breakthrough in AD research. Such groups have, at best, partial connections with other research groups, including, quite often, groups in the same basic science discipline as their own. Even at major conferences experts usually attend only those sessions relevant to their specific discipline. The move toward prevention has not eliminated a degree of isolation or the tensions evident among researchers; quite the contrary, it appears to have exacerbated the situation in some cases, as we shall see, particularly when concerns about funding come to the fore. Nevertheless, several events, including the flurry of publicity in connection with the move to AD prevention, the number of sessions devoted to its implementation at major conferences, and the number of articles published that bear upon this shift, have ensured that all AD experts are alert and ready to participate in discussions about the implication and implementation of this move.

Many experts wear more than one cap; most are clinician/researchers who have their working lives divided between these two activities. A second group consists of basic scientists who devote their entire time to laboratory research, for the purposes of developing drugs, modeling dementia in animals, doing protein research and crystallography, researching cell signaling, performing genome studies, and so on. Increasingly, many of these researchers are likely to be employed by or to have their research projects at least partially funded by pharmaceutical companies even when they hold university appointments. Some specialists are forensic scientists, and many of these individuals have links to governmental departments. Yet others are family and general practitioners who care daily for patients and families dealing with dementia. Another cluster of experts, including epidemiologists and population geneticists, usually do little or no clinical work, and focus their research endeavors primarily on delineating the causes and distribution of AD among human populations.

Whatever their specialty, the majority of researchers willingly acknowledge today that environmental variables, social and physical, may well be implicated in AD causation, and this is usually expressed with reference to "gene/environment" interactions. Many, but not all, then go on to argue that this raises a degree of complexity that cannot readily be dealt with. Hence, it is argued, a search for molecular changes internal to the body is the best place to start. Such a search is for the very beginnings of a final common pathway assumed to bring about AD pathology, no matter what may be previously implicated "upstream" at a larger scale either internal or external to the body. But other researchers take very different approaches, and among them are those who argue strongly for the entanglement theory; their research is driven less by a search for a cure and more by an effort to find out who among us are at increased risk for AD on the basis, largely, of social, political, and environmental factors.

Clinicians who deal primarily with patients but have no treatment to offer them often become impatient with epidemiologists who insist that enumeration of AD cases, and hence of risk estimates, are inaccurate because AD diagnoses are not accurate in the first place. Clinicians working in specialty clinics usually insist that today their diagnoses are highly reliable. But for many clinicians working in primary care, because no effective medication is available, their attention is focused on family support rather than pushing for further tests in the hope of obtaining greater diagnostic accuracy. The tensions among experts alluded to here will become readily apparent in the chapters that follow.

It is also the case that in connection with AD, recent findings have made it increasingly hard for any one research collective to complacently argue that closure is just around the corner on the basis of the most recent "truth claims." At present, very many interventions, many exquisitely designed, are producing results that challenge assumptions long held among groups of researchers. Recent work on biomarkers and on genetics, for example, suggests a degree of complexity involved, even at the molecular level, that has heretofore not been fully envisioned. The uncertainty stakes have been raised significantly, and in this situation resultant tensions may well become a rich incentive, perhaps a painful stimulus, to backtrack, reconsider, reluctantly give up cherished assumptions, and forge ahead in modified directions, sometimes by exchanging ideas at great length across disciplines. One other largely unspoken matter links together researchers in the Alzheimer's world: like vast numbers of people everywhere, many researchers come from families that have firsthand experience with dementia, giving enormous incentive to press forward in order to hound out this scourge.

Turning briefly to the social sciences, the British anthropologist Marilyn Strathern, writing about the eternal difficulty in anthropology of how best to represent ethnographic findings, especially those accrued in cross-cultural settings, makes use of the concept of "partial connections," an idea that bears some resemblance to that of Ludwik Fleck's styles of thought. Most anthropologists are acutely aware that their writing can never adequately portray what their informants have told them, in particular the unexamined assumptions embedded in what was said; nor can they ever satisfactorily describe the lived experience of others. Furthermore, inevitably the ethnographer's assumptions color the manner in which research findings are written up. Although some of the researchers I have interviewed in this project have become friends, in one or two instances close friends, at best I have partial connections with the majority of researchers whom I have interviewed and had conversations with while doing this research. The story I tell in the following chapters is necessarily incomplete, and no doubt leaves much to be desired in the minds of certain experts, but the timing is appropriate because many AD specialists believe that their approach is currently undergoing a shift that has a sense of urgency about it, and is attracting considerable attention due to global concern about aging populations.

Data Collection and Outline of Chapters

The data on which this book is founded are based on more than 80 extensive interviews and conversations with dementia experts from various specialities in the United States, Canada, and the United Kingdom. Of these individuals, 12 are key informants with whom I have met more than once, and I have met repeatedly with 3 or 4 of them. I have also observed outpatient care in clinical settings. In addition, I have attended numerous conferences and meetings large and small, including six gatherings of the major annual forum for dissemination of Alzheimer research, newly renamed the Alzheimer's Association International Conference. I have benefited from responses and questions posed by listeners to papers I have presented at professional meetings and delivered as invited lectures. In addition, I have given several presentations at gatherings designed to educate the public about Alzheimer's—the questions raised after such talks have been among the most difficult to answer. My message is one of neither certainty nor imminent breakthroughs, and this leaves many affected family members and certain involved researchers and clinicians with considerable disquiet, especially those who continue to hope in vain, in my opinion, that a cure for AD may be shortly forthcoming.

It is impossible to review the entire corpus of literature on Alzheimer disease and dementia, composed of several thousands of publications each year in English alone, but I have covered a substantial sampling of these research findings and editorial commentaries. My arguments in this book are illustrated throughout by means of excerpts taken from both interviews and written sources. The opinions of the experts date from the time of the interview with them, ranging between 2008 and 2012, and certain of these individuals may since have revised their thinking in light of recent findings. I refer to written sources up to the end of 2012, at which time this volume went into final production.

The chapter following this introduction is devoted to the "discovery" of Alzheimer disease and a somewhat truncated genealogy of its history to the present time. Emphasis is given to the virtual disappearance of AD for over four decades after its initial identification, followed by its rediscovery in the late 1960s in association with government and medical recognition of aging populations and their impending burden on society. The consolidation of what has been the dominant research paradigm in AD research for the past four decades—the amyloid cascade hypothesis, grounded in localization theory—is discussed. Throughout this chapter, difficulties in attempting to unravel the entanglement of "normal" aging from dementia, evident from Alois Alzheimer's time on, are pointed out.

Chapter 2 sets out by briefly considering repeated attempts at diagnostic refinement and standardization of AD carried out over the years. The difficulties of reconciling repeated mismatches between a clinical and a neuropathological diagnosis of AD are discussed, as are the discrepancies in diagnoses between specialized memory clinics and general and family practice settings. This is followed in chapter 3 with a discussion of the adoption in the 1980s of the clinical diagnosis known as "mild cognitive impairment." Formal recognition of MCI, the value of which continues to be debated in some circles, is the result of an exerted effort to systematically identify incipient AD in the clinic and is closely associated with the founding of specialist memory clinics. Ethnographic findings are presented in this chapter of interviews with individuals who have been diagnosed with MCI. Chapter 4 presents an account of the shift, commencing in the late 1980s, to the molecularization of AD, and the attempt to identify significant bodily changes as much as 20 years before behavioral changes can be diagnosed in individuals. The rationale for efforts to formulate a "prodromal," diagnosis before behavioral symptoms or memory loss are detected is considered, followed by a presentation of the involved molecular diagnostic tools—biomarkers—with emphasis on spinal taps, neuroimaging, and genetic testing. The significance of the first two of these biomarkers is attributed to their apparent ability to detect the onset of the amyloid cascade process. The anomalies and uncertainties associated with biomarker testing are discussed.

The first four chapters make it clear that the ontological status of AD is open to debate and, moreover, that very many cases of assumed AD never come to medical attention, with the result that estimates of the population prevalence of AD are not reliable. This situation provides a link with the four chapters that follow, in which the genetics of AD and findings based on individual testing for a specific susceptibility gene, APOEε4, associated with late-onset AD, are presented. In order to estimate individual risk for a condition on the basis of the presence of a specific gene, population databases composed of reliably diagnosed cases are indispensible. Hence, the value of genetic testing for AD, other than as a biomarker for ongoing research, is open to question.

Alzheimer genetics is introduced in chapter 5, including an account of the genes associated with rare, familial, early-onset Alzheimer disease. Discussion is included about why such patients are thought of as excellent research subjects in the search for a "cure" not only for early-onset AD but also for the much more common form of late-onset Alzheimer's. Recruitment of a group of Colombian research subjects who come from families with early-onset AD (now known among experts as dominantly inherited AD) is examined. I then turn to the susceptibility gene, APOE, the ε4 variation of which is associated with increased risk for late-onset AD, although, as yet, under poorly understood circumstances. Epidemiological research makes it clear that the way in which the APOE genotype functions is elusive because its effects are modified by the presence of other genes and by environmental variables. This raises profound difficulties for estimating individual risk estimates for AD on the basis of APOE genotyping.

Findings from the newly developed technology of genome-wide association studies (GWAS) being applied to the investigation of AD, primarily in the United States, United Kingdom, and France are discussed in chapter 6. These linked research projects make use of many thousands of DNA samples procured from individuals diagnosed with AD, which are then assessed using high-speed throughput technology and compared with control samples, in an attempt to find out what combinations of genes put individuals at increased risk. To date, these enormously expensive projects have provided few if any startling new insights, and many researchers are highly skeptical as to their value. However, others believe that GWAS is a first step toward a more sophisticated way of understanding the interrelated pathways of the numerous genes that appear to be implicated in AD.

Chapter 7 opens with a brief consideration of the anthropological literature on divination. This is followed by a condensed discussion composed primarily of ethnographic research findings about the responses of individuals and their families to genetic testing for single gene disorders such as Huntington disease and genetic testing for breast cancer. The concepts of "genetic body," "geneticization," and "biosociality" are introduced. The greater part of this chapter is taken up with an account of a National Institutes of Health randomized trial carried out with individuals whose families have one or more members diagnosed with AD, and who have undergone genotyping for the susceptibility gene APOE. The findings presented in this chapter represent the first extensive analysis of the responses of individuals undergoing testing for a susceptibility gene. The manner in which the trial was conceived and put into practice is set out, drawing on interviews with the principal investigators and genetic counselors. Much of the chapter is devoted to an extensive presentation of interview responses of nearly 80 individuals who had participated in the trial as subjects. The findings are in striking contrast to responses to genetic testing carried out to date for other conditions associated with single known genes. This chapter also presents findings from interviews with people who come from AD families but who have not been genotyped—the narrative accounts given by these individuals do not differ substantially from those of individuals who have undergone genetic testing. In closing, a brief account is given of the official position taken by AD societies in connection with genetic testing, and also an analysis of media reporting about the genetics of AD.

The concluding chapter sets out with a reader-friendly discussion about the shift in orientation by molecular biologists away from genes per se, into the wider fields of genomics and epigenetics—that is, to the contextualization of genes. Emphasis is given by many of these researchers to the dynamic interaction of DNA with both micro-environments internal to the body and macro-environments external to the body, and their interrelationship. This approach takes into account individual developmental and life course experiences that may bring about long-lasting modification to or even permanent changes in gene function. These emerging epigenetic models embrace complexity and acknowledge irresolvable uncertainty. The effects of historical, political, and social environments on individual biology and lived experience are also recognized, but thus far only spottily researched.

The fundamental tensions present in the AD world discussed in the opening Orientations chapter are taken up once again in closing the book. The concept of a "boundary-traversing mind" is introduced as a mediating entity between the brain and environments exterior to the body. The entanglement of aging and dementia and the related matter of the disputed ontology of AD is visited again on the basis of recent empirical findings, and the nature/nurture dichotomy is considered in conclusion, opening with a discussion of recent research into the epigenetics of AD and culminating in a discussion of "embedded bodies." The inseparable interrelationship of these three tensions is brought to the fore in this discussion.

This final chapter closes with a return to the global concern about aging societies, and the so-called epidemic of aging. It is argued that a public health approach to aging and Alzheimer's will have a much greater effect in reducing the incidence of AD worldwide than will the technologically oriented molecular approach currently being heralded as a paradigm shift. Should such an approach be effective, and there is little evidence to date to be optimistic that this will be the case, the extent of investment in advanced medical facilities and highly trained expertise required to put it in place is not realistic beyond wealthy segments of the world, especially given the global economy of the present and the increasing gaps between rich and poor.

To date, much of what has been written about Alzheimer disease by social scientists has focused on caregiving and on what is so often characterized as a "loss of self" or disappearance of the "real" person. This research is invaluable and has brought about many important insights.[52] *The Alzheimer Conundrum* takes a very different tack and is designed to be complementary to the extant body of work that has produced moving accounts about the subjective experience of dementia; documented the social burden, particularly on families, of dealing with demented relatives; and set out the political and cultural ramifications of the management of demented persons across the globe. However, a short Afterword is included in this present volume based on a brief discussion of the works of the deceased artist William Utermohlen, who made an agreement with his psychiatrist to keep working after being informed that he had Alzheimer's disease. These poignant images, several of which are reproduced here, together with relevant details from an analysis in the medical journal *The Lancet*, remind readers why it is imperative that the challenging endeavor of confronting Alzheimer's must be a global priority.

Chapter 1
Making And Remaking
Alzheimer Disease

I . . . now see my reluctance to apply the term Alzheimer's to my father
as a way of protecting the specificity of Earl Franzen from the gen-
erality of a named condition. Conditions have symptoms; symptoms
point to the organic basis of everything we are. They point to the brain
as meat. And, where I ought to recognize that, yes, the brain is meat, I
seem instead to maintain a blind spot across which I then interpolate
stories that emphasize the more soul-like aspects of the self.

—Jonathan Franzen, "My Father's Brain"[1]

In January 2011 I attended a lecture delivered in an engaging manner in a
Montréal hospital about pathbreaking basic science research in connection
with amyloid plaques and neurofibrillary tangles, long thought to be diagnostic of Alzheimer disease. The lecture was given by an invited guest from Harvard University, Bradley Hyman, and took place in a room packed with young
molecular biologists from diverse countries of origin, interspersed here and
there with a few clinicians. In striking contrast to the majority of basic science
lecturers who launch immediately into their specialized subject making use
of the obligatory PowerPoint presentation, Dr. Hyman started out by making
two general statements. First, he noted the estimated number of people living with AD worldwide at the present time, and the projected number of more
than 115 million by 2050, to which he added the wry comment that his audience had clearly chosen to be in the right field. Second, he referred briefly to
the history of AD, noting that it was the demonstration of plaques and tangles
in the brains of demented people that had allowed the condition to be definitively identified by Alois Alzheimer over 100 years ago. Hyman then reminded
his audience that here they were, assembled in 2011, still struggling to understand the reasons for the formation and significance of plaques and tangles.
He went on to state that his team is now able to observe the production of a
single plaque or tangle in a mouse's brain and track its growth over the ensuing
hours. The audience viewed this remarkable feat on video—an innovation that
may, perhaps, move us one step closer to solving part of the molecular aspects
of this stubborn Alzheimer puzzle.

During his talk Bradley Hyman had not mentioned that the significance of amyloid plaques is currently being debated in the AD world, due largely to repeated failures of clinical trials designed to target the removal of amyloid plaques in the brain. Nor did he note that a second reason causing dispute is irrefutable evidence showing that among individuals whose brains exhibit plaques (whether demonstrated in vivo by means of neuroimaging, or at autopsy), a good number do not exhibit the behavioral changes associated with dementia. The difficulties posed in attempting to rigorously delineate normal from pathological aging are made starkly evident by these findings. When prodded a little by a clinician during the question period, Dr. Hyman agreed that certain people are apparently able to "maintain homeostasis" in their brains, even in the presence of amyloid plaques, acknowledging, somewhat belatedly, an entanglement of "normal" aging and dementia.

This chapter opens with a brief discussion of the "discovery" of Alzheimer disease early in the 20th century, at a time when the significance of neuropathology as causal of mental illness began to be firmly established. The history of the disease is then tracked throughout the 20th century, showing how the question of the relationship of "normal" aging to dementia has never been satisfactorily resolved. Arguments revolve around interpretations of the significance of specific neuropathological changes associated with both aging and dementia and are of immediate relevance in determining what directions to take with respect to drug development designed to limit the ravages of AD. At a more fundamental level, such debates raise questions about the ontology of AD and what exactly will count as its defining pathological signs. An obverse question then follows: what "protects" those many individuals who never become demented during life but who harbor what is believed to be definitive Alzheimer pathology in their brains? This second question, although cursorily posed quite often these days, has never been systematically examined.

Is Aging a Disease?

The evolution of senile dementia has traditionally been considered to represent an aspect of senescence which, in turn, is the normal final phase of human performance that occurs as a prelude to death. Yet there has always been vast disagreement regarding the meaning of this statement.

—Richard M. Torack, *The Pathologic Physiology of Dementia*[2]

The idea that many people become demented in old age has a very long history well documented in the major literate traditions. In *As You Like It*, Shakespeare's character Jacques, a professional melancholic, tells his audience how the elderly are perceived as they pass through the last of the seven stages of human life:

Last scene of all
That ends this strange eventful history [of humankind]
Is second childishness and mere oblivion,
Sans teeth, sans eyes, sans taste, sans everything.

(Shakespeare, *As You Like It*, Act 2, Scene 7)

Given that the play is a comedy, it is perhaps tempting to interpret the gruesome characterization of old age as recounted by Jacques as ironic.[3] But, during Shakespeare's time, the dominant idea was one of decline and decay, not very different from our own, and irony was not in play, although it was also recognized that by no means everyone becomes demented, even when very old.[4] However, until well into the 20th century, in contrast to the present day, what was characterized as senile dementia was rarely regarded as pathology, but simply as part of aging itself.

No doubt most people coped as best they could by keeping their affected relatives at home. If families simply could not or did not want to deal with dependent elderly, then some were placed in poor houses and others taken to asylums. Alternatively, they were left to wander the streets to beg and scavenge—a situation that remains all too evident in many parts of the world today. The epigraph above by the neurologist Richard Torack suggests that it was generally assumed that everyone would eventually become demented—senile dementia, one's second childhood, was a "natural" end to life, although this might take place at a great age, by which time most people would have succumbed to some other condition.

Somatikers and Psychikers

Although senile dementia among older people was widely regarded as normal aging by the medical profession in the 18th and 19th centuries, other types of dementia that affected people of all ages were subsumed under the overarching concept of "mental derangement." The majority of the patients housed in asylums suffered from an extreme form of dementia associated with tertiary syphilis, and epileptics too were commonly shut away in these institutions. When care was no longer provided by families, cases of senile dementia were also usually housed in these custodial asylums. Typically, these and other patients were kept in prison-like conditions, and more often than not they were constantly shackled. Alois Alzheimer worked as a psychiatrist in such an establishment at the end of the 19th century in Frankfurt, but he then moved in 1903 to commence work in Heidelberg, and then later in Munich and Bresleau, in newly founded university clinics designed for the purposes of teaching and research, where patients stayed a relatively short time before being removed to asylums once it was clear that nothing further could be done for them. A notable feature of Alzheimer's career, then, is that it "traversed both of these psychiatric cultures."[5] Cases of senile dementia were most likely to be housed in the custodial asylums.

The numerous accounts about the "discovery" of Alzheimer disease rarely relate how Alzheimer was deeply involved in clinical care throughout his career, that he spent a great deal of time trying to converse with patients and took exceptionally comprehensive medical histories. He worked at a time when humanistic reforms were beginning to be implemented and was directly responsible for implementing nonrestraint practices together with other reforms in the large Frankfurt hospital where he was first employed, including regular bathing of patients not simply for reasons of hygiene, but also to calm and soothe them.

However, Alzheimer and his close colleague Bielschowsky declared that their stated mission was to move psychiatry forward with "the assistance of the microscope,"[6] and Alzheimer had always shown a predilection from the time he did research as a medical student for drawing remarkable pathohistological diagrams.[7] But, at heart, he was clearly also a caring, dedicated clinician. As his biographers state, "Alzheimer was an obsessed doctor and scientist."[8] His purpose was not to "reduce" the condition of dementia entirely to neuropathology, but to establish irrefutable links between clinical changes and pathology seen at autopsy.

Above all, Alzheimer wanted the medical world to recognize that mental illnesses have an undeniable material component. There was an obvious political reason for taking such a position because it could then be established that dementia-like conditions are not part of the spiritual/theological domain, but undeniably biological in origin and therefore not attributable with moral implications. A related reason, linked to the first, is that Alzheimer was an early adherent of the idea of "cortical localization," and hence was classed by his contemporaries as a "somaticizer." On the other hand, given the attention that he paid to his patients, and his predilection for improving their care, it is not unreasonable to surmise that Alzheimer was also seeking to reduce the stigma so often associated with the mentally ill and the inhumane treatment that was their lot. Common thinking of the day drew on the concept of "degeneracy" in which it was assumed that certain people, notably the poor, were predisposed to hereditary degenerative disorders, including mental derangement, exacerbated by alcoholism, sexual excess, venereal disease, and other "immoral" behavior, and it seems that Alzheimer questioned this demeaning thesis.

In order to better understand the "discovery" of Alzheimer disease, it is necessary to briefly touch on theories about causation in the early days of psychiatry. A tension has been evident throughout the history of Western medicine between accounts that favor somatic origins of mental illness and those that privilege social and individual behavioral causes including, on occasion, narratives about retribution by otherworldly entities. By the turn of the 19th century, when the earliest signs of the reform movement in the care of the mentally ill began to take shape, a "moral therapy" movement emerged. One of its pioneers was Philippe Pinel, who directed attention for the first time to the patient's "story" in an effort to discern what might have taken place in the life of the patient to precipitate her or his illness. Pinel asserted that because no organic lesions could be seen in the brain, the belief that mental illness has material origins must be a false premise. His ideas were extraordinarily influential among physicians in both Europe and North America, but it was not long before the somaticists "struck back."[9] Phrenologists of the late 18th century postulated the idea of localization of "faculties" in different parts of the cortex of the human brain, and over the ensuing decades, debate moved toward the somatic end of the spectrum. This move was bolstered by experimentation carried out by Broca, Ferrier, and others. In 19th-century Germany tensions between so-called localists and holists were most apparent, culminating in an attempt at a synthesis by Wilhelm Griesinger, whose arguments were well known to Alzheimer. Griesinger's position, described as "multifactorial," postulated predisposing factors including "psychical causes" that bring about a state of "intense irritation of the brain."[10]

By the late 19th century, the subspecialties of "organic neuropsychiatry" and "brain pathology" took form as new, hardened articulations for somaticism. As Michel Foucault was to note when commenting on this shift, "Disease is an autopsy in the darkness of the body,"[11] and findings from numerous autopsies conducted by Jean-Martin Charcot, known today as the "father" of modern neurology, and his colleagues at the Paris asylums of Saltpêtrière and Bicêtre anchored these new specialties. These institutions housed 3,000 to 4,000 patients and hence were a rich source of material for brain dissections. In 1899, Charles Hughes, the editor of *The Alienist and Neurologist* and a follower of organic neuropsychiatry, wrote,

> There is no such thing as insanity without disease . . . involving the brain. . . . There is no expression of mental derangement without a substratum of cortex disease, either in the neuron, in the enveloping membranes of the brain, in the nourishing blood supply, in the behavior of the vaso-motor system mechanism.[12]

But when dealing with elderly patients, a pathologically oriented position was more difficult to sustain because of the continuing widespread acceptance of an inevitable, "natural" decline among the aged, bringing about the onset of a second childhood, as Jacques in *As You Like It* reminds us.

Senility in Old Age

From classical times in Europe until early in the 19th century, influenced by ideas that originally emanated from the Middle East, the life cycle was conceptualized as rather clearly defined epochs based on age. Certain illnesses and conditions were associated with specific epochs in the life cycle thought to bring about instability in the nervous system, notably at pubescence and the climacteric (believed to be common to both men and women). The term "dementia" was assigned to people of any age and was used to indicate "any state of psychological dilapidation associated with chronic brain disease."[13] However, among the elderly it was assumed that an inevitable depletion of the vitality necessary for life was the primary cause of dementia, and from the beginning of the 19th century "senile dementia" was used almost exclusively when referring to the elderly—a condition of normal aging, in which, as some medical practitioners noted, memory loss rather than the florid symptoms associated with "derangement" was usually the primary symptom. Clearly, delineating normal from pathological has never been easy; "they implode," as Cohen puts it.[14]

It was in this milieu, in which neuropathology was being recognized as the key to understanding the class of conditions known as dementia and, paradoxically, senile dementia continued to be widely recognized as "normal," that Alzheimer began in 1901 to write a remarkably extensive clinical history of the 51-year-old woman whose case fascinated him, and who came to be known in the medical literature as Auguste D. When Alzheimer first met her in the Asylum for the Insane and Epileptic in Frankfurt am Main, where he was employed as a senior physician, Auguste Deter had just had a lunch of cauliflower and pork. Alzheimer asked her,

> "What are you eating?"
> "Spinach."
> She chewed the meat.
> "What are you eating now?"
> "First I eat the potatoes and then the horseradish."

Two days later Alzheimer noted that Auguste D was "constantly fearful and at a loss. She said, over and over, 'I won't let myself be cut,' acted as if she were blind, and when walking about groped the faces of other patients, and was often struck by them in return. When asked what she was doing she said: 'I have to tidy up.'"[15] For months, until he left the hospital in Frankfurt, Alzheimer saw his patient virtually every day, making extensive notes of his attempted conversations with her and of her moods and behavior. Three months after he first met Auguste D, she was neither able to converse with Alzheimer nor answer his questions. Her behavior, Alzheimer wrote, was now hostile, and she lashed out when he tried to examine her. He noted that she often screamed spontaneously for hours on end and wandered about aimlessly, sometimes having paroxysmal fits that lasted for several hours. After he left Frankfurt, Alzheimer continued to inquire about his former patient, and when she died in April 1906, five years after first being admitted to hospital, Alzheimer immediately requested that he be given her brain for autopsy. His biographers note, "Alzheimer believed that behind the clinical symptoms, marked by forgetfulness and jealous fantasies, a new 'peculiar disease,' as he put it, could be found in Auguste D."[16]

Seduced by Plaques and Tangles

The innovation that many agree brought about the "discovery" of Alzheimer disease was a silver precipitation technique first developed by the Italian scientist Camillo Golgi in 1873, and shortly thereafter modified by the Spanish neurologist and photographer Santiago Ramón y Cajal. This novel technique radically transformed the emerging discipline of the neurosciences, and was regarded as such an important breakthrough that these two scientists were jointly awarded the Nobel Prize for medicine. Prior to the development of effective staining techniques it had already been postulated by the somaticists that psychiatric disorder must be intimately associated with specific demonstrable changes in the gross anatomy of the brain, but this could not be proven beyond noting anatomical changes detectable by eye, the significance of which was speculative. Nevertheless, efforts were made as early as the 1850s to create subcategories of mental illness based on gross neuroanatomical lesions found at autopsy. Commencing in the 1880s, microscopic examinations (figure 1.1) were carried out in earnest and, with improved preservation and fixing and the development of new stains, it was possible to discern in closer detail the material changes assumed to be causal of psychiatric disorder and senility.[17] As the neurologist Peter Whitehouse has noted,

> The staining process is a centrally important one that, literally, brings the lesion into the medical gaze. As part of a process that made a thing visible, it thereby made it (appear) real. The staining process asserted a commonality of the pathologic lesions.[18]

The silver precipitation technique was refined in 1902 by the German scientist Max Bielschowsky and shortly thereafter made use of by Alois Alzheimer to stain sections of the brain tissue he had procured after the death of Auguste D together with that from several other patients diagnosed with the disease soon to become his eponym. The histological slides allowed Alzheimer to see with the aid of a microscope what had not before been evident, namely the clumped structures he labeled as neurofibrillary tangles that would thereafter be recognized as one of the key autopsy signifiers of Alzheimer disease.[19]

Neurofibrils are present in normal cells and had been observed well before Alzheimer's time, but the stain clearly showed the extent to which they had accumulated excessively to form abnormally dense twisted fibers inside the nerve cells. A second signifier of Alzheimer disease, amyloid plaques, found between the nerve cells (neurons), were also visible when he applied a different stain, methyl blue-eosin, but plaques had already been definitively

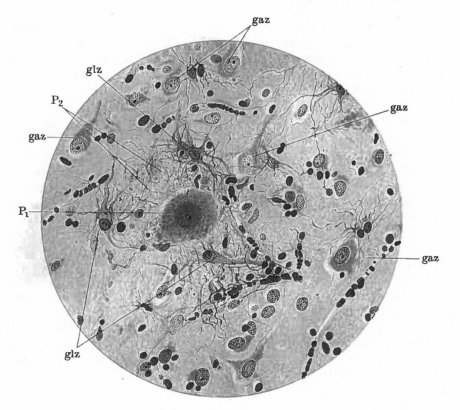

Figure 1.1.

Alois Alzheimer, 1911. Neuritic plaque, in Javier DeFelipe,
Cajal's Butterflies of the Soul: Science and Art (2010). Reproduced with
the permission of Oxford University Press p.278

described by several other medical researchers working at the end of the 19th century.

Both Alzheimer and Bielschowsky had spent a great deal of time dealing with patients with the dementia-like condition associated with tertiary syphilis; they had dissected the brains of many patients who had died of this disease and of epilepsy—the condition that more than any other would lead to an acceptance of localization theory.[20] Alzheimer had also studied anatomical changes in the brains of older demented patients where syphilis was not implicated. He and his colleagues were already convinced that lesions visible to the naked eye in the autopsied brains of demented patients, even without the aid of a microscope, were important signifiers of mental illness; what they saw when looking down the microscope at preparations of brain tissue only confirmed their beliefs.

Alzheimer gave a report at a meeting of the South West German Alienists on November 4, 1906, about the case of Auguste D. He informed his audience that she had "presented" with progressive cognitive impairment, hallucinations, delusions, and "marked psychosocial incompetence." Alzheimer added that on postmortem he had found brain atrophy, arteriosclerotic changes, senile plaques, and neurofibrillary tangles.[21] He concluded his presentation with the following comment:

> Taken in all, we clearly have a distinct disease process before us. Such processes have been discovered in great numbers in recent years. This observation suggests to us that we should not be content to locate any clinically unclear cases of illness in one of the familiar categories of disease known to us to save ourselves the effort of understanding them. There are undoubtedly far more mental illnesses than are listed in our textbooks. In many such cases, a later histologic examination will allow us to elucidate the case.[22]

Alzheimer's report of Auguste D's case received a poor reception at the meeting. No one present answered the chair's call for questions, and the paper was rebuffed by the so-called anti-Kraepelinians present, who argued that "there can be no talk of nosological specificity." They were adamantly opposed to the possibility that specific pathological anatomy accounted for named mental illnesses.[23] Even though Alzheimer's reputation was well established, and the leading psychiatrist of the day with whom Alzheimer worked, Emil Kraepelin, ably defended him, the assembled group moved on to discuss a paper delivered by a disciple of Sigmund Freud.

While Alzheimer was working in the Frankfurt asylum he was dealing with a large number of cases of senile dementia, but in the ensuing years, after he moved to the teaching hospitals, he was no longer responsible for such patients on an ongoing basis because they were relatively quickly transferred to asylums. In the years following his presentation of the case of Auguste D, Alzheimer and his colleagues reported only eight similar cases; at least two of these had already been reported elsewhere, and, furthermore, their authenticity was questionable. These patients were not elderly and therefore not "typically" senile, and there were serious doubts as to whether the cases represented anything significantly new that warranted revising the current taxonomy. What is more, the second key case written up by one of Alzheimer's colleagues, that of a 56-year-old laborer, Johann F, showed no signs of neurofibrillary tangles at postmortem.[24]

But a move toward recognition of localization theory was very much in the air despite the existence of vocal anti-Kraepelinians. L. W. Weber, a senior physician practicing in Göttingen, noted in 1905, "It is a fact now scarcely contested, that all mental disorders depend on pathological processes in the brain . . . in this sense every mental disorder may be termed a brain disease." And in 1910, Emil Kraepelin, in the eighth edition of his extraordinarily influential revised textbook on psychiatry, made a cautious but nevertheless clear distinction between conditions he described as presenile and senile dementia. The former, presenile dementia, he named Alzheimer disease, but he noted, "The clinical interpretation of Alzheimer disease is still unclear at the moment."[25] The question remains as to why Kraepelin took this position on the basis of what appears to be fragile evidence. The historian German Berrios argues that the re-reporting of several of the cases suggests that there may have been great pressure in the laboratory where Alzheimer worked alongside Kraepelin to find evidence for a "new" disease, but it has never been satisfactorily established whether or not Alzheimer in the end agreed with Kraepelin's decision to name this disease, tagged or not, with his name. The comments of one of Alzheimer's close colleagues, Gaetano Perusini, strongly suggest that Alzheimer continued to believe that he had documented nothing but a few atypical forms of senile dementia in which age was the sole distinguishing feature, but vacillation on his part is also evident.[26] It seems undeniable that the institutional constraints imposed by teaching and research hospitals in which Alzheimer worked for most of his life, in which long-term care was not provided, ensured that he could not observe his patients in a manner that he had carried out so successfully while caring for Auguste D in an asylum.[27]

Among the reasons put forward by historians for Kraepelin's apparently hasty move to name Alzheimer disease are the following: Kraepelin did so for scientific reasons because he himself was convinced that the cases being reported demanded taxonomical revisions. But evidence for the veracity of this reason remains slim. The existence of a rival neurological department in Prague headed up by Arnold Pick, who would shortly have another form of dementia named after him, has been given as a second reason. But the most commonly offered reason is that Kraepelin was feeling threatened by Sigmund Freud and the increasing interest being shown in a psychoanalytic approach to the interpretation and management of mental illness. Hence, it is argued, Kraepelin was experiencing a sense of urgency to document the pathological substrates of mental conditions in order to put them on a sound footing. However, Berrios argues strongly against this position and reminds us that in principle Freud, himself a neurologist, had no antipathy to recognition of organic, localized foundations of dementia.[28]

It is also pertinent to note that Kraepelin was not a hard-nosed somaticist. He had traveled to many parts of the world, and strongly believed that it was important to establish a comparative psychiatry. In common with many like-minded thinkers of his day, and profoundly influenced by his trip to Southeast Asia in 1904, Kraepelin postulated the "psychic character" of peoples who live in similar environments, and following in the footsteps of his teacher Wilhelm Wundt, he called for the formation of a discipline of comparative ethnopsychology.[29] In short, a satisfactory explanation for Kraepelin's apparent rush to name a new disease is still wanting, but clearly the idea that localized changes in the brain cause mental illness was sufficiently well established for Kraepelin to feel justified in principle to make his move, even though the supporting evidence was rather slim.

The slides on which Alzheimer's initial observations were made were lost for many years. Remarkably, they came to light in 1998. Upon reexamination it was agreed that Alzheimer's conclusions had been entirely accurate by present-day standards. Atrophied cells and plaques were clearly present in the preparations made of brain tissue from both Auguste D and Johann F, but a massive number of neurofibrillary tangles appeared only on the slides of tissue taken from Auguste.

The Partial Eclipse of Alzheimer Disease

Despite its baptism by Kraepelin, Alzheimer disease lost its way for four decades. Numerous scientists and clinicians of the day disagreed with Kraepelin's designation of a new disease, and no systematic follow-up was undertaken to consolidate its recognition. It was already well known at the time, largely on the basis of the autopsied brains of syphilitics and epileptics, that amyloid plaques and neurofibrillary tangles occur in individuals other than those diagnosed with Alzheimer's. This effectively weakened any argument made for a relationship between these specific neuropathological findings and the behavioral changes seen in patients diagnosed with presenile dementia.[30] A second reason is that the outbreak of the First World War ensured that resources were diverted away from basic laboratory work to the war effort. And a further reason was that the energies of the drug companies of the day were directed toward what was termed arteriosclerotic or vascular dementia, on the assumption that pharmacological agents would soon be found to combat this problem. Senile dementia continued to be associated primarily with aging itself, and the matter of early age of onset alone did not prove to be sufficiently convincing evidence for presenile dementia to be accepted as a distinct disease. This was the case even though Alzheimer's close colleagues of the day started to use the label, and Alzheimer himself continued to insist, as he had done since his presentation of Auguste Deter's case, "We must reach a stage in which the vast well-known disease groups must be subdivided into many smaller groups, each with its own clinical and anatomical characteristics."[31]

Although Alzheimer spoke up for recognition of links between localized neuroanatomy and behavioral changes, scholars who know the literature of the time well and have facility with German have made comments such as the following: "Alzheimer himself could be counted among the 'doubters' who did not necessarily believe that AD represented anything but a precocious form of senile dementia."[32] Even so, for Alzheimer, both presenile dementia and senile dementia were not "normal" aging but rather irreversible conditions, the material reality of which could be located in the brain. When plaques were relabeled in 1910 as "senile plaques," this confounded the matter further—plaques were now specifically associated with aging, but they were also understood as the prime signifier of dementia as a disease.[33]

During the 1920s more cases of presenile dementia were documented, but hesitation about its validity as an isolable disease persisted, leaving the problem of Alzheimer disease and its relationship to aging unresolved. Another line of thought further muddied the waters: a good number of experts of the day believed that cerebral arteriosclerosis was the primary cause of dementia. Meantime it was reported in the 1930s by a German researcher, on the basis of autopsy findings, that "84% of persons dying over the age of 65 had 'senile plaques' in their brains," suggesting that plaques are, in effect, a normal part of aging.[34] The upshot of the uncertainty was that the dementias were "seriously neglected" for many years, in part because they fell into a no-man's land, neither fish nor fowl, neither neurological nor psychiatric disorder, and geriatrics did not as yet exist as a specialty.[35] This situation was to persist until the 1970s.[36]

Medicalization of Aging

During the latter part of the 19th century medical concern about the plight of the elderly increased, resulting in the formation of what would become the specialty of geriatric medicine and hence the beginnings of the medicalization of old age. In his *Lectures on Senile and Chronic Diseases*, published in 1867, Charcot noted that the importance of studying diseases of old age was no longer contested.[37] France, Germany, and Great Britain were at the center of this move that then spread to North America. Early on, certain of the key figures in this emerging field started to describe old age itself as a disease-like condition.[38] Emphasis was given to the slowing of body activity, loss of sociability, becoming bedridden, and "senile degeneration."[39] As the historian Martha Holstein has noted, "[I]n what may seem contradictory to modern readers, turn of the century investigators often described 'normal' aging somewhat quixotically as pathological or at least as 'a quasi-pathological process of cell and tissue degeneration.'"[40] Thus dementia could be simultaneously "normal" and "pathological." Ignatz Nascher, the founder of gerontology in the United States, argued that "senile changes" were "deviations in degree . . . usually permanent, progressive, and uncontrollable; rarely remissive or changeable."[41] But the historian Jesse Ballenger points out that despite a widespread negative stereotype about the ravages of senility there was a large popular literature available at the time informing its readers how to avoid senility by paying "careful attention to hygiene, exercising, seeking out the companionship of the young, and many other stratagems."[42]

Ballenger insists that what is known as the "dark ages" of dementia research, from about the 1920s until 1970, is not an accurate depiction. He points out that between the mid-1930s and the 1950s a surge of interest in senile dementia took place, notably among American psychiatrists. This was in part stimulated by what was described as a "demographic avalanche of aging"[43] under way, with the result that state mental institutions in the United States were becoming clogged with elderly senile patients, thus detracting from the professional authority of psychiatry. At the same time, the profession of gerontology argued forcefully for recognition of and provision for diseases that affect older patients in disproportionate numbers. Among the U.S. psychiatrists of the day who turned to a consideration of senile dementia, David Rothschild is perhaps the best known.

Rothschild argued that because plaques and tangles are found in several medical conditions and not simply in senile dementia, these formations should be understood as a generalized tissue reaction in response to a number of biological factors. But Rothschild was not content with this observation alone; together with his colleagues he began to examine the relationship of emotional and personality disturbances to Alzheimer's. Rothschild argued explicitly that the usual approach to senile dementia is reductionistic, noting that there is "too exclusive a preoccupation with cerebral pathology." He went on, "The changes occur in living, mentally functioning persons who may react to a given situation, including an organic one, in various ways."[44] Rothschild argued that people have differing abilities to compensate for organic lesions and that this should be investigated. To reinforce his position, Rothschild emphasized the contradictory findings shown repeatedly since Alzheimer's time, namely that the presence and degree of dementia in a living patient often show discrepancies with the presence and degree of pathological structures at autopsy. Ballenger suggests that Rothschild's publications were so influential that the therapeutic nihilism usual among psychiatrists when confronted by senile patients began to break down.

Moreover, Rothschild's arguments were drawn on to create a highly influential literature of the day in which normative societal attitudes toward aging and management of the elderly were extensively criticized. The locus of senile mental deterioration should no longer be located in the aging brain, it was now argued; rather, society strips the older person of a reason for living.[45] The psychodynamic variation on this position also aired at this time was that the forgetfulness associated with senility is a form of repression, a protection against the fear and frustration associated with growing old in the mid-20th century. By the 1970s discussion highlighting systematic discrimination against the elderly was common, but at the same time it was acknowledged that a minority do indeed suffer from severe organic problems that require investigation.

The Politicization of Alzheimer Disease

Ballenger points out that during these decades a limited organic approach to senility never entirely disappeared and that, commencing from the 1960s, five innovative changes took place that once again put Alzheimer disease firmly on the map. First, in the 1960s, a large series of brain autopsies of patients diagnosed with dementia were carried out in England. It was asserted that the vast majority of these brains showed relatively few signs of arteriosclerotic changes (the pathology that had occupied both drug companies and neuroscientists for some time), but that they exhibited numerous plaques and tangles as described decades earlier by Alzheimer.[46] These clinicians argued explicitly that the amount of damage caused by the plaques and tangles corresponded closely to the severity of the behavioral changes in patients. However, the researchers were also obliged to acknowledge that in a considerable proportion of their nondemented control group, plaques, tangles, and brain shrinkage were markedly evident at autopsy.

A second significant change was the development of the electron microscope, permitting important refinements in classification of neuropathologies. For the first time, specific biochemical changes were associated with structural changes made visible by this powerful new instrument. It was possible to distinguish clearly between the composition of plaques and tangles, and arguments began to take place about the primacy of plaques or tangles in disease causation. A psychological approach to dementia was pushed to one side, to be replaced by a "new agenda for pathological research."[47]

Third, in the 1970s, the influential neurologist Robert Katzman publicly insisted that received wisdom of the time, namely that aging inevitably results in senility, be abandoned. Several commentators pointed out that the word "senile" had in effect become a term of abuse and, furthermore, that its use perpetuated discrimination against the elderly.[48] Katzman, following a suggestion first made in 1948 by R. D. Newton on the basis of 150 autopsies carried out at Middlesex Hospital in London, declared that both "presenile dementia" and all cases of senility should be recognized as pathology, be labeled as Alzheimer disease, and understood as entirely distinct from normal aging.[49] Katzman's position has been described as a "neo-Kraepelin" concept of Alzheimer disease,[50] and it culminated in what has been characterized as the "rediscovery" of this condition.[51] As Ballenger notes, "[B]y the end of the 1970s, if senility had not been eradicated, as an earlier generation of gerontologic activists had dreamed, it had at least been thoroughly disciplined—relegated by biomedical scientists to various discrete, well-defined disease entities that, at least in theory, no longer contaminated the entire experience of aging."[52]

It was at this time that Alzheimer disease began to be billed by Robert Katzman and others as the fourth or fifth leading cause of death in the United States. Writing Katzman's obituary in the *New York Times* in 2008, Roger Segelken argued that a major transformation had come about in the view of the medical community after the publication of an editorial by Katzman in 1976 in the *Archives of Neurology*, followed in 1977 by a conference that he organized. In the editorial Katzman described AD as a "major killer" and discussed its "malignancy." Prior to its publication, fewer than 150 articles in all had been published on Alzheimer disease. From the publication of the editorial until the time of the publication of Katzman's obituary in 2008, over 45,000 articles had been published,[53] a number that continues to rise exponentially.

The fourth innovation in the 1970s was the formation in the United States of the National Institute on Aging (NIA), founded specifically to foster a comprehensive research program on aging. And fifth was the emergence of an incipient Alzheimer's movement, and its consolidation commencing from 1977 as the Alzheimer's Disease and Related Disorders Association (ADRDA), with strong support from the first director of the NIA, the gerontologist and psychiatrist Robert Butler. Activities of the ADRDA in turn increased the legitimacy of the NIA, in large part because it lobbied government and also set about raising money for research into Alzheimer disease.[54] At this time the language associated with an epidemic, "a ticking time bomb," began to be evident, together with concerns about looming skyrocketing medical and social costs.[55] However, Whitehouse and colleagues have pointed out that emphasis was given from the outset to finding effective treatments and ultimately a means of prevention of AD, rather than paying attention to the desperate needs of caregivers.

It was argued by advocacy groups that widespread recognition of Alzheimer's as a disease was first essential and that, when making a case to the NIA, an emphasis on curative medicine would be the best way of addressing not only the search for a cure, but also the social and personal burdens imposed by the disease. Not surprisingly, little money was set aside to specifically deal with caregiving.[56] Similarly, in Great Britain, the Medical Research Council denoted research into dementia and Alzheimer disease as a priority area.[57] The sociologist Patrick Fox points out that although clearly efforts to treat and cure AD are admirable, it must be kept in mind that this endeavor is a business that involves powerful economic interests centered on "the marketplace of disease diagnosis and treatment."[58] It is also fostered by a "discourse of hope" sustained by involved families.[59] This characterization of AD continues to be relevant and accounts in part for the consolidation of what came to be the dominant AD paradigm known as the "amyloid cascade hypothesis," believed to be causal of a buildup of plaques in the brain (see the following chapter).

Fox notes that interest in the newly formed organization began to explode in 1980 after a letter from a family member of an Alzheimer disease victim was published in the nationally syndicated column "Dear Abby." Following this publication, the ADRDA received more than 30,000 letters, precipitating interest among the public in AD as nothing else before it had done.[60] It rapidly became clear that many families desperately hoped for assistance with caregiving, but, of equal importance, they wanted senility recognized as a disease of the brain for which a cure should be sought out, thus challenging the stigma associated with dementia, and at the same time shifting the moral burden for the occurrence of the disease away from themselves. These families were adamant that they had little time for psychosocial or psychodynamic models of senility. Fox and colleagues have recently shown how this two-edged sword—the tension between cure and care—has been exacerbated over time, particularly following the Reagan years, so that support for care of AD patients has been reduced today to little more than tax credits for most affected families in the United States.[61]

When I talked with the neurogeneticist John Hardy in 2008, he stressed that consolidation of the concept of Alzheimer disease as a singular condition "was just a political maneuver to get funding, and some people then actually came to believe that this is the case." Writing in the early 1990s about the rise of a medical model to tackle the problem of dementia, the eminent psychiatrist William Lishman, associated with the Institute of Psychiatry at the Maudsley Hospital in London, sounds a cautionary note: "Certain observations remain obstinately to remind us that aging and Alzheimer's dementia are interlinked closely."[62] Lishman notes that many of the clinical features of Alzheimer's, notably memory loss, are strongly associated simply with aging. And so too are changes within the brain, above all, the presence of plaques and tangles in the hippocampus. He points out other biological changes in common between the aging process and Alzheimer's, and concludes with the important observation that, as individuals move into old age, it becomes increasingly difficult to distinguish Alzheimer's and old age because of their common features. Lishman insists that the possibility must be entertained that Alzheimer disease is, after all, "simply brain aging." He argues that if "a marker is found that is indisputably associated with the condition labeled as Alzheimer's alone, a marker never seen in healthy aging individuals," then the idea that Alzheimer's and neural aging are indistinguishable must be dismissed.[63] Lishman concludes that the aging process itself may be under strong genetic and environmental influences so that in certain circumstances it will be "accelerated and intensified," in effect becoming a form of "precocious aging."[64] And he adds, "In all persons the shift from so-called 'normality' to becoming 'a case of Alzheimer's dementia' will occur when the process has passed a certain threshold in its development."[65]

The recently deceased British psychiatrist Martin Roth argued in the 1990s for research into what he described as "reserve capacity," by which he meant the likelihood that certain individuals have an abundance of neurons that permit them to stay above a crucial threshold in which mental activity is preserved throughout life. In contrast to that of Lishman, the position that Roth took is one of a marked discontinuity between aging and dementia; he argued that when a crucial threshold is passed, the involved molecular pathways diverge, even though some overlap may be possible. As far as Roth was concerned, "an aging person either does or does not have AD," and those without AD may have some memory loss, but they do not lose their sense of identity, ability to retain any information, or capacity for reasoning. Nor do they die as a result of being demented. Roth was adamant: "AD cannot be accounted for in terms of a continuous and predictable extension of normal mental aging."[66] But he nevertheless acknowledged that genetics and environment no doubt play a role in accounting for why some individuals are more vulnerable to AD than are others.

A Brief History of Normal

Michel Foucault's work on the "archaeology of medical perception" illustrates how the rise of anatomical pathology in the 19th century helped create the standardized body of modern, scientific medicine.[67] He argued that representations of the body in premodern medicine were in effect "mute," a blank slate on which the timeless truth of disease was to be deciphered through a parsing of signs. Gradually, the "truth" of disease was displaced into the body: "[O]pen up a few cadavers," exhorted the great Parisian anatomist Bichat, "and you will see disappear the obscurity that observation alone could not dissipate."[68] The body could now "speak," but only once the differences there observed could be understood as pathological deviations from a healthy norm. This marked a shift from the earlier notion that diseases are pure "essences" that mark the body from within.

Foucault extended his argument to note that it became possible to "see" differences among abnormalities made visible when bodies were dissected. This variation was interpreted either as different diseases or as different stages in the unfolding of the same disease, that could then be linked back to the signs and symptoms experienced and spoken about by patients that were at times visible on the surface of their bodies. The anatomized body became an invariant, a standardized measure of disease mechanism. Extrapolating from the cadaver to ailing patients made it possible to imagine the symptoms that afflicted them as signifiers of invisible processes deep within the body; thus was "localization theory" solidified.

Until well into the 19th century use of the term "normal" was virtually limited to the fields of mathematics and physics. It was not until an internalizing approach to the body based on anatomy took hold that arguments about the relationship between normal and abnormal biological states were seriously debated for the first time. Auguste Comte, writing in 1851, noted a major shift in conceptualization that had taken place when the physician Broussais argued in the 1820s that the phenomenon of disease is of essentially the same kind as that of health and, thus, health and disease differ from each other only in "intensity."[69] Application of epidemiological methods to the study of populations allows this continuum to be mapped, and the results can then be made use of in clinical settings, with the assumption that all human bodies are biologically equivalent. Conversely, findings about individual bodily conditions such as those observed in clinical studies may be extrapolated to entire human populations, although, given that such findings are not representative of any given population, they often introduce major problems in connection with interpretation as to their significance.[70] Broussais postulated not only that normality should be understood as being on a continuum with pathology but also, furthermore, that deviation must be understood with reference to a "normal" state.[71] This theme was taken up and expanded upon by several influential thinkers during the course of the 19th century, among them Auguste Comte and Claude Bernard.

In the 1960s, the philosopher and physician Georges Canguilhem, Foucault's teacher, in writing a synthesis of the work of the previous century in connection with normality, noted that "strictly speaking . . . there is no biological science of the normal. There is a science of biological situations and conditions called normal." Canguilhem concluded that normality can be understood only in context, "as situated in action," and moreover, diversity does not infer sickness, nor does "objective" pathology exist outside of the laboratories, clinics, and operating theatres where it is made visible.[72] Canguilhem's argument, contra Broussais, was that the "normal" and the "pathological" are two fundamentally different states that cannot logically be placed on the same continuum. Their reconciliation along a biological continuum, Canguilhem argued, was in fact an artifact of decontextualized clinical and laboratory methods used in biomedical research, and "normalization" can lead to the mistaken assumption that what is statistically "abnormal" is inevitably pathological or, alternatively, that no pathology lies in what is statistically "normal." Discussion in contemporary Alzheimer circles about "exceeding a threshold," "maintaining homeostasis," "reserve capacity," and continuities and discontinuities between normal aging and dementia reflect this much older debate about disease, and the relationship of normal and pathological. The confusion about this relationship is most often expressed in current scientific articles about Alzheimer disease in terms of the oxymoron "normal subjects exhibit neuropathology," or some close equivalent.

When Is Pathology Normal?

The brain, to a greater extent than other major organs, has long presented a formidable barrier for medical diagnosticians. Until very recently no technologies existed that allowed one to "see" into the living brain, and for this reason, from Alzheimer's time on, two diagnoses have been made use of: the first is a clinical diagnosis based on an apparent decline over time in cognitive function, determined by means of psychological testing and clinical judgment, and the second is a postmortem diagnosis based on autopsy findings. Clinicians I have talked to in specialty clinics usually insist that when a diagnosis is made by an experienced physician, one who has tracked a patient for a good number of months or even years, the diagnosis is accurate in the great majority of cases; the Alzheimer's is "real." During this time, extended conversations have taken place at each meeting with the patient and a caregiver who has been requested to accompany the patient to the clinic, and various psychological tests have also been repeatedly administered. It is clinicians, of course, who must deal with the complaints, concerns, and the assessment of "functional deficits" associated with dementia. But they are required to write a diagnosis of "possible" or "probable" AD on patients' charts, on the assumption that only at autopsy will the truth be revealed.

In turn, neuropathologists I have interviewed insist that pathological findings represent the "true" diagnosis of AD, even though they acknowledge that their findings do not always correspond precisely with those of other pathologists. Clinicians and pathologists agree that clinical and neuropathological diagnoses are, for the most part, in concurrence, but acknowledge that this is not always the case. These discrepancies have been evident since Alzheimer's time, as we have seen, and were highlighted yet again when findings from the so-called Nun study were first published. This research commenced in 1986 and involved 678 Catholic sisters who belonged to an order called the School Sisters of Notre Dame located in seven regions of the United States. Statements written by these nuns when they were young women about why they wanted to enter the order, carefully stored for decades, were matched with neuropsychological test results administered throughout the latter part of their lives from age 75 on, and then subsequently linked to autopsy findings after death (every nun had agreed when she entered the project to donate her brain for autopsy).

It is argued on the basis of this study that those individuals who showed imagination and complexity in their thinking while young (that is, exhibited "high idea density," in the language of the researchers) were less likely to succumb to Alzheimer disease when they grew older.[73] This finding was not related to number of years of formal education, and was borne out as the autopsy results gradually accumulated: 90% of those nuns whose brains exhibited extensive neuropathology had shown "low idea density" as 20-year-olds. This research gave an enormous boost to what came to be known as the "cerebral reserve" hypothesis—such reserve being laid down commencing in utero.

Of even greater interest for the present discussion was the finding that a small proportion of the nuns who coped very ably with the neuropsychological battery of tests turned out at autopsy to have extensive signs of plaques and tangles, once again confirming observations made repeatedly over the past 80 years. Conversely, it was also clear that a few individuals whose autopsied brains revealed a relatively small number of anatomical changes exhibited all the behavioral signs of dementia while alive.[74]

David Snowdon, the principal investigator in this study, made the following comment about one of the nuns whom he came to know well:

Sister Mary, the gold standard for the Nun study, was a remarkable woman who had high cognitive test scores before her death at 101 years of age. What is more remarkable is that she maintained this high status despite having abundant neurofibrillary tangles and senile plaques, the classic lesions of Alzheimer disease.[75]

When asked to comment on the findings of the Nun's study, a Montréal neuropathologist had this to say:

> Well, if I find little neuropathology and the clinician says there's "probable AD," then I have to find a reason why. And if I don't think the quantitative changes of what I'm looking at in the brain is Alzheimer disease, then I have to find another reason. Is it vascular dementia? Is it another neurodegenerative disease? So, I mean, when the patient is clinically demented, there's got to be something in the brain to explain that. The difficult thing is, we know that there's no direct correlation between the density of plaques and tangles, and the clinical presentation. So we can have a brain that's full of neuropathology, but clinically the patient was not very demented. We can't categorize and say this is "moderate" Alzheimer's, this is "severe" Alzheimer's just by histological findings. That means nothing. All we can say is that there are sufficient tangles to confirm the clinical diagnosis, but there's no direct relation so as you can say that if you have more than 20 plaques in this area, or so many tangles, that's a severe case of the disease. So there's no correlation between clinical severity and histological severity.

Of course, neuropathology is the expression par excellence of localization theory in practice. Pathologists often know relatively little about the patient whose brain tissue they examine under the microscope, and all they have in front of them may be the clinician's final diagnostic report of "probable dementia." Their task is one of verification or falsification and, if the latter, to account for the death using a different diagnosis. We begin to get a sense here of just how uncertain it is as to what should count as an AD "case"—making the creation of robust figures for use in public health arenas and in creating population databases highly problematic. Moreover, the way in which the person is, in effect, "disappeared," to be transformed into a thoroughly decontextualized AD case, one reduced entirely to neuropathology, is strikingly evident. This, in contrast to autopsy findings in connection with many other conditions, the pathological report does not provide definitive causal information—rather, it confirms that Alzheimer's is present in the brain. The situation cannot be compared with a laboratory demonstration of the syphilis spirochete or the exposure of a brain tumor at autopsy. The intricacies involved were highlighted for me when an experienced neuropathologist remarked, without irony, "I've never seen two human brains in which the pathological signs of dementia are the same."

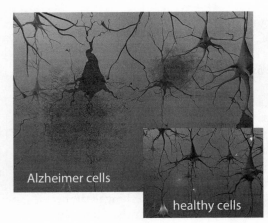

Alzheimer's tissue has many fewer nerve cells and synapses than a healthy brain.

Plaques, abnormal clusters of protein fragments, build up between nerve cells.

Dead and dying nerve cells contain *tangles*, which are made up of twisted strands of another protein.

Scientists are not absolutely sure what causes cell death and tissue loss in the Alzheimer's brain, but plaques and tangles are prime suspects.

Figure 1.2.

Under the microscope: Plaques and tangles. Reproduced with the permission of the Alzheimer's Association, from Inside the Brain: An Interactive Tour, Alzheimer's Association, http://www.alz.org/alzheimers_disease_4719.asp. This tour is available in 14 different languages.

Another epidemiological project, one in which the population was carefully delineated, assessed the neuropathological status at death of 456 individuals who had agreed to donate their brains for research and had previously been tested clinically for AD. This study showed, as expected, that the neuropathological features assumed to be diagnostic of Alzheimer disease and dementia, including plaques, tangles, and cell loss (figure 1.2), are more evident in older than in younger subjects. However, although the correlation between postmortem neuropathology and clinically diagnosed AD and dementia was strong in the group aged 60 to 75, this finding was shown to be less and less strong with increasing age. The researchers concluded that a considerable overlap exists in the so-called neuropathological features present at postmortem of individuals aged 75 and older who were clinically diagnosed with dementia while alive, and those without dementia.[76]

It was argued that additional factors other than the simple presence of Alzheimer neuropathology must determine the clinical expression of dementia, especially in those aged 75 years and older. The researchers insist, "[T]he pathological basis of dementia should be considered as an interaction among pathological changes, compensatory mechanisms, and underlying synaptic dysfunction" and, "cognitive dysfunction in later life is a life-span issue and is affected by genetic, developmental, and lifestyle factors, accumulated neural insults, innate and acquired cerebral reserve and compensatory mechanisms, and age-related decline."[77]

One member of this research team, the British epidemiologist Carol Brayne, has been arguing for years that research into dementia should move away from a model that poses as its basic question "has he got it?" to one of "how much of it has he got?" and "why?"[78] And cumulative research in recent years based on population research has confirmed that so-called mixed dementias—usually of vascular dementia and Alzheimer disease—are the most common; some findings suggest that the vascular changes may initiate the dementia.[79] Another study found that "dementia of unknown etiology" accounted for 5% of all cases of dementia among patients dying in their 70s, 21% of patients dying in their 80s, and 48% of patients dying in their 90s. It was concluded that a significant percentage of demented patients older than 80 years do not meet pathological criteria for either AD or a second relatively common kind of dementia (Lewy body dementia).[80]

Almost as an afterthought, one of the most highly recognized and experienced of AD experts, John Morris, suggests along with his coauthor in the concluding sentence of a 2006 article, "An additional research focus, of course, should determine why many older adults with neuropathologic evidence of AD manage to remain cognitively normal despite numerous AD lesions littering their brains."[81]

A Diffuse Clinical Syndrome

Since the beginning of the 20th century two major ontological matters have persisted in the AD world. The first, as a recent article title in the *British Medical Journal* put it, is, "What do we mean by Alzheimer disease?"[82] The second, linked to the first, is the question of whether or not AD is an inevitable part of "normal" aging, or is it a bona fide neuropathological disease, entirely different from aging? Throughout the 20th century repeated efforts were made to have AD recognized as a disease, not merely to aid in the search for a cure but also, as we have seen, for powerful social and political reasons. However, in the first decade of the 21st century emerging research, largely resulting from epidemiological population-based data, has shown repeatedly that the question of what exactly constitutes AD, and who might be at risk for it, remains unanswered. These findings tip the scales toward an argument for the inextricable entanglement of dementia with aging, but, even so, an approach grounded in localization theory remains dominant, in large part because pharmaceutical companies support it.

Two researchers have recently called for the condition glossed as Alzheimer disease to be recognized as a "diffuse clinical syndrome," one that reflects the gradual accumulation of "multiple pathologies, arising from multiple interlocking risk factors over the life course."[83] Alzheimer's is not a "yes-no clinical diagnostic category," they argue, nor can one tell when a symptom gradient associated with AD-like symptoms commences, because it is increasingly becoming clear that events that happen in utero and in early life may well be implicated. Richards and Brayne insist that if indeed Alzheimer's is a diffuse clinical syndrome, as they argue, then a therapeutic "silver bullet" is unlikely to be forthcoming, and the focus should instead be on better management of the numerous factors throughout the life cycle linked to increased risk for becoming demented. These authors conclude,

> No straightforward correspondence exists between higher mental function and the burden of lesions in the aging brain. If this shifts the focus away from detailed diagnostic classification made on the basis of assumed clinical-pathological correlation and towards a global pragmatic approach to the needs of patients and carers, and to modifiable lifetime risk factors, then the apparent loss of scientific precision is a gain to clinical practice.[84]

This article highlights inherent, incommensurable tensions among the four principal sets of expert actors in the Alzheimer world: basic science researchers, often with drug company support; academically based clinicians; physicians whose specialties are gerontology or family medicine and general practitioners working in primary care; and those researchers in epidemiology and public health. Basic scientists and academic clinicians (very many of whom work in specialist memory clinics) constitute the heart of the biomedical endeavor that favors localization theory.

The sociologists Cambrosio and Keating characterize contemporary biomedicine as a "bio-clinical" hybrid in which tight working links are established between what is termed in the medical world "bench and bedside." The bodies of patients in tertiary care settings (whether hospitalized or ambulatory) are a source of endless research material, in the form of specimens, imaging data, medical records, and so on, deposited in computerized repositories that may be drawn on in order to both refine extant knowledge and generate new knowledge that will then, in turn, be standardized and made available for use in clinical practice. Clinical trials using human subjects also take place in tertiary care settings, most of which are designed with the purpose of developing new drugs. In effect, patients in such settings become hybrids—patient/research subjects—the moment they enter the hospital. In laboratory investigations, the disease or condition takes priority over the actual condition of human research subjects. But this is not to suggest that the clinical care given to hospitalized tertiary care patients who are also research subjects is wanting in any way, although on occasion this may the case, depending on the location.

John Le Carré's novel *The Constant Gardener* depicts an appalling exploitative research situation that is not complete fiction in certain poor countries where drug company interests are at work.

Aging as a Continuum

In an illuminating article, the sociologist/historian team of Tiago Moreira and Paolo Palladino has written about the role that gerontologists have played in recent years in both the United States and the United Kingdom in bringing about a substantial transformation in how diseases of aging should be managed. This transformation is described as an era moving toward "health promotion and disease prevention in the 21st century."[85] They cite one of the last articles written by Robert Butler together with colleagues, in which it is argued that to persist with the customary anatomical divisions of the body characteristic of 19th- and 20th-century medicine on which medical departments and clinical care are based is inadequate to address the numerous chronic, long-term illnesses of the late 20th and early 21st centuries. Criticism from several sources, including that of the U.K. House of Lords, is summarized by Moreira and Palladino as follows:

> The protracted temporal unfolding of these illnesses is so nearly coterminous with ageing that it unsettles the epistemic pairing of the "normal" and the "pathological" that underpins the clinical perspective on ageing. Furthermore, this pairing assumes that the two states can be situated proximally and intervened upon directly, but this obscures the understanding of the diverse and complex processes involved.[86]

The argument made by the Science and Technology Committee of the House of Lords called for research into the "generic" process of aging.[87] This committee in effect charged that the current organization of biomedicine no doubt serves clinicians and researchers well, but fails with respect to the elderly themselves.[88] It is also the case, since the time in the 1970s when Alzheimer's was fully recognized as a disease, that the U.S. government approach to aging has been similarly grounded in a disease-specific program that, paradoxically, was initially headed up by Robert Butler.

Today, gerontologists, whose numbers continue to be relatively small, especially in North America, together with certain primary care physicians, are arguing for a new approach, one that sets out from a theoretical stance similar to that taken by the evolutionary biologist Tom Kirkwood: "aging is a continuum, affecting all of us all the time." Kirkwood elaborates by pointing out that "there are scientific connections between birth, early years, childhood and adolescence that have major impacts on health and quality of life in middle and old age."[89] His position, in common with several of the medical researchers cited above, is that aging is inextricably entangled with conditions such as dementia that become manifest in later life, and recognition of this entanglement should become the cornerstone in a move toward their prevention.

Nor surprisingly, many basic science researchers such as Dr. Hyman, whose work tracking the formation of plaques in mice was mentioned at the beginning of this chapter, continue to put their energies into unraveling the molecular pathways thought to be uniquely associated with the onset of Alzheimer disease. Such research receives the bulk of available corporate and government funding because if a cure for this condition is to be found, then it will be based on results obtained from this kind of research. Basic science research is also strongly encouraged by advocacy groups, notably the AD societies and individuals working for them, whose fund-raising activities depend to a great extent on promotion of the idea that a cure for this devastating condition is just over the horizon. The tension between localization and entanglement stances has, if anything, become more aggravated than ever in an aging world where no cure for AD is in sight.

In the following chapter I turn to a discussion of repeated efforts to standardize an Alzheimer diagnosis—a task that continues to be extraordinarily difficult to achieve, in turn suggesting that the AD phenomenon is indeed heterogeneous, and simultaneously putting into question claims about the numbers of AD cases.

Chapter 2
Striving to Standardize Alzheimer Disease

Depending on where you set your sights, Alzheimer disease is a scientific puzzle, a medical whodunit, a psychosocial tragedy, a financial disaster or an ethical, legal and political dilemma.

—Zaven Khachaturian, "Plundered Memories"[1]

The historian of medicine Charles Rosenberg writes, "[T]he modern history of diagnosis is inextricably related to disease specificity, to the notion that diseases can and should be thought of as entities existing outside the unique manifestations of illness in particular men and women."[2] Rosenberg adds, "Diagnosis labels, defines, and predicts and, in doing so, helps constitute and legitimate the reality that it discerns."[3] In elaborating on this "specificity revolution," Rosenberg highlights, as have many other researchers writing about changes since the late 19th century, the way in which the medical world has with ever-increasing acceleration developed an array of technologies and "objective" tools designed to reliably replicate and standardize diagnostic procedures.[4] These days it is common practice to have a clinical diagnosis confirmed by laboratory examinations of specimens procured from patients. If the laboratory findings do not support the clinical assessment, then, usually, the clinician's diagnosis is trumped, and clinician and patient must adjust their understanding of what has gone amiss. But entities understood as irrefutable signs of "disease" also come to light for the first time with laboratory testing, usually because they cause no symptoms, and may never do so. Hepatitis C is one obvious example and HIV another. In both these instances the virus can be measurably present in the body, but the individual remains perfectly healthy. Multiple sclerosis is a further example, one that signals diagnostic problems similar to those documented in connection with the dementias. Neuroimaging has shown that the characteristic lesions associated with this condition—multiple sclerosis plaques—may be present in individuals who are entirely unaffected by the disease.[5] Conditions such as these make the task of delineating disease specificity extraordinarily difficult,[6] and it is abundantly evident that simply demonstrating the presence of so-called pathological entities or features in a body is not sufficient, and may actually be very misleading, as the previous chapter made clear. Questions of ontology and epistemology come to the fore rather dramatically in these instances: When is a "disease" not a disease? When is "pathology" normal? Under what conditions does a "natural" entity become "pathological"?

The first section of this chapter briefly reviews the literature that points out inconsistencies in both the clinical and neuropathological diagnosis of AD despite repeated efforts over the years to standardize them. A diagnosis of AD involves the demonstration at autopsy of neuritic amyloid plaques, neurofibrillary tangles, and also cell loss or shrinkage of brain tissue. For more than half a century, notably once the electron microscope was made available, efforts have been made to understand why and how plaques and tangles form in the brain, at times to excess, and what exactly they consist of. For two decades the preeminent model to account for plaque buildup has been the "amyloid cascade hypothesis" that, it is argued, initiates the eventual formation of tangles and other neuropathological changes. In the second part of this chapter I discuss this paradigm, on the foundation of which the majority of efforts to develop drugs to combat AD have been based. The model is currently being questioned by an increasing number of key researchers, but has by no means been overturned, and continues to be a driving force, even as the entire Alzheimer enterprise moves to include prevention as a major goal. The third section of this chapter returns to the question of dual diagnostics—symptomatic and neuropathology—and to current efforts to restandardize the criteria yet one more time in both Europe and North America. In the United States a new clinical diagnosis, made public in the spring of 2011 by the National Institute on Aging (NIA) and the Alzheimer's Association of the United States, is distributed along a time dimension that extends backward through the life span of individuals and has three phases: a presymptomatic phase, an early symptomatic, predementia phase, and a full-blown dementia phase. This most recent round of standardization is explicitly designed to move the management of Alzheimer's to a point in life well before the assumed amyloid cascade would have started to gather momentum.

Standardizing a Diagnosis of Alzheimer Disease

The dementias present a special challenge because no blood, urine, or other laboratory test of human secretions or fluids can, as yet, affirm or refute a clinical diagnosis in the living patient. It has long been agreed that confirmation of an AD diagnosis can be obtained only at autopsy. Hence, verification of accuracy is necessarily after the fact of death. Confirmed autopsy results are of use in creating medical records and for epidemiological, public health, and basic science research, but such neuropathologically confirmed diagnoses have little or no direct relevance for the clinic. However, families often wish to be informed about autopsy reports. Confounding matters, a "shortage" of neuropathologists exists in the United States, Canada, and the United Kingdom, and no doubt elsewhere; relatively few cases of dementia ultimately go to autopsy, and families may have to wait up to three years to obtain verification of the diagnosis of their deceased relative.[7] Furthermore, as one American neuropathologist noted, autopsy costs exceed financial returns to the hospital. Establishing AD specificity in individual cases is, therefore, exceptionally challenging: because autopsies are not routinely carried out and, when they are conducted, the findings do not always provide evidence that is congruent with the initial clinical diagnosis. As we saw in the previous chapter, what exactly constitutes an "AD brain" remains in question, especially because research makes it increasingly evident the extent to which normal and pathological aging are entangled.

Despite these difficulties, since the 1970s, AD has steadily become an uncontestable reality, in effect, a social actor in its own right[8]—an entity in association with which a specific language, images, estimates of incidence and prevalence, and concerns about a looming epidemic are mobilized. The concept of Alzheimer's is an artifact of value, one which has taken on a life of its own among medical practitioners, the public, and advocacy groups alike. And yet, every medical specialist with whom I have talked readily agrees that accurately estimating the number of Alzheimer cases leaves a great deal to be desired. Some believe that a dramatic underdiagnosis is probably the case, particularly when one keeps in mind the global situation,[9] while others, focusing on findings about the heterogeneity associated with AD, believe that what passes for "pure AD" may well be overdiagnosed, and others note that the presence of rarer forms of dementia, such as frontotemporal dementia, can rather easily be overlooked.

Given that no decisively effective medication exists for AD, there is little need for primary care physicians to be overly concerned at present about the accuracy of their diagnoses. Moreover, certain clinicians have told me that it is well known that some doctors in Canada may diagnose a patient as having AD based on neuropsychological testing scores in which the figures are "fudged" to make them fall under the cutoff point below which individuals are eligible for cheaper medication as part of the socialized health care system. On occasion families may actually push doctors to confirm an AD diagnosis even when the physician continues to have doubts. Likely Canada is not alone in this practice. Evidently, as a result of these confounding factors, national and global estimates of AD cases cannot possibly provide an accurate estimate of the number of individuals at increased risk because statistically reliable population databases do not exist. Those few communities that have been consistently researched over the years using well-designed epidemiological methods are an exception. In these instances the resultant data are much more reliable,[10] but strictly speaking such findings apply only to the studied communities, making extrapolations from them about the nationwide number of elderly people at risk for AD inappropriate, unless it is assumed that age and sex are the only significant variables implicated in AD incidence.

A brief purview of the massive literature on AD diagnoses over the past two and a half decades highlights two things. First, there have been repeated efforts over the years to refine the standardization of the diagnosis both for use in the clinic and for epidemiological purposes. And, second, in practice, the criteria for making a diagnosis, whether clinical or based on neuropathology, are not entirely uniform, despite the effort put into standardization. This situation makes very clear just how challenging is the pinning down of the AD phenomenon, with enormous consequences for estimations of AD prevalence and incidence.[11]

At a research workshop sponsored in part by the NIA in 1984, the organizer, neurologist Dr. Zaven Khachaturian, made the following comment:

> We are still uncertain whether AD is a specific, discrete, *qualitative* disorder such as an infectious process, endogenous or exogenous toxic disorder, or biochemical deficiency, or whether it is a *quantitative* disorder, in which an exaggeration and acceleration of the normal aging processes occur and dementia appears when neural reserves are exhausted and compensatory mechanisms fail.[12]

Khachaturian noted that incorrect diagnosis of AD is "quite common" and may be as high as 30% in the general population. The goal of workshop participants, then, was to better define the criteria for AD, which would, it was assumed, improve diagnostic validity. Perhaps not surprisingly, there was no discussion about the ontological status of AD—the material "facticity" of AD was assumed to be unassailable, if only the diagnosis could be improved.

Later that same year, what has since become the most commonly used criteria for the clinical diagnosis of Alzheimer's were put out jointly by the National Institute of Neurological and Communicative Disorders and Stroke (NINCDS) and the Alzheimer's Disease and Related Disorders Association (ADRDA; abbreviated shortly thereafter simply to the Alzheimer's Association), both in the United States. These criteria, known as NINCDS-ADRDA criteria, were extensively revised between 2007 and 2011. But there are several other widely used criteria that have been available for many years, among the best known in the English-speaking world being the American Psychiatric Association's *Diagnostic and Statistical Manual of Mental Disorders (DSM)*, which has also undergone repeated revisions; the World Health Organization's International Classification of Diseases (ICD), which too has been revised; and the Cambridge Examination for Mental Disorders of the Elderly (CAMDEX). These criteria all require that evidence of cognitive impairment indicating dementia be established by neuropsychological testing. The most commonly used screening test for cognitive impairment is the 10-minute Mini-Mental State Examination (MMSE),[13] supplemented by further more comprehensive tests of cognitive function requiring up to several hours to administer, when a clinician seeks to confirm the diagnosis. Repeat testing over the course of months or even years then follows, during which time possible comorbidities are ruled out and a steady, irreversible progression of cognitive decline is carefully documented, following which a clinical diagnosis of probable Alzheimer's is made.

The usual assumption has been that these various diagnostic classifications are essentially interchangeable. But one project carried out in 1997 involved an examination of 1879 men and women aged 65 years or older who were originally among more than 10,000 individuals enrolled in the Canadian Study of Health and Aging, a project designed to ascertain the prevalence of all types of dementia across Canada in both community and institutional settings.[14] The results showed that the criteria used for diagnosis differed by a factor of 10 in the number of subjects classified as having dementia; this depended on which among six of the standardized diagnostic criteria were used.

Additional problems have been brought to light as a result of systematic research; prominent among these are the very numerous cases of so-called mixed dementias, atypical dementias, and "cognitive impairment no dementia," all of which conditions have been highlighted as confounding factors in diagnostic standardization.[15] Current neuropathological findings indicate strongly that 30% or more of dementias may be a mixture of vascular dementia and Alzheimer disease, but in the clinic, up to a decade ago, it was the case that mixed dementias were not usually recognized. Interrater reliability has also been shown to be a problem—neuropsychologists and neurologists do not always agree on diagnoses.[16]

A clinical diagnosis of AD is frequently described as one of exclusion, that is, in clinical practice AD has been and often continues to be the diagnosis that remains after all other known forms of dementia and many other conditions, including hypothyroidism, stroke, vitamin B12 deficiency, folic acid deficiency, head trauma, Parkinson's disease, Huntington disease, and several further conditions, are first ruled out.[17] A diagnosis of AD has often been described to me as a "wastebasket" category by general practitioners. When I met with David Bennett, a Chicago-based neurologist, and inquired what he thought about this idiom, he replied,

> What I would say is that historically, Alzheimer disease was truly a wastebasket. So if you go back a few decades, Alzheimer disease in the textbooks was a diagnosis of exclusion. You would take somebody with dementia, and you would say, they don't have this, they don't have that, they don't have the other thing, and therefore they must have Alzheimer disease. But, significantly, the wastebasket didn't have all the Alzheimer disease in it because—well, strokes don't prevent you from getting Alzheimer disease. So if you throw out your strokes, okay, very many of those people have Alzheimer disease, but are missed. And if B12 deficiency and hyperthyroidism are thrown out, you often throw out AD as well.

So basically these researchers were left with their dementia cases after they'd thrown out all the comorbidities. And of course they didn't throw out these conditions in their controls who weren't demented. And then they compared these cases and controls, violating every epidemiologic principle. So this wastebasket was a mess. And so what's happened over the years with the recognition of mixed dementias is the change from a diagnosis of exclusion to a diagnosis of inclusion.

When I asked exactly what he meant by this, Bennett explained,

Well, today when we train people, they learn that Alzheimer disease coexists with other common diseases. And these diseases probably actually contribute to dementia and make it *more* likely. We train people to ask themselves, does the person have the characteristic temporal and spatial pattern of a cognitive deficit? And so if you have somebody who moves from a progressive amnesia [memory loss], to profound amnesia, followed by problems of orientation and so on, I mean that's a pattern that's just so unmistakable, that 90 plus percent—there are some things that can mimic it, but there's a pattern that 90 percent plus most of the time—these people have Alzheimer disease. We see this pattern just from cognitive testing. In our research here we keep ourselves blinded at each year of follow-up and, even so, 90 plus percent of them prove to have the AD pathology when they come to autopsy. So this approach to diagnosis is not a wastebasket.

But a problem remains, and the reason why I agree with you, is that along with the Alzheimer's pathology comes all kinds of other things. So, for example, infarcts [local tissue death from stroke], Lewy bodies [which characterize a rare form of dementia known as Lewy body dementia], and other things, are contributing to whether someone tips over the edge. Now we've moved beyond the wastebasket and we've got an Alzheimer's bucket and in that Alzheimer's bucket is mostly Alzheimer's—and we're pretty good at that—but in that bucket are all kinds of other crap that also tags along. And the difficulty is that, even if you can separate these things out, the value of doing this from a research point of view is dubious at best, and probably introduces bias. So from a research point of view, you don't want to be studying "pure" Alzheimer disease because you're going to find some really weird things. But people still do this.

Some of these people are friends of mine and they take their brain tissue, for example, and they say, I'm going to throw out everyone who had infarcts and Lewy bodies and blah blah blah, and I'm just going to study my pure Alzheimer disease. And it's kind of like, it's not the way it works in the world. So they find weird things. They find sometimes the opposite of what we find when we keep everyone in, and it's probably because of the bias that they've introduced by throwing stuff out. So you can find diabetes cases who have infarcts, but if you throw out all your infarcts and you just try to understand the relationship of diabetes to AD pathology, it's not clear what you're studying anymore. (April 2010)

Studies have shown that in Alzheimer disease research centers and in memory clinics—frequently, but not exclusively attached to teaching hospitals—AD diagnoses are confirmed at autopsy in over 90% of the cases, as David Bennett points out is the situation for his unit.[18] But, clearly, for research purposes, attempting to isolate "pure" AD introduces difficulties because a "normal" aging brain is full of "crap" (a word used surprisingly often by AD scientists). But, so-called brain "crap" or "junk" is a reality of aging, particularly in old age, and should be dealt with systematically in research protocols rather than ignored. Cases of "pure" AD neuropathology are, in effect, researcher-created artifacts that can introduce serious difficulties into studies designed to investigate risk for Alzheimer's, as Bennett's comments make clear.

A further major difficulty is that by far the majority of patients with dementia are never seen at memory clinics or similar specialist research centers. One Montréal family practitioner commented,

Do I send people to memory clinics? No, I don't initiate referrals to memory clinics. I have only done it on one or two occasions where it was an atypical presentation that I wasn't clear about. I don't initiate, but I do respond to families asking about it. And when families ask, what I reply is that there's an eight month to a year waiting time; that the clinics are—for the most part—funded and designed to gather research data; that the treatment modalities that the patient will be offered would likely be no different from what I would do here, and that the choice of medication might be a little bit better informed at a memory clinic but that it probably wouldn't make any difference. And then I say, "you make your choice." I give them all that I think they need to know, and in my experience, the vast majority doesn't go to the memory clinic. And my patients are mostly solidly middle class.

The ones who I've seen go and who hang on do it because they believe in research. They do it because it gives them something to do that makes them feel a bit better. They tend to have slightly higher IQs and therefore intellectually, they're interested in what goes on in the memory clinic, and it gives them a sense of control that they may feel would be lacking if they were only followed by a family doctor. And often it's the spouse who encourages this rather than the patient himself. I take a lot of time listening to the family as whole; talking about daily life, following up on them carefully over the years, and it seems most patients really appreciate this, even if it's clear you can't cure the problem. (May 2011)

This practitioner and other primary care practitioners whom I interviewed (12 in all) are concerned that once patients go to specialty clinics they essentially become research subjects, and that their quality of life and that of their families may not be attended to adequately. A good number of these practitioners have little faith in the medications currently on the market and administer them sparingly. Should a highly effective drug become available, this situation would no doubt change—under these circumstances these clinicians would ensure that a finely tuned differential diagnosis be carried out.

Another Montréal-based family practitioner, Robert Diez d'Aux, who recently returned to the city after working for nearly 25 years in a small town in Tennessee, said,

So far all we've really got to offer are platitudes, more or less. You know: get lots of exercise, good nutrition, do crosswords (which I do myself!), and watch your weight. My position is, when the family comes to see me, if they don't identify a problem and they are managing well at home, then I don't want to upset the apple cart. So, I certainly don't rush into a diagnosis of AD, because, even if the patient doesn't do too well on the animal fluency test, for example [naming as many four-legged animals as one can in one minute], or on the Mini-Mental, I don't really know if they have some dementia until I've followed them for months. I very, very rarely sent families to the Memory Clinic when I was in Tennessee, because they would have had to drive for two and a half hours to Vanderbilt in Nashville. I thought it was better to support the family rather than trigger a consultation cascade. I can live with uncertainty, especially if the family isn't pushing for a diagnosis of the "A" word. It makes little difference to patient well-being. (May 2011)

As part of his practice, Tom Dening, formerly a Cambridge University–based old-age psychiatrist, routinely makes house calls in the villages and communities that surround Cambridge. He agreed to take me along with him on two half days when he was visiting patients and their families in their homes. Dening states that the majority of people referred to his clinic for memory impairment are older than 80 years of age. "Some of them are quite a bit more than that," he adds, "it just so happens that I'm going to see someone who is 95 this morning." Dening is emphatic that when an individual is very old it is impossible to be precise when making a diagnosis because, inevitably, one is confronted with a mixture of clinical features that may or may not contribute to dementia: "So many factors are in play that my most common diagnosis for people age 80 and over is one of mixed dementia." Moreover, he adds,

> The dementia concept is quite rocky with old people because there's a psycho-social component that must be taken into consideration— one has to ask, is this person "normal" for their age, or not? Even normal Olympic athlete types who are 95 years old are not going to score 30 on an MMSE. It's important to recognize that there may be very, very few 95-year-olds who are *not* MCI [mild cognitive impairment]. What *is* a normal baseline at this age? We really have little idea, and the situation is that the condition is so heterogeneous it's impossible to have a gold standard for [diagnostic] sensitivity and specificity—at least among the numerous older patients one sees these days. (September 2010)

A very busy general practitioner working in a relatively economically deprived part of Montréal emphasized that in his opinion among his patients aged 75 and older he is usually dealing with individuals who have what he describes as a "syndrome." Even with a socialized medical system in which no fees for service are charged, David Dunn finds that his patients come to see him "very late in the game," whatever their problem (unless it is an emergency), at which time it is virtually impossible to help them change lifelong habits that bring about poor health. Such patients may well have dementia, he stated, but in addition they have diabetes, high blood pressure, and other problems. Among this mixed patient population, including recent immigrants and some refugees from Eastern Europe, Russia, China, Senegal, the DRC, and many other locations, few of them, says Dunn, have good social supports, and a few are in hiding from authorities or members of their own "ethnic group." When they come to see a doctor their purpose is very often to obtain medication and little else.

An alert practitioner might well suspect early signs of dementia among some of these older patients quite frequently, as does Dunn, but when he attempts to administer a modified version of the MMSE he finds that he is confronted with language barriers that even his trilingualism cannot surmount and, in addition, often suspicion and resentment on the part of some of his patients at being made to do a "mental test." Dunn notes that a good number of his patients do not return for follow-up visits, making it impossible to diagnose anything other than "possible dementia"—a diagnosis that may well not be correct, especially if the patient is depressed. Under the circumstances, Dunn is most likely to write a diagnosis of diabetes or another common problem that he can verify on the clinical chart, and leave it at that.

Trained in epidemiology in addition to general medicine, David Dunn agreed with me without hesitation that official statistics on the incidence and prevalence of AD are without doubt highly inaccurate. But when he makes a differential diagnosis in his practice, his intention is not to discard everything he finds until he exposes "pure AD." In primary care, there can be no waste-basket for leftover "real" cases, Dunn argues, because comprehensive treatment of patients and families is the primary objective, and clinicians note all the significant observations they have made in their examination, otherwise they would be failing in their care of patients. Dunn sees some positive signs of change with the formation of group practices made up physicians, nurses, and other health care practitioners, all of whom are specialists in geriatrics. Primary care practitioners can refer their patients to these practices specifically for assessment of dementia and other common disorders among the elderly. But, inevitably, not all patients are willing to go.

Neuropathology as the Gold Standard

As we have seen, clinical diagnoses of AD can be confirmed only at autopsy—neuropathological findings have been the gold standard for over a century. But here again, problems associated with accuracy and replication abound. Not only are autopsies infrequently carried out, inconsistencies have been shown in the findings, notably in mild to moderate cases, and especially for older subjects. One study in the early 1990s showed that quantitative estimates of the number of plaques and tangles in the same fixed slices of cortical tissue sent to 12 different neuropathological laboratories varied by a factor of 10.[19] Inaccuracies have also been demonstrated where autopsies show that mixed dementias are involved.[20]

The British neuropsychologist C. J. Gilleard summed up the matter in 2000 as follows:

> The clinical diagnosis of Alzheimer's assumes a distinct neuropathology to "confirm the diagnosis." But it is apparent that the rather stylized "types of dementia" beloved by textbook writers are expressed more clearly in text than in tissue.[21]

Gilleard charges, "[T]here is confusion over what constitutes the common disease process termed Alzheimer's," and he goes on to argue that a degree of ambiguity and arbitrariness has been tolerated in the field.[22] Specialists working in memory clinics are liable to insist that in recent years this kind of uncertainty has been eliminated, but epidemiologists in particular challenge this position, claiming that the complexity of the situation continues to be underestimated.

John Morris, one of the doyens of Alzheimer researchers in the United States, in a carefully crafted review article published in 1995, cautioned,

> The fundamental relationship . . . of plaques and tangles to AD phenotype is only incompletely understood. The presence of these lesions in the neocortex remains the cardinal feature for both the qualitative and quantitative assessment of AD, but neither lesion is pathgnomonic [that is, a conclusive diagnosis cannot be made from such findings].[23]

And John Breitner, an epidemiologist and psychiatrist, has pointed out how the *DSM* (both *DSM-III* and *DSM-IV*) and the ICD use criteria for establishing dementia "cases" that conflate clinical signs with what are assumed to be signs of irreversible brain pathology. In other words, in Breitner's estimation, criteria are used that are "at once syndromic and neuropathological." He continues, "It is an important goal in dementia research, as in all of medicine, that we characterize the underlying pathological or etiologic entity responsible for the clinical picture."[24] Before this can be done, he argues, a reliable, replicable understanding of what exactly constitutes the dementia syndrome is needed. A 2010 draft of *DSM-5* proposed that the term "dementia" of the Alzheimer's type should be dropped entirely, in large part due to its stigmatizing qualities, and be replaced by the term "major cognitive disorder"—a euphemistic phrase, now formally incorporated into *DSM-5*, that will do little to clarify matters.[25]

The research cited in the previous chapter by Carol Brayne and colleagues suggests that in the case of AD-type dementia we are not dealing with one syndrome, but with at least two and possibly several types, each of which is closely associated with aging, an unavoidable process that Brayne insists is "the elephant in the room." It is patently clear that aging completely dwarfs all other variables associated with risk for AD and dementia (aside from rare, early-onset AD).[26] Furthermore, as more and more research findings emerge, including a good number that demonstrate the remarkable plasticity of the human brain, the difficulty of characterizing even a "diffuse clinical syndrome" on the basis of a high degree of concordance between standardized clinical and neuropathological diagnoses may not, ultimately, be possible. Current research findings repeatedly destabilize the situation and challenge dogged assumptions anchored in a lingering belief about localization theory that goes unquestioned by many researchers. In recent years, for example, an abnormal form of a protein known as TDP43 associated with ALS (Lou Gehrig's disease) and also with the relatively rare form of dementia known as frontotemporal dementia, from which patients with Lou Gehrig's disease may also suffer, has been found in up to 20% of AD patients.[27] It has been suggested in this case that AD diagnoses may be "masking" frontotemporal dementia.

In summary, in specialty clinics, an assumption is often made that it is possible to diagnose "pure" AD in the clinic with a high degree of specificity and sensitivity. This diagnosis is based on observed behavioral changes associated with a decline of cognitive capacities in individuals who are monitored over months and years, and who continue to decline steadily until a rather dramatic downturn frequently takes place in the final months before death. As Zaven Khachaturian puts it,

> The disease quietly loots the brain, nerve cell by nerve cell, a burglar returning to the same house each night. Typically the first symptom to appear is forgetfulness. Then comes more severe memory loss, followed by confusion, garbled speech and movements, hallucinations, personality changes and moods that can swing from anger to anxiety to depression. As the brain loses mass the rest of the body gradually shuts down.[28]

No one who has cared for a relative afflicted by AD could take issue with this description, even if some might be uncomfortable with the use of dramatic metaphors in scientific papers. Khachaturian admits that the speed of this deterioration can differ greatly among patients, and that death can take place between 2 and 20 years after the first symptoms appear, suggesting that molecular complexity is at work that varies greatly from one patient to another. And Khachaturian also acknowledges that mixed dementia is very prevalent. Epidemiological and basic science findings both add support to an argument that a diagnosis of "pure" AD is an artifact of attempts at standardization. Recently Carol Brayne has reiterated the position she has taken for many years, namely that the dominant assumption about neuropathology as the cause of AD had resulted in "a reification of particular diagnostic entities without the necessary empirical evidence from relevant populations."[29] Should such robust evidence ever be amassed, even then, it may ultimately prove to be impossible to consistently link clinical diagnoses of AD with standardized postmortem neuropathological findings due to the very complexity, plasticity, and adaptability evident in aging brains. Such an argument does not, of course, diminish the reality of dementia.

A well-known Cleveland neurologist, Peter Whitehouse, with 30 years of specialized clinical practice and neurological research under his belt, feeling increasingly ill at ease with the situation in the AD world, published a book in 2008 titled *The Myth of Alzheimer's: What You Aren't Being Told about Today's Most Dreaded Diagnosis*.[30] When I talked with him, Whitehouse made it clear that he continues to diagnose his patients using what he describes as "an American Academy of Neurology standard approach." He quickly notes that he does not use tests such as spinal taps (as is the case for many neurologists (see chapter 4), but he does use CT scans, and occasionally an MRI (magnetic resonance image). He adds,

> It sounds a little strange, but most people come to me—about 70%
> I would say—having seen somebody else and certainly having read

something about memory problems, so I use the first part of the examination to kind of get a sense of what they think about the issues. I usually spend the time with the patient and caregiver together, offering them an opportunity for separate time if they wish it. I spend a lot of time talking about language. I tell the patient and family that we still don't understand very much about dementia. I believe that the most accurate statement you can make to an older person coming to a clinic with dementia is that they have a "mixed dementia." I also use the word "aging" a lot, and "brain aging" and talk about multiple factors that affect brain aging. It's trivial just to say that AD is heterogeneous—no disease is ever exactly the same in two bodies; what you really have to ask is what is the relationship of dementia to aging in the particular person in front of you.

When people ask if they might have Alzheimer's, I reply that we don't really know how Alzheimer's can be separated out from aging. So I try to undermine the concept—to get them away from thinking that they are beginning the "funeral that never ends," or that they've got a progressively degenerative disease. There are things one can do, as I write in my book, to make aging much more "successful" even when someone in the family has dementia. I think we need to focus much more on brain fitness.

Of course, the national Alzheimer's Association criticized my book, particularly because of its title. In the public forum they try to scare people to raise money, and my book is a threat. I think this is particularly so in America, and the links with drug companies push things in this direction. The amount of money involved is a powerful incentive to keep advocating for an Alzheimer cure and not to look around for new ways of approaching this massive problem. (June 2010)

I have asked several of Peter Whitehouse's colleagues what they think of his book. Predictably the answers have been varied, ranging from "Peter has lost his marbles" to "I haven't time to look at that kind of thing." But the most common answer has been that a new vision is needed in the dementia world, and that a good shake-up by Peter Whitehouse is not at all a bad thing—"[H]e's pushing the boundaries, challenging people and making them think," replied one eminent neurologist.

When I asked William Thies, until recently the medical and scientific officer of the Alzheimer's Association, what he thought of *The Myth of Alzheimer's*, he responded,

Let me say at once that I know what Peter has written and I disagree with it. It's perfectly clear to me that there are people who have Alzheimer disease and there are people who don't. And there's a pathological process involved. Now, is it a single monolithic disease? We'd be ignoring some of the most famous lessons of history if we thought it was such, and I think there's interesting work being done now that suggests that many of the people that show dementia actually have mixed dementia and what you probably have is a continuum. Just to put two points on the continuum (there will be more) but just to put two, it would be amyloid pathology and vascular disease. And you are going to have some people who are at one end of the continuum and others at the other end, but most will be in the middle. But we don't really know what the distribution curve looks like yet. So I don't have any problem with the idea that if you look across the whole population of people who are currently labeled as Alzheimer disease, that there will be some sub groups in there and, if I had to make a guess, I'd say that therapy for AD will end up looking like therapy for hypertension . . . it will be tailored to the individual.

In response to my question about AD being a wastebasket category, Thies said,

It's no more a diagnosis of exclusion than many other transient diagnoses are. The first day one comes up with a diagnosis for anything it may prove to be wrong, but I think we're getting better and better at identifying AD and the pathological studies really do bear that out, as long as good clinicians are involved, . . . further down the road I wouldn't be at all surprised if we found it best to create subcategories. Ultimately I think therapies will be the best tools for allowing us to make distinctions into subsets of Alzheimer disease. In the end, one only gets confirmation from therapeutic outcomes. (April 2010)

It is understandable that William Thies, as scientific officer of the Alzheimer's Association, whose working life was closely associated with raising money to defeat Alzheimer's disease, would be very cautious about entertaining the idea that AD might not be "real." But his willingness to accept the likelihood of emerging subcategories on the basis of drug development is not at odds with his position in the Association. Even if the Alzheimer category fragments, the overarching concept will continue to be used to promote fund-raising, political activities, and policy making.

To my inquiry into what kind of research the Alzheimer's Association generally funds, Thies replied, "a fair amount of amyloid science—well, that's kind of the state of the science, you know," but he stressed that efforts are made to fund across the spectrum of research applications, and to give young researchers with innovative ideas a chance to get established. Thies emphasized that rigorous Alzheimer research has a very short history of only 30 or 35 years to date, and he insists that it remains "an open question" as to whether "intervention into amyloid" is likely to be effective. He then added a further proviso: "and another thing that complicates matters is that our cognitive tools are so dull that we can't really tell how cognitively intact people are," adding greatly to diagnostic problems.

Summarizing, despite inherent problems with standardization of both clinical and neuropathological AD diagnoses, and explicit doubts expressed by certain researchers about the very concept of AD, the basic science paradigm that has held sway for over 20 years in the Alzheimer's research world continues to be dominant—it is a model in which amyloid deposition in the form of plaques is for all intents and purposes regarded as the master key.

The Amyloid Mafia and the Prevailing Paradigm of Alzheimer Disease

Arguably the most influential article associated with Alzheimer research appeared in 1992 in *Science*, one author of which was the well-known British neurogeneticist John Hardy. Hardy is today the most cited Alzheimer researcher in the United Kingdom and has on occasion appeared in a Hawaiian shirt and other unconventional dress to deliver plenary addresses to audiences sometimes of well over 1,000 people. The article, coauthored with Gerald Higgins, a neurobiologist formerly associated with the NIA, is subtitled "The Amyloid Cascade Hypothesis,"[31] and the theory it postulated quickly became the dominant paradigm for understanding Alzheimer disease causation at the molecular level. The authors emphasized at the outset, "Alzheimer's is characterized by various pathological markers in the brain—large numbers of plaques surrounded by neurons containing neurofibrillary tangles, vascular damage from extensive plaque deposition, and neuronal cell loss."[32] The second paragraph then set out the cascade hypothesis: "deposition of amyloid β protein (AβP), the main component of the plaques, is the causative agent of Alzheimer's pathology and . . . the neurofibrillary tangles, cell loss, vascular damage and dementia follow as a direct result of this deposition."[33]

The cascade hypothesis states that what is known as the amyloid precursor protein (AβPP), when "cleaved" by a specific enzyme, y-secretase, itself composed of four proteins, can result in an excess amount of toxic Aβ protein, resulting in a cascade of negative effects in the brain. The authors generated this hypothesis on the basis of what were at the time newly discovered findings in connection with the genetics of early-onset Alzheimer disease in which mutations of the APP gene are implicated (see chapter 5). They also noted that people with Down syndrome, once adults, are highly susceptible to Alzheimer disease because the region of chromosome 21 that contains the APP gene is involved in that condition. Hardy and Higgins acknowledged that several upstream pathways no doubt exist that can set off processes contributing to AD, but their position when the hypothesis was first formulated was that the final common AD pathway is precipitated by the toxic effects of a buildup of Aβ protein, and it is this buildup that causes the disease.

Since the publication of this article, a vast amount of research has been directed toward solving the question of what exactly leads to excess precipitation of the Aβ protein, and by far the majority of drug trials in connection with Alzheimer's, costing billions of dollars, have been designed to eliminate excess plaques in the brains of experimental mice and humans—with no real success to date.[34]

In 2000, Mark Smith, a Cleveland-based neuropathologist, and three of his colleagues published a letter in the prestigious medical journal *The Lancet*, titled "Amyloid-β Junkies." The letter started out, "Evidence indicates that, in the very near future, by either snorting or injecting antibodies to amyloid-β, it will be possible to remove amyloid-β-laden senile plaques from the brains of individuals with Alzheimer disease." These letter writers were referring to a 1999 article claiming that by means of amyloid-beta immunization Alzheimer-disease-like pathology had been attenuated in mice. Smith and colleagues ran through all the reasons why they were certain that such an immunization might do more harm than good in humans, including the fact that animals with massive doses of amyloid-β more extensive than that found in human AD do not replicate the full range of AD pathology, nor the dementia associated with Alzheimer disease. They concluded that attempting to remove amyloid-β is playing a dangerous game; in their opinion, amyloid may well play a role in "defense of the aged brain."[35]

In contrast to Smith's position, in 2002, John Hardy together with Dennis Selkoe, a Harvard-based neuronal cell biologist, robustly defended the amyloid cascade hypothesis in an article cited to date well over 5,000 times. They argued that basic science findings in the 1980s that had provided a genetic framework that enabled the postulation of the hypothesis "marked the beginning of the modern era of research" in connection with AD.[36] However, it was also noted in this article that research had shown that the number of amyloid plaques in the brain does not correlate well with "the degree of cognitive impairment." It was also pointed out that cell culture systems on which most of the research has been carried out may well not adequately reflect "the complexity of Aβ economy in the human brain." Two summary points were made by these authors: first, although gaps in knowledge are evident, none of the currently perceived weaknesses of the amyloid hypothesis provide good enough reasons to abandon it; and, second, drug development based on the cascade hypothesis to date has not slowed cognitive decline in patients. Even so, it was argued, given that no other alternative hypothesis had been proposed, the amyloid cascade hypothesis should not be abandoned. Numerous research projects are designed to unravel some of the persistent problems associated with amyloid analyses, notably by further molecularizing Aβ into specific subtypes and researching their independent effects on cell death in the brain.[37]

Scientists whose focus of investigation is tangle formation have long felt themselves overshadowed by the so-called amyloid mafia, but on the whole such in-house debates have remained reasonably friendly. However, as Mudher and Lovestone, neuroscientists at the Institute of Psychiatry in London, have pointed out, these debates "assumed religious overtones" when those whose investigations concentrate on amyloid activity were dubbed "baptists" (β-amyloid protein) and those whose efforts are devoted to an explication of the tangles became known as "tauists," because excess production of a hyper-phosphorylated form of the protein tau is associated with tangle buildup.[38] Certain tauists have argued that researchers devoted to the amyloid cascade hypothesis have been inclined to think of neurofibrillary tangles as an "epiphenomenon," and take issue with this position.[39] They point out that it is clear that tau-related pathology alone can cause neurodegeneration in the brain; that plaque formation by itself does not induce tau pathology; and that in transgenic mouse models the cascade hypothesis does not apply because plaques alone are produced in mice, and the brains of these animals exhibit no neuronal loss, no tau pathology, and no tangles.[40] It is also of note that it is human APP that is inserted into the mice, and that mouse APP is not toxic to mice.

An article in *Science* titled "Tauists and Baptists United—Well Almost!" reminds readers that β-amyloid plaques have never been proven to cause neurodegeneration—rather, an association has simply been established.[41] However, research findings may have "cooled down" the argument because two research groups working with new models of transgenic mice have shown what had not previously be demonstrated—that plaque and tangle pathways are themselves "entangled." This suggests that too much emphasis may indeed have been given to amyloid alone.[42]

Criticisms about the amyloid mafia have been in print since the early part of this century, as *The Lancet* letter makes clear, but have gone largely unheeded until very recently. A set of brief articles published in 2006 in *Nature Medicine* are the product of a call by the journal editors to 34 leading scientists to set out what they believe are the most important research findings in AD research since 2003. The editors summarize, "Notably, of the papers our experts chose as the most important, 95% are related to the processing or pathogenesis of amyloid-β." They continue, "Does this mean that amyloid-β is accepted by all as the primary causative factor in Alzheimer disease pathogenesis . . . ? Perhaps, but there are some indications from experts that the time may be ripe for the influx of new ideas into Alzheimer disease research."[43]

The lead article in this collection, written by the science writer Apoorva Mandavilli, is titled, "The Amyloid Code." This article initially airs the idea of an amyloid conspiracy, but quickly drops this suggestion to embrace a slightly more jocular tone by embarking on a discussion about the "Church of the Holy Amyloid," an expression coined by Mark Smith. Mandavilli points out that for well over a decade every aspect of Alzheimer research has been dominated by the amyloid hypothesis, ranging from the search for a cure to animal modeling, most funded grants, and the most-cited articles.[44]

Mandavilli notes that many of the scientists to whom she talked who are associated with the nature/genetics collection are less passionate in their criticisms of amyloid than is Smith, but even so they repeatedly made comments about the dangers of "putting all of your eggs in one basket." The interviewed scientists warn that with such a complex disease "it is foolhardy, perhaps even dangerous, to focus exclusively on one theory." One eminent researcher concludes, somewhat paradoxically, that evidence for the amyloid hypothesis is overwhelming but, even so, "[t]he whole field is governed by an old boy's network that is not very positive. It needs new blood, new movement, new ideas."[45] The work of Mark Smith and his colleagues highlights new directions that, in fact, are not so new,[46] but are currently being given more coverage than previously. Smith and colleagues insist that investigators should move upstream and consider more generalized bodily changes such as oxidative stress, inflammation, mitochondrial changes that are part of aging, and cell cycle dysfunction—all factors that may set off a chain of events leading eventually to amyloid and tau deposition.

Indirectly in support of Mark Smith's position, a Finnish researcher, C.P.J. Maury, published an article in 2009 in the *Journal of Internal Medicine* in which he argued that although amyloid has usually been thought of as a "pathological structure," growing evidence indicates that amyloid may well contribute to "normal physiology." Maury points out that β-amyloid is "highly conserved," meaning that it evolved long ago, and is present in numerous organisms ranging from bacteria to mammals and spans all aspects of cellular life. For proteins such as β-amyloid to be biologically active they must be appropriately folded into a three-dimensional structure, and in vivo are subject to strict "quality control" for this purpose. Maury goes on to describe in detail the diverse and essential positive functions of amyloid deposition in organisms of all kinds, including the preservation of homeostasis in mammalian cells. While acknowledging that amyloid may well be a sign of disease, his argument, like that of Smith and others, is that without doubt it also has indispensible positive functions.[47]

A 2010 article by Smith and coauthors, published in the *Journal of Alzheimer's Disease* and cited as the best article of that year, comments, "[T]he continued targeting of end stage lesions in the face of repeated failure, or worse, is a losing proposition."[48] When I talked with Mark Smith shortly before his death at age 45 as the result of a shocking hit-and-run car accident, he suggested that amyloid should be understood as "scar tissue" and that it could be best thought of positively—as a sign of healing. Plaques are not the "cause" of AD, Smith insisted; rather, they are a sign of the body fending off troubles that commence further upstream.

Smith understood AD as being on a continuum with normal aging, and under certain circumstances plaques and tangles become irrefutable evidence

healthy   advanced

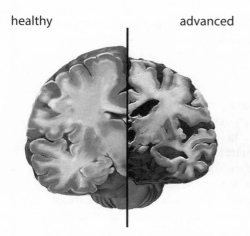

In the Alzheimer's brain:

The **cortex shrivels up**, damaging areas involved in thinking, planning and remembering.

Shrinkage is especially severe in the **hippocampus**, an area of the cortex that plays a key role in the formation of new memories.

Ventricles (fluid-filled spaces within the brain) grow larger.

Figure 2.1.

Cell loss in a brain diagnosed with Alzheimer's disease. Reproduced with the permission of the Alzheimer's Association, from Inside the Brain: An Inter-active Tour, Alzheimer's Association, http://www.alz.org/alzheimers _disease_4719.asp. This tour is available in 14 different languages.

that a threshold has been crossed into pathology. However, he insisted, given that plaques and tangles are not limited to AD alone, they simply cannot be re-garded as the defining features of AD. Smith was in no doubt that the neuropa-thology he observed routinely at autopsies is evidence of age-related disease, but, he emphasized, as do other neuroscientists, that the number of neurons present and signs of cell shrinkage (figure 2.1) are of more significance than are the mere presence of plaques and tangles. His estimation as an experi-enced neuropathologist was that "educated people" can lose up to 50% of their brain tissue and continue to function reasonably adequately. Almost as an af-terthought, Smith added, "[I]n older people a 'normal' brain looks very like an AD brain at autopsy," and in common with perhaps the majority of researchers today, Smith noted that "mixed dementias" are most frequently found at au-topsies of older individuals.

Rudolph Tanzi, a neurogeneticist, holds an endowed chair at Harvard and is director of the Genetics and Aging Unit associated with the Massachusetts General Hospital. Tanzi is billed on the Internet as a "rock star of science," and is one of the best known and most outspoken Alzheimer specialists of the day. His position on the role of amyloid is not unlike that of Mark Smith, and when I met him in early 2010 in his research unit he had this to say:

> [T]o this day, plaques don't confirm dementia. What *is* replicated over and over again is that if you grind up part of the brain at autopsy and measure the amyloid-β load—that always correlates with dementia.

But plaques that a neuropathologist can sit there and count using a microscope and extrapolate from—that doesn't correlate. Why? Because the plaque's not the problem!

I think of plaques as the end-game. I call them brain pearls because, just like an oyster, when you put a piece of sand in, it makes a pearl to entrap the sand to stop the pathogens. An oyster makes a pearl not because the sand is an irritant. The sand is covered with microbes and pathogens and the oyster protects its gut from them by producing a pearl to entrap that piece of sand and then you get a vicious cycle and it keeps going. You should leave plaques alone because they are sequestering the amyloid-β away into a neat little ball that cannot get into the synapse. If Aβ oligomers get in the synapse the synapse dies, the dendrites start to prune back and the nerve cell body dies. Plaque stops that happening as best it can.

It doesn't bother me when somebody shows me a case with a lot of plaques. I am interested though, genetically, in understanding how someone with a lot of plaques comes through cognitively intact. What protects them?

Tanzi's thoughtful, occasionally provocative comments on the current situation in connection with AD genetics will be discussed further in later chapters.

Anti-amyloid Therapeutics

The historian Jesse Ballenger has recounted the way in which efforts to develop antidementia drugs have played a crucial role over the years in the transformation of old age. His has researched the American situation, but his comments apply equally to the majority of European countries, Australia, New Zealand, Japan, several countries in Central and South America, and increasingly elsewhere. The first drugs designed to treat dementia began to appear in the mid-1950s in the United States and the United Kingdom, in an era when psychiatrists and social gerontologists were the specialists primarily concerned with dementia. These drugs were designed with the hope that patients would be able to continue an active social life and avoid further mental deterioration. As Ballenger notes, "By the early 1980s, the social and cultural transformation of aging had worked seismic changes in the landscape of dementia."[49] The very term "senility" was under attack, and Robert Butler coined the phrase "ageism" in making a powerful argument that broad social, political, and medical changes were needed to deal with the aging society. It was at this time that AD was rediscovered as a discrete entity and real investment began to be made in finding a cure for this condition that was increasingly termed in the professional literature as SDAT (senile dementia of the Alzheimer type), making it clearly discontinuous with normal aging.

Ballenger emphasizes that the labeling of AD as a discrete entity was politically effective and allowed the medical world and advocacy groups to stress that a hunt for a cure should be launched forthwith. Even if a cure was not found, a delay in onset of the disease by five years or so would result in enormous economic savings (and incidentally reduce considerable suffering).[50] The sociologist Tiago Moreira argues that efforts to standardize an AD diagnosis were intimately related to drug development, and that drug development in turn, from the 1980s on, has been managed within a regulatory framework in which free enterprise, scientific innovation, and the interests of insurance companies have all played a part.[51]

By the mid-1970s it had been shown by several research groups that the brains of patients diagnosed with Alzheimer's exhibited a pronounced deficit in a neurotransmitter known as acetylcholine, a deficit associated with memory loss.[52] In the early 1980s Peter Whitehouse and colleagues at Case Western Reserve University discovered a "cholinergic center" in the cerebral cortex in which a marked decrease in acetylcholine was associated with cell loss and was presumed responsible for much of the symptomatology of AD including amyloid plaques. This finding became formulated as the "cholinergic hypothesis" of AD symptomatology—a model for the proximate cause of AD.[53] Since that time, the majority of drugs that have shown sufficient efficacy to be marketed to combat AD have been known as cholinesterase inhibitors. These drugs, although they improve some clinical symptoms of AD, have mixed results at best, are effective for only a short time in some patients, do not halt the progression of the condition, and are often accompanied by unwanted side reactions.

The story of Alzheimer drug development is complex (for a comprehensive discussion, see works by Ballenger, Moriera, and Whitehouse, among others),[54] but one thing is strikingly evident: despite widespread recognition for many years by a good number of eminent researchers that AD is heterogeneous, the amyloid cascade hypothesis has dominated the search for a drug to "cure" this condition. This search has been spurred on by a remarkable sustenance of hope with repeated predictions that Alzheimer's would be defeated within five years. Only in the past several years has this language been toned down as numerous drug trials costing "untold millions" have consistently resulted in failure.[55]

In the summer of 2009 Forbes reporter Robert Langreth published an interview with Mark Smith, in which he labeled him as a "renegade" because Smith had argued that anti-amyloid therapies could actually harm patients by precipitating them into a downward spiral. But, by 2010, shortly before Mark's death, the tone of reporting had changed. Numerous drug companies, including Pfizer, Elan, Bristol-Myers Squibb, and Eli Lilly, had all been working on drugs designed to counter the buildup of β-amyloid plaque. In 2010 two clinical trials were stopped in the same week. One carried out by Eli Lilly and Co. was hastily withdrawn when it was shown that patients taking the drug were having more cognitive problems than were members of the control group taking a placebo. The drug also appeared to increase the risk of skin cancer. John Hardy was asked to comment online about the Eli Lilly failure. He stated that there are two key questions that have to be asked: First, did the drug hit the target but still not work? If this was so, then this suggests that the entire class of drugs is simply not appropriate. Second, Hardy questioned whether the scientists had neuroimaging data to show that the amyloid had actually been cleared in the trial subjects. If the amyloid was cleared and the patients became worse, this would be a very bad sign and clearly would put the amyloid cascade hypothesis into question.[56] Chapter 5 includes a discussion of several trials about to get under way that are in part designed to answer the latter of Hardy's queries.

The second failed trial was designed to create a vaccine that would teach the immune system to recognize amyloid as though it is a foreign body. The researchers saw no measurable clinical gains in the treated group; the immunized patients did not live longer, nor did they take longer to reach a stage of severe dementia than did members of the control group. Two patients, even though their amyloid plaques were virtually cleared from their brains, died of profound end-stage dementia. These findings are based on a small number of patients, but nevertheless the researchers concluded that "the presence of plaques does not seem to be a prerequisite for progressive cognitive impairment in AD."[57] Other vaccine trials caused encephalitis, resulting in death in a few subjects. Despite these findings further such trials are in progress.

The FDA has recently set up a Coalition Against Major Diseases in the United States. The chief executive of this endeavor stated with respect to Alzheimer's and Parkinson's diseases, "We really believe drugs are failing because we honestly don't understand the disease." The hope is to remedy this situation by pooling findings obtained from drug companies and academic institutions, to form a massive computerized database.

The cholinergic hypothesis had a second, quite different impact on the approach to AD, in addition to emphasis on amyloid removal from the brain. Due to the commercial availability over the past fifteen years of medications available for symptomatic therapy, a more "precise" clinical diagnosis is necessary, and physicians have been taught to base the diagnosis largely on the presence and extent of measurable memory loss that is far more easy to document than are changes in personality, social behavior, and emotional tone. Hence, demonstrated memory loss has in effect become a synecdoche for AD itself, thus reducing the concept to one associated largely with memory function alone.[58] Furthermore, clinical testing for AD has become heavily dependent on just one aspect of memory—that of recall—rather than giving equal attention to the other well-recognized forms of memory function.[59]

The Move to Prevention: Finding a New Lexicon

In 2005, 15 international dementia experts were brought together to attend a workshop in Florence designed to discuss the need for revised diagnostic criteria for AD. The resultant 19-author article appeared in 2007 in *The Lancet Neurology*. This article was the first definitive sign of a sea change in connection with Alzheimer diagnostics. The article, the senior author of which is Bruno Dubois of the Hôpital de la Salpêtrière in Paris, sets out by stressing that in the clinic no definitive "diagnostic biomarker" exists for AD, and readers are reminded that a reliable diagnosis can only be made by means of histopathological confirmation. The article claims that in recent years the clinical diagnosis for AD has improved, is "no longer described in exclusionary terms," and can now be taken further and "characterized more definitively on a phenotypic basis."[60] Dubois and colleagues argue that potential patients can be assessed for "distinct markers," including structural brain changes assumed to be definitive signs of prodromal, presymptomatic AD. They insist that prodromal AD "must be distinguished within the broad and heterogenous state of cognitive functioning that falls outside normal aging."[61]

The proposal in this article is to retain core diagnostic criteria for probable AD as they are currently set out, in which patients or family report a gradual and progressive change in memory function for the worse for at least six months. This complaint must be confirmed by psychological testing. In addition, at least one "support feature" is called for, including positive results from one of the following: an MRI scan, cerebrospinal fluid analysis, positron emission tomography (PET) neuroimaging (see chapter 4), and/or demonstration of an autosomal dominant gene mutation for AD. These four "support features" are described as "biological footprints of the disease." It is emphasized that because the "clinical phenotype of AD is better known than its biological phenotype," for the time being at least, the clinical diagnosis should take priority. Validation studies of the biomarkers will be needed and can be applied retrospectively to large cohorts of diagnosed cases that are already part of ongoing research. Validation should also be carried out prospectively on people with and without dementia, who will then be followed to postmortem. The scope of this proposed research enterprise takes away one's breath, and the need for extensive financial resources and technical skills is acknowledged.[62] The authors argue that these criteria "represent a cultural shift requiring a more biologically focused work up," one that is justified because AD diagnosed in the clinic has already progressed so far along an irreversible pathway, no pharmacological intervention will ever be effective.

Many adherents of the amyloid cascade hypothesis think of the move to an earlier diagnosis as positive. An article in *Nature* in 2008 insists, "Researchers have very strong reasons for believing that plaques and tangles are the cause of Alzheimer's symptoms, rather than just the markers of it," and the author cites well-known research findings to back up this assertion.[63] Simon Love-stone, even though a committed tauist, is quoted as saying, "It could well be that there has been too much focus on early events of the amyloid cascade. . . . But let's not throw out the baby with the bathwater—there is just too much convincing preclinical research that supports the amyloid-cascade hypothesis."[64] More than 30 trials were under way in 2008 for treatments based on the amyloid cascade hypothesis, and the assumption is that if "surrogate" biomarkers ("biological footprints") are tested for, rather than waiting until the disease is diagnosable in the clinic, clinical trials will be of shorter duration and cheaper.[65] The conclusion of the *Nature* article is that the biggest tests of the amyloid hypothesis are yet to come.

Perhaps this is indeed the case if the 2011 remarks John Hardy made at a symposium sponsored by the pharmaceutical industry are anything to go by. The symposium was titled "The Amyloid Cascade Hypothesis Has Misled the Pharmaceutical Industry," a title deliberately selected by Hardy to generate debate among several AD experts. As part of the introductory talks at this gathering, Hardy said, "I am convinced that, for early onset [AD] of the familial type [referring to the rare single gene form of AD—see chapter 5], the [amyloid cascade] hypothesis is proven. The situation remains less clear for . . . late-onset disease [the "late-onset" condition commonly diagnosed as AD]." During the debate Simon Lovestone asked John Hardy, "Why don't mice get AD?" And Hardy replied, "I don't know. The system that worries me most with respect to . . . AD is the vascular system. . . . The vascular system of a mouse [is] very different to that of humans and therefore this may have a strong effect on why mice don't get AD."[66]

In the following two chapters I discuss at greater length the move to transform the clinical diagnosis of Alzheimer's into one to be made at the "prodromal" stage, with the ultimate objective of directing a substantial amount research toward the goal of prevention. A new round of standardization involving the condition known as MCI (mild cognitive impairment) and the so-called biological footprints of prodromal AD, including those found in cerebrospinal fluid and by means of PET scanning, is under way. For the foreseeable future these investigations will be limited to research settings alone. This new approach is attractive to many because it enables detection of amyloid changes in cerebrospinal fluid, and the assumption is that if amyloid changes are spotted very early, then its deposition may well be reversible, or at least its accumulation may be slowed down. In other words, the amyloid cascade hypothesis continues to be the center of attention. But, as the article in *Nature* suggests, and as a large number of the researchers whom I interviewed indicated, this approach as far as they are concerned may well be mired in outmoded thinking.

The move to identification of biomarkers is likely to compound the difficulty of disassociating normal and pathological aging. As we will see in the next chapter, biomarkers do not reliably identify AD, but merely indicate probabilities for becoming demented. If one assumes that aging and AD are on a continuum and acknowledges that AD cannot be characterized by one or more features unique to it, then the paradox of entanglement cannot be resolved by means of biomarker testing. Given this situation, we may well finally be forced to confront the question repeatedly set to one side throughout the history of Alzheimer disease: Why are certain individuals "protected" from the ravages that amyloid deposition and/or other intimately related molecular changes apparently precipitate in many others?

Chapter 3
Paths to Alzheimer Prevention

Although it was not until two or three years later that the Alzheimer symptoms became fully apparent, I have sometimes wondered if Iris knew that her own career as a novelist was nearly over.

—John Bayley, *Iris: A Memoir of Iris Murdoch*[1]

The idea of being "predisposed" to an illness stems from the early 18th century, when discussion among medical practitioners and lay people alike about the "constitution" of individuals and its relation to disease vulnerability began to take place.[2] However, François Ewald, a political scientist, argues that the "philosophy of risk" as we understand it today is very much a product of contemporary society—a radical epistemological transformation involving a "mutation" in attitudes toward justice, responsibility, time, causality, and destiny.[3] This transformation was part of a shift toward a secularized approach to life where "the ills that befall us lose their old providential meaning." In a world without God, control of events is left entirely in human hands—a logical outcome when life is made into a rational enterprise.[4]

The anthropologist Mary Douglas argued that use of the word "risk" in this restricted sense, rather than its former usage as a synonym for "danger" or "hazard," has the rhetorical effect of creating an aura of neutrality, of cloaking the concept in scientific legitimacy. Paradoxically, this permits statements about risk to be readily associated with moral approbation. Danger, reworded as risk, is removed from the sphere of the unpredictable, the supernatural, the divine, and is placed squarely at the feet of responsible individuals. Risk, in Douglas's words, becomes "a forensic resource" whereby people can be held accountable.[5] As the French sociologist Robert Castel puts it, mobilization of the concept of risk becomes a novel mode of surveillance—self-surveillance—a component of the microphysics of power that Foucault noted with reference to emerging neoliberal society.[6] Castel warns that such "hyper-rationalism" comes with a cost, and that there may be "iatrogenic aspects" to this new form of prevention among which chronic anxiety can be prominent.[7]

A moment's reflection confirms that everyday life is riddled with warnings about risks to our health—above all about what will happen if we do not eat balanced diets, exercise regularly, and avoid smoking. Self-surveillance is expected even when poverty and inhuman working conditions make such behavior virtually impossible. In addition are risks posed by toxic and stressful environments, variables over which it is usually possible for individuals to exert only minimal control. A third form of risk—"corporeal" or "embodied" risk[8]—is the topic of discussion in this chapter. Perhaps the most familiar examples of embodied risk are cholesterol levels and blood pressure readings. In addition, millions of people routinely undergo pap smears, PSA tests, mammograms, bone scans, and other forms of routinized screening designed to indicate who is at risk for cervical cancer, prostate cancer, breast cancer, osteoporosis, and so on. A large body of recent publications about these screening programs highlights the uncertainties and discrepancies that arise when interpreting test results associated with this kind of surveillance of individual patients.[9]

The move toward the prevention of AD represents a shift in which it is assumed that embodied risk can be made manifest in the form of biomarkers.[10] This shift is to be accomplished by researching, standardizing, and gradually routinizing the use of several biomarkers believed to put individuals at increased risk for AD. In contrast to lifestyle and environmental risks, and to many types of embodied biomarkers such as cholesterol levels, individual accountability cannot be associated with AD biomarkers. Individuals are not responsible for their own genes,[11] nor can they be held responsible for the condition of their cerebrospinal fluid or for deposits of plaque in their brains, because the cause or causes of AD are simply not understood, and no irrefutable associations among everyday life, behavior, and biomarkers for AD have ever been shown. What is more, people cannot be held accountable for aging itself (not yet!).

Before elaborating on AD biomarkers and their detection in individual bodies, a discussion of the clinical diagnosis known as mild cognitive impairment (MCI) is necessary. Standardization of this diagnosis was the first formal move toward distinguishing incipient AD from so-called normal aging. The second part of the chapter turns to the role of the media in fueling public interest in a revitalized approach to AD, including its newly marked emphasis on early detection. A proposed revised lexicon for the definition of AD, one that unhooks the century-old clinical and postmortem diagnoses from each other in setting out the move to the prevention of AD, is then briefly outlined. This proposal, initially spearheaded by European researchers, argues that an "in vivo" diagnosis of AD neuropathology involving biomarker detection should now become the gold standard of AD diagnosis, and the concept of MCI will henceforth have use only in clinical settings where other more stringent tests cannot be carried out.

Circumscribing Mild Cognitive Impairment

I was complaining to my doctor, my family doctor, for quite a while, that I can't remember what happened yesterday. And he kept saying it's nothing, nothing, and he gave me all kinds of tests, and he kept saying, it's nothing. And then he finally said, look, if you insist, I'll send you to the memory clinic.

—Josef Hoffman, 85, born in Berlin[12]

My wife kept saying: "John, your forgetting this, and forgetting that." When I retired I wasn't interested in anything. I don't know what money I have in the bank, I don't know anything any more and my wife went with me to see our doctor, and she refused, and refused, and refused to do anything. She said you do not have Alzheimer's, even when I said I think I'm losing my memory. Finally she gave me a referral to this memory clinic.

—Duneth Bandjar, 66, born in Suriname

The term "benign senescent forgetfulness" was coined in 1958 by a Czech physician trained in both neurology and psychiatry, initially in Prague and then in other locations in Eastern Europe. Voijtech Adalbert Kral and his family were interned from 1942 to 1945 in a concentration camp at Theresienstadt, in what was then Czechoslovakia, where he was allowed to practice as a physician. After the war, the Kral family emigrated to Montréal, where, while employed at what was then known as the Protestant Insane Asylum,[13] Kral began to undertake research in connection with memory disturbances and dementia. In the 1950s he was made director of the Gerontologic Unit at the Allan Memorial Hospital, affiliated with McGill University, the first hospital with such a specialty in North America. In 1955, Kral became a consultant at the Montréal Hebrew Old People's and Sheltering Home, where he conducted survey research on memory disorders with the elderly residents, the majority of whom were older than 70 years of age. His assignment was to make recommendations to improve conditions in the home.[14] In doing this study Kral became one of the first people to make use of standardized psychological scales to assess the cognitive condition of research subjects. Among the 162 residents, Kral and his co-researcher Wigdor found that 77% of them not only had memory disturbances but also had signs of psychosis or major neurosis. When he carried out a second survey with a subset of 52 of the residents whom Kral described as "well preserved," he found that nearly half of them were without any memory deficit or other impairment. A quarter of the remaining subjects were labeled as having benign senescent forgetfulness, and another smaller group were diagnosed with "incipient amnestic syndrome." These early efforts at creating boundaries among normal cognitive aging, impaired memory function, and dementia were followed by repeated efforts at refinement and resulted in a proliferation of proposed diagnostic subdivisions.

By the 1980s two psychological rating scales were being widely used—the Global Deterioration Scale and the Clinical Dementia Rating Scale—both of which were created to recognize subjects without dementia "who nevertheless exhibited some evidence for cognitive dysfunction." Based on their test responses, subjects could be subdivided into cases considered as "questionable," "borderline," and "preclinical AD."[15] This last category was applied to individuals whose condition was regarded as intermediary between "normal" and "demented." The introduction in 1984 of standardized criteria for the diagnosis of AD ensured that persons whose test scores indicated that they had "mild memory problems" were not included. One unanticipated result of this change was a proliferation of labels indicating incipient memory problems including "cognitive impairment no dementia," "isolated memory loss," "mild cognitive disorder," and others. The label that came to dominate and is used to this day to signify memory loss with no dementia is that of mild cognitive impairment (MCI). First coined in the late 1980s, the term became closely associated with the name of Ronald Petersen, a neurologist at the Mayo Clinic in Rochester, Minnesota. In 1995, Petersen and colleagues first used MCI as an independent clinical diagnostic category and applied it to research subjects defined as "cognitively normal" who could live entirely independently, but who had subjective memory complaints.[16] When given age-adjusted memory tests, these individuals scored relatively poorly. Petersen's group argued that MCI "deserved recognition" because "as preventive treatments for AD become available, it will become incumbent on clinicians to identify persons at risk of AD and those with the earliest signs of clinical impairment."[17] In connection with research, notably clinical trials, it was quickly recognized that there would be a distinct advantage in having access to samples of nondemented individuals who were quite possibly on the path to dementia, and people diagnosed with MCI, usually on the basis of subjective complaints, rapidly became a target "at-risk" population. Use of this diagnosis has always been more prevalent in the United States and Canada than elsewhere in the world.

In 2001 requests were made to the U.S. Food and Drug Administration (FDA) to have MCI formally recognized as a medical diagnosis for which drugs to alleviate memory loss could be developed (to date no such drugs are available). In response, a panel was convened to provide answers to several questions, including these: Can MCI be clearly defined in a clinical setting? Can MCI be distinguished from AD and other causes of dementia? Disagreements among AD experts were laid bare when this panel met, particularly in connection with deciding at what point along a continuum of cognitive decline, drugs, once available, should be prescribed for MCI.

Ron Petersen was quick to agree that MCI could not be readily diagnosed in primary care settings, but he insisted that in specialist clinics it could be reliably identified. However, this comment did not put everyone at ease. The sociologist Tiago Moreira who attended the hearings, cites the comments of Dr. van Belle, a member of the committee:

I think we have defined a mystery. We have sort of put a fence around the mystery but we have really heard many ways of defining MCI today and I am not sure that there is a consistent operational entity that we can deal with at a relatively simple level . . . in terms of really having a clinical entity, I just haven't seen the evidence yet.[18]

It was eventually agreed that the clinical utility of MCI derived primarily from the fact that this category produced measurable effects in clinical trials.[19] Significantly, Ron Petersen published a monograph in 2003 in which he argued that "ultimately MCI is likely to be a heuristic concept. . . . At some point the term will be discarded and another will take its place."[20] Moreira and colleagues conclude that these hearings are an example of the way in which "the production and temporary stabilization of biomedicine's knowledge and entities requires continuous 'uncertainty work' in the clinic, laboratory, and regulatory forum."[21]

Between 1999 and 2004 the number of publications on MCI increased sixfold, to over 300 articles,[22] and since that time many more have appeared. A consensus report on MCI published in 2004 concluded that one of the criteria for this condition is that "the individual is neither normal nor demented."[23] The entanglement of aging and dementia could not have been made more evident. Today, individuals diagnosed with MCI are research subjects not only in clinical trials, but also in neuroimaging initiatives, and for tracing and tracking biomarkers associated with presymptomatic AD (see chapter 4). Inevitably, they become "hybrid bioclinical entities" because they are at once patients and research subjects.[24] Such a situation is common in tertiary care hospitals today in connection with many conditions, but it is particularly troubling for people diagnosed with MCI, as we will see shortly.

In 2005 the International Psychogeriatric Association in Bethesda, Maryland, convened an "Expert Conference" of professionals drawn from the AD world. The work assigned to this 40 strong international group was to "clarify the diagnosis and management of mild cognitive impairment," having first reviewed all the literature relevant to their assigned topics. The objective was not to reach a consensus about MCI, but to "offer an expert opinion on where mild cognitive impairment stands as a clinical entity."[25] The 2006 article based on this Expert Conference made it clear that diagnoses of MCI vary considerably among medical settings but added, "The syndrome of mild cognitive impairment as a transition state between normal aging and dementia has increased awareness that memory complaints in elderly people, particularly when accompanied by subtle cognitive performance difficulties, should be assessed in a systematic way by clinicians."[26]

In 2007 *Nature Clinical Practice Neurology* published two short articles, one written by Ronald Petersen and the other by Peter Whitehouse. These articles were published as a pair under the heading "Viewpoint." The contrasting opinions they express highlight the continuing debate among experts on the subject of MCI. Petersen's article, with a subtitle "What Do We Tell Our Patients?" reminds his readers that, because it is a degenerative disorder, by definition

AD has a gradual onset and progression. Hence, early on in the process, there is inevitably a time when individuals begin to exhibit some, but not all, of the clinical features of the disorder. Petersen adds that this phase of incomplete AD symptomatology, now known as "amnestic mild cognitive impairment,"[27] is a stage in which "an affected individual has memory impairment beyond what would be expected of their age,"[28] but he is quick to cite data demonstrating to his satisfaction that although "clinical acumen" with respect to MCI is good, "it is not, at the amnestic MCI stage, sufficiently specific to warrant a clinical diagnosis of AD."[29] Petersen comments on findings showing that people with amnestic MCI who are seen in specialty clinics progress to a diagnosis of "probable AD" at a rate of 10% to 15% per year, markedly higher than the general population (a finding also highlighted in the Expert Conference publication noted above), and he uses these findings to bolster his argument for the clinical usefulness of the MCI construct.

The gist of Petersen's Viewpoint essay is that because, as yet, there is no unequivocal biomarker for AD, care must be taken to avoid prematurely diagnosing individuals with this condition, above all because of the social consequences that are likely to follow, including loss of a driver's license, loss of health care insurance (in the United States), and difficulties in keeping employment. His position is that an AD diagnosis "had better be correct," and that usage of the MCI category gives everyone—patients, families, and physicians alike—a space in which to come to terms with the strong possibility of approaching AD, while making allowances for the inevitable uncertainties involved, including the fact that some people so diagnosed will not progress to dementia. In an earlier article, Petersen had argued that the cognitive and behavioral changes associated with MCI are devastating and should not simply be dismissed as part of aging. Even so, he stressed, individuals should not be actively recruited into research projects only then to be told that they have MCI; rather, individuals who come independently to clinics seeking out care and who, as a result, are diagnosed with MCI should be counseled about the uncertainties associated with the diagnosis.[30] Presumably such individuals are then free of their own volition to become research subjects.

In his essay, Peter Whitehouse notes that an email survey sent to participants immediately following the 2005 Expert Conference revealed that only 57% of the participants thought that MCI should be used clinically as a diagnosis. Whitehouse writes emphatically that "AD is arguably not one entity, but comprises several conditions associated with processes that are part of normal brain aging. . . . [I]t is time we seriously think about the words we use to label the continuum of normal aging, and the hardening of categories from which the field of dementia studies currently suffers."[31]

In early 2007 I interviewed Ronald Petersen at the Mayo Clinic and challenged him about a comment I had heard him make at a workshop I attended in which he stated that MCI is essentially an interim category, one that may well be replaced by a diagnosis based on the detection of significant biomarkers in patients. Petersen, a thoughtful interviewee with a calm demeanor, had this to say in response:

> I think that there are biomarkers that will inform us more definitively as to what's going to happen with the outcome, but I'm not sure that I'm ready to retire the [MCI] construct yet because what will happen is our diagnostic threshold for AD will probably move back to earlier stages . . . as that moves, I suspect the threshold for MCI might also recede a bit. . . . [W]here you draw these lines clinically now is arbitrary . . . and I'm not certain there will ever be a definitive biomarker for Alzheimer disease . . . you know, yes or no . . . I don't think so, because aging is so inextricably interwoven here that, er . . . I know this is almost a heresy—but its almost like there's nothing that is absolutely unique to AD. There's a strong aging influence involved . . . and we always ask the question, if you live to a hundred and twenty won't everyone get Alzheimer's? But this doesn't mean that we shouldn't try to do all the things we can to diagnose and ameliorate symptoms. But where do you draw the line between a little bit of forgetfulness and something more serious? How much is too much? The further back in time, the more blurred that gets. I definitely think that MCI is a construct that will probably be replaced down the road. But we have to remember that the clinical symptoms are not going to be the first symptoms—there will be signs of amyloid before there are clinical symptoms, IF amyloid is the culprit. We don't even know that. But if it is, not everyone with amyloid is going to get AD, so I'd certainly hang on to the importance of early clinical symptoms, I'm not abandoning ship yet. (March 2007)

A Montréal neurologist, Howard Chertkow, notes that MCI can be characterized and diagnosed by physicians with a very high specificity and sensitivity when the diagnosis is made using accepted clinical criteria and the internationally recognized screening tool, the Montréal Cognitive Assessment Instrument (MoCA), (figure 3.1) created in 1996.[32] He adds that a clinical history and further testing is needed to confirm a diagnosis but, even so, he describes MCI as "a measure of clinical uncertainty" for the very reason that it cannot be ascertained who among diagnosed patients will "convert" to AD. Furthermore, it is well recognized that some patients diagnosed with MCI later "rement" and once again become "normal."

Given the uncertainties associated with MCI expressed by the experts, an obvious question to ask is what the experiences are of people who have actively sought out referrals to memory clinics because they or their partners are worried about increasing memory difficulties. With this in mind, in 2005, two research assistants and myself interviewed 31 people attending a memory clinic established in 1991 in the Jewish General Hospital, a teaching hospital that is

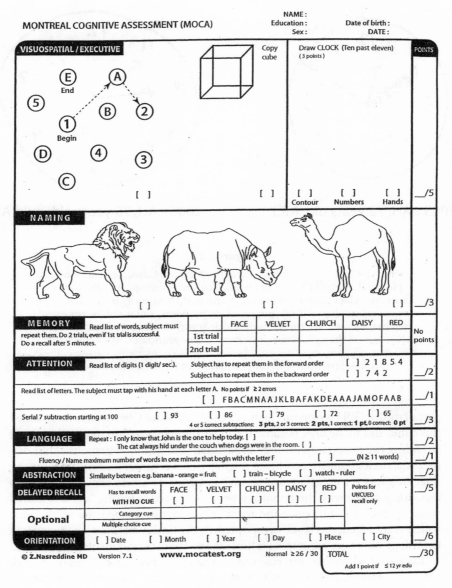

Figure 3.1.

Montréal Cognitive Assessment instrument. Reproduced with the permission of Ziad Nasreddine MD.

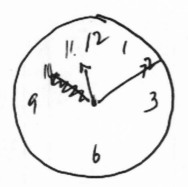

2003 - AH, 72 years old, MCI   2006 - AH, 75 years old, mild AD

Figure 3.2.

Draw-a-Clock test, administered at the Memory Clinic, Jewish General
Hospital, McGill University Health Center, Montréal. This illustration
shows two clocks drawn by the same patient three years apart. Each time he
was asked to "draw a clock, put on all the numbers, and make the time read
ten past eleven." On the left is the clock drawn after being diagnosed with
MCI, and on the right is the clock drawn three years later, when this patient
was clinically diagnosed with mild dementia due to Alzheimer disease. Illustration reproduced with the permission of Howard Chertkow MD.

part of the McGill University Health Center.[33] Howard Chertkow, who founded
this clinic, did so, he says, because it had become clear to him that many people
attending the hospital neurology department did not have dementia, but neither was their cognitive condition normal in his estimation. From the outset,
the clinic was set up as a research unit designed to assess semantic memory
loss, and patients with this condition were diagnosed as having "age-associated
memory decline." But once the term "mild cognitive impairment" became
widely recognized, this label was adopted. At the time, Chertkow's colleagues in
neurology were very eager to have him take on these time-consuming patients.

Everyone whom we interviewed in the clinic had been diagnosed with
MCI. The sample consisted of 15 women and 16 men; their mean age was approaching 78, with the youngest being 64 and the oldest 89. Interviews were
conducted in either English or French, depending on the preferred language
of the interviewee.[34] Potential patients cannot make an appointment to be
seen at the clinic without a referral from their general or family practitioner, or
a physician of another specialty. A nurse carries out the first assessment over
the phone, at which time she talks to the potential patient's closest relative or
partner, someone who, preferably, is living in the same household. This phone
call can last for 30 minutes or more.

In recent years the waiting time for a first appointment has averaged between six and eight months, although efforts are being made to cut this back.
Patients are asked to attend the clinic together with a family member or close
friend, but many come alone. At the first visit a resident spends up to one and
a half hours taking a full history and doing a lengthy physical and neurological
examination that includes short nonstandardized tests of language, memory,
attention, and decision making, among other items (figure 3.2). This is followed by administration of the psychological scale (MoCA)[35] developed at the
clinic, the results of which are then reviewed together with a consultant. This
first visit is followed up with blood work and CT scans to exclude various illnesses that can produce memory loss. Patients are asked to return after a two-
month interval, at which time those among them diagnosed with MCI are
educated about what is known about this condition and given a handout. This
five-page document states clearly that the patient has been diagnosed with
MCI and explains what is currently understood about the causes of this condition, including details about certain prescription medications that produce
MCI-like symptoms. It is explained that a few individuals are likely to improve
with time; some will continue to have MCI, but never become demonstrably
worse, and between one-third and two-thirds of MCI patients will progress
to dementia. It is made clear in the handout that clinicians cannot judge into
which category indivduals are likely to fall in the future. Plenty of time is given
for patients and their relatives to ask questions, and it is during this second
visit that a discussion takes place to ascertain if the patient would agree to become a research subject, which involves submitting to blood tests, brain scans,
and genotyping of the APOE gene. Research subjects understand when they
sign informed consent that they will not be informed of their APOE results.

A good number of people are not eligible to become research subjects, including those who have pacemakers in place (in which case they cannot undergo an MRI scan), have been diagnosed with other serious illnesses, are very
frail, or have no family support. Among those who are accepted, the dropout
rate of patients over the years is high, at about 50%, including those who die.
At the clinic, in addition to patients with dementia and MCI, patients are seen
who came initially complaining of memory loss, and on assessment are diagnosed only with "subjective memory impairment." Some of these individuals
will eventually progress to MCI, but a good number do not. Once diagnosed,
patients with MCI and subjective memory complaints are usually followed up
at one-year intervals.

The term "mild cognitive impairment" appears in the media much less frequently than does "Alzheimer's" and is clearly a cumbersome term for nonmed-
ical people. Before carrying out interviews, we had been reassured that everyone to whom we would be talking had been told on their first visit that they were
being assessed for "mild cognitive impairment," and then at the second visit
they had been informed of their diagnosis and had received the handout about
MCI that was discussed with them. The individuals to whom we talked had all

been attending the clinic for several years, some for four or five years or longer. Accordingly, after introductions and some general questions, we then asked by way of confirmation, "I understand that your diagnosis is one of mild cognitive impairment, is that right?" A few answers were clearly in the affirmative:

> Yes, I was told I have mild cognitive impairment. Which I agree with, because, you know, I don't forget everything. (Marilena Nistor, 73, born in Romania)

Diane Blacker, 64 years old, has mixed feelings about her condition:

> I was told I have mild cognitive impairment. That's all I learned. And now they keep doing tests—that's it. And then sometimes it shows I'm at the beginnings of Alzheimer's. But I was given the sense that, yeah, the last time I was here, that I hadn't deteriorated . . . he just said I'm not really much worse. . . . I think in a way it's good to find out . . . it's helpful you know, if you don't come and check it out, then you don't know if you're worse or better, you know. But better it's not. But is it worse? I don't know. I don't really get many answers here because Dr. S is very busy, but I guess it's good that I'm not one of his priorities. Then I'd be more nervous. So, I don't really know what's happening from the last test to the next test, you know.

Others appeared to be relatively comforted by the idea that they do not have Alzheimer's, but did not make use of the term MCI:

> Well, they told me that, uh, even though I'm very forgetful, and I think that it's getting worse, that it doesn't necessarily mean that I have Alzheimer's. (Susan Bestor, 83)

Several more patients were less informative. Duneth Bandjar, a 66-year-old from Suriname who said that he had been coming to the clinic for three years or so, when asked about mild cognitive impairment, simply said, "They can't find anything." And 79-year-old Eileen Grey, when asked what the doctor had told her about her condition, insisted,

> Nothing really. Nothing at all. I mean I've taken all the tests and everything else, and they've always said, "Oh, that's fine." All I know is that it's fine!

One person insisted that she is a volunteer at the clinic and not a patient, and said she has only marginal concerns about her memory. And several more simply answered abruptly, "no." But when Charles Bregner, aged 81, stated that he had not heard of the term MCI, his wife, who was present, contradicted him, saying, "[T]hey told you you've got 'mild cognitive impairment.'" Her husband then said, "Mild cognitive whatever, but it's getting less and less mild I think . . . but when I ask them, they say it's about the same."

Claude Jolivet, 74, agreed that his doctor might have told him something about his condition, but added, "I can't recall anything." A second man of 74, Thomas Bestor, insisted that he had never heard about mild cognitive impairment, but added that he had been given some papers by his doctor: "I started reading them, but it gets too hard for me to understand. So I put them away."

When Alan Salter, 75, gave the following reply, we were concerned that perhaps we had been mistaken in our understanding that everyone had been told of their diagnosis:

> Well, now that you mention it, that's the first time I've gotten a diagnosis, ah, directly. The only thing they've said, there is some lack of memory, you know? This is really interesting.

Following this experience, we changed our opening gambit to "what have you been told about your condition?" After this, responses such as the following from Jeremy Poland were frequent:

> I don't think Dr. C has ever told me about my condition. I only spoke with him for an hour, I guess, at the most. He asks me some of those questions I can't answer. He knows I'm not going to answer them! I don't think I ever received a diagnosis.

At this point his wife interrupted: "It's just mild memory loss." He added, "mild to her but bad to me, 'cause I get mad at myself, and the madder I get, the worse I get."

Some people made it clear that they feel rather upset with their experience at the clinic, having come away repeatedly with the impression that they have been put through a series of tests of benefit only to the doctor: "I have to show them what I know, but they haven't told me anything about what I can do at home to stop getting worse."

Several others said that they were hoping for some practical guidance to stop their condition from getting any worse, but grumbled that this had not happened, or that they had learned nothing new:

> The new doctor I have places a lot of importance on nutrition. He's written articles about it. He gave me concrete suggestions, but there are a lot of things I was already doing, like eating salmon—that kind of thing. And vegetables, fruits, blueberries . . . antioxidants. (Pauline Vigeant, 64)

As noted above, one cannot attend the clinic unless referred; even so, three people stated emphatically that they had just walked in and been accepted. Two of these people had probably initially volunteered to be "normal" controls for ongoing research at the clinic, but after assessment were judged to have MCI. They were then asked if they would like to come to the clinic as patients and had accepted, but had apparently forgotten that this was the case. The third person was adamant that she had walked in off the street.

Among the people whom we interviewed only three spouses were present at the time, and answers such as those above made it abundantly clear the extent to which many of these patients are troubled by forgetfulness; the distress they were experiencing as a result of their memory loss was striking. Some people laughed in an embarrassed way when talking with us because their ability to recall things was so poor; others worried that their memories were getting worse even when they had been reassured on the basis of the testing they had undergone that this was not the case. Yet others adamantly denied the extent of their problem and sat oddly mute, not attempting to answer our questions about their lives. On reviewing the interview data it was clear that the majority of people to whom we had talked were on the brink of no longer being able to manage their own lives satisfactorily.

Interpreting these interview findings is exceptionally difficult because of the degree to which memory loss is implicated. It is abundantly evident why people are asked to have a spouse or friend accompany them to the clinic, but most do not comply, and among those who do, it is the spouse who has usually initiated visits to the clinic and then insisted on attending the follow-up visits as requested. In one case, 80-year-old Jan Reslowski, born in Poland, hardly said a word throughout our interview, and his wife took it upon herself to answer the questions. This man, a baker throughout his working life, has since been diagnosed with Alzheimer disease. His wife and daughter came to the clinic together with him when I met him for a second time five years after his first interview. They were clearly very concerned because Mr. Reslowski could no longer remember how to bake bread at home and was increasingly unable to manage independently at all. It was also apparent that he was unable to speak more than a word or two in English, although this had not formerly been the case, and he had no detectable difficulty when the family spoke Polish together.

A great deal of research shows that large numbers of patients, regardless of what kind of medical problem they have, forget or misunderstand what they are told in doctor's offices. But, of course, memory loss makes it particularly difficult for patients to recall what they were told, even when they have been given the specifics of their case in writing, and their diagnosis is repeatedly mentioned on their visits to the clinic. In addition, language problems are implicated in a cosmopolitan multilingual city such as Montréal. Ten people whom we interviewed have neither of the official languages, French and English, as their first language, and were born outside Canada. It is well known that as AD progresses a gradual loss of access to languages other than the first language takes place.[36] The odds are high that several more of these patients, even though they were born in Montréal, would not have spoken English or

French until they went to day care or primary school. Under these circumstances, visits to the doctor are exceptionally stressful as patients become increasingly cognizant of their growing language deficiencies.

Another point that became apparent when talking with these individuals for an hour or longer was that being given a technical term to gloss their condition of memory loss was of little importance to most. Even though nearly three-quarters of the people with whom we spoke made it very clear that memory difficulties was the reason they had come to the clinic, many then added that, above all, they wanted to be reassured that they did not have AD. Equally important, perhaps, is that the term "mild cognitive impairment" is easily confused in the minds of many patients and their families from "mild memory loss," and this is the phrase people use most often when talking about their difficulties.

Once subjective memory difficulties become seriously troubling, in contrast to numerous people with other medical conditions, MCI patients do not avidly search out everything that has been written on their condition. One interviewee said, "I think I read something in a paper, or something like that. I don't read so much . . . I can't remember that much." Another said, "I have a memory book, my nephew sent me one, and if I would remember to read it, you know, it might help, but I forget!" A few more thought they may have read something, possibly what they were given at the clinic, but they could not recall if this was really so or not, and yet others claimed that reading about memory loss would simply make matters worse: "I try to stay away from learning too much 'cos the more I read the more nervous I get."

When asked about their principal concerns at the present time, by far the majority responded that it was a fear about progressing to Alzheimer disease. An 85-year-old man born in Berlin said, "My primary concern is that I'm afraid that I'm getting Alzheimer's. That's what I'm afraid of . . . because mild usually becomes less mild." Aniela Bakula, a 78-year-old holocaust survivor who lost nine relatives in Auschwitz, laughed nervously when asked what were her major concerns:

> I'm more worried about getting Alzheimer's than dying. It scares me because, you know that's . . . you're not a person at that point and you can't do anything for yourself. I'm very independent; I've been working for 35 years and I raised four kids, no housekeepers, no maids. To have to have someone look after me, it's something I wouldn't want to be around for. My mother-in-law had it, my girlfriend's husband had it, and I've seen it and it's worse than being dead. You know, if you're dead you're at least gone and you don't require people to wait on you and feel bad for you.

Moshe Weinberger, a great-grandfather of 83, is the sole caregiver of his wife who has advanced Alzheimer's, and he himself has recently been diagnosed with MCI. He spoke of his wife at length during the interview, and preferred not to speak of his own condition:

> It's called the Golden Years; believe me, it's not golden. Sorry to scare you.... Let me ask you this ... why is it that my wife has such strength? If we're walking and I'm holding her hand as I always do, and if she wants to go another way she'll pull me. You can't stop her. And she doesn't talk. When she does talk she doesn't talk in English. She talks in two languages: Yiddish and gibberish. We're sitting downstairs and a big fat woman walks by and she says out loud "look at that big fat ass on her," and the woman is standing right in front of her. And she was so gentle, my wife, so terrific. I'd rather shoot myself than end up like my wife. And she always said to me, "if I end up like my father, you can kill me."

Several people expressed strong concerns about losing their independence. An 83-year-old divorcee, Julie Vaillant said,

> I'm resigned to the fact that my memory is going to get much worse, and I'll probably have to get special help of some kind. I'll probably end up like my siblings who are all in homes. I don't want to be a burden on my children, so it's a worry. And, of course, there's always that nagging feeling that I'm supposedly, in quotation marks, "normal" now, but I'll probably end up with Alzheimer's.

Alin Baboescu, a 72-year-old with nine grandchildren who was born in Romania, is also very worried about the burden he may place on his family in the near future:

> My concern is not to fall on the shoulders of my family—that's what I'm afraid of. I'm not afraid of dying of sickness . . . but now I can't sleep at night and I find myself . . . I wish it were over. A friend of mine just died two weeks ago, and he died peacefully, and I said, gee, this is nice. I mean, I don't like thinking like this, but this is what I thought.

Maya Szabo, 89, is a widow of Hungarian origin who lives on her own. She says,

> I have nobody to turn to, nobody. My brother is sick with Alzheimer's and I'm watching him, he's fading away. My mother and my cousin both had this too.

She feels she can no longer handle banking or pay her bills and wonders how she can "organize" her life from now on.

Being unable to do banking was a common complaint of many of the patients, together with an enormous frustration about not being able to keep track of items and appointments, as well as finding themselves asking the same questions repeatedly because they were unable to remember the answer they had already given. In addition were worries, expressed most strongly by women, about no longer being able to be the caregiver for their spouse. It is clear that these interviewees, in contrast to individuals who have progressed to a condition that is undeniably one of dementia, are acutely aware of what they can no longer deny—memory loss that is beginning to interfere seriously in their everyday lives—and the majority are convinced that they will progress to Alzheimer's before too long.

These patients find themselves in a liminal state, to use Victor Turner's apt term—"betwixt and between"—neither "normal nor demented" as certain AD experts put it. Turner's analysis of Ndembu ritual was primarily about rites of passage, notably life cycle transitions and the rituals performed in public to ensure that they are completed successfully. Many such transitions are joyous events, in which people are ceremoniously returned to everyday life as socially matured individuals, often having endured deprivation during the ritual. But rituals associated with the end of life, in which fears about contagion and pollution must be rigorously controlled, are carefully bounded by practices of separation. Most of the patients diagnosed with MCI to whom we talked made clear that they are overwhelmingly concerned about an approaching separation—a social death—involving a departure from their former, everyday lives, a concern shared visibly by their spouses when present. This pending loss of self is exceptionally fraught because so many have witnessed family members and friends succumb to and die of Alzheimer's, and they recall all too well the ambivalent, often negative terms that they themselves quietly used when talking about these fading people.

Although the subjective judgment of the patients is often that their memory is indeed slowly becoming worse and worse, many are reassured when they are told on their return visits to the clinic that the tests do not suggest a decline and, above all, to learn that they do not have AD—the future can be postponed, and some quietly embrace the hope that they will be among those diagnosed with MCI who never "convert" to AD. All but six of the people whom we interviewed were suffering from other conditions, often more than one, including heart problems, asthma, diabetes, cancer, and so on. As one person put it, "I'm always in between experts who try to identify my problems, but I think they've decided that my memory is not that serious, mainly it's the age."

Virtually all of these patients had agreed to become research subjects at the clinic when asked. When they sign the consent form they relinquish the right to learn the results of tests carried on them out as part of the research protocol. Fulfilling the basic requirements of "informed" consent is, inevitably, a difficulty given the extent of memory problems that many of these patients are already experiencing by the time they arrive in the clinic, and their partners in effect become legal substitutes. Moreover, their level of confusion and uncertainty is very often heightened once they become research subjects. Aline Beaubien, 64, when asked what she had been told about her condition, said,

> Um, they sent me to the research project. I got an MRI at the Neurological Institute and some . . . some blood tests . . . two research projects, right? But they haven't told me anything. The neurological tests lasted for an hour . . . that's long eh, an hour? I was curious to know the results, I'm not sure if they are going to tell Dr. R in the Memory Clinic . . . maybe I don't remember well, but maybe they did tell me.

Even after enrollment as research subjects, these patients receive no medication because nothing has as yet been approved specifically for MCI. The interview data suggest that patients are often disappointed because no drugs are prescribed, and one or two complained that they feel like guinea pigs.

John Morris, one of the most experienced Alzheimer neurologists in the United States, has been part the Alzheimer's Disease Research Center at Washington University in St. Louis since 1997, a group that he currently heads up. When I talked with him he said, "[P]eople don't like to have terms like MCI thrown at them. We predicted this from our clinical experience. It's not a disease. People say, 'so I've got to wait to see if I dement?'" Morris added, "No one wants to be told they've got Alzheimer's but it's better to know what one's dealing with and develop some management plan than go into some nether world." A diagnosis of early Alzheimer disease is given to patients seen by the Morris group, and the term "MCI" is avoided altogether. When asked how many of these patients "rement," John Morris said, "Oh, I think out of 640 people 7 reverted. We are pretty good at what we do." In contrast to many other AD research centers, this group does not compare the results of the battery of neuropsychological tests that they give their clients to standardized norms to assess the significance of the test results. Instead, clinicians interview and test people repeatedly over many years, often 5 or even 10 years, and the previous test results are used as the standard against which to assess possible deterioration. In order to do this effectively, a family member or companion is also consulted.

The Washington University group has the advantage of working closely with families with whom they have built ongoing liaisons as research subjects over the years, many of whom live relatively close to the hospital. Few city-based tertiary care hospitals and memory clinics have this advantage—patients usually come from far and wide. William Thies, the chief medical and scientific officer of the Alzheimer's Association, made it clear why not knowing something about the life history of patients seen in a memory clinic can be a problem:

> One of the big complicating factors at the moment is that our cognitive tools are so dull that we really are not good at telling how

cognitively intact people are. So, with both the low end and the high end of cognitive function, these tools are not very useful because there's a population in the world that's MCI from birth . . . some people just aren't blessed with much in the way of tools for making their way through life. So they're going to have problems that are going to trigger any one of the tests that we currently use, almost at any age. If you see these people at 70 they probably won't be as good as they were at 25 and you'd clearly label them with cognitive problems. On the other hand, they're probably not going to progress the way people with Alzheimer's do because they may well not have it. And on the other end of the scale, people with great cognitive abilities cope for a long time, even with AD.

It is possible that the responses set out above given by individuals diagnosed with MCI are a portent of widespread concerns that may well erupt among the public at large, if and when testing for dementia biomarkers becomes routinized. Will the "chronic anxiety" that Robert Castel associates with risk predictions indeed become embedded to a greater extent in everyday life than is already the case, when people pause to contemplate their aging and death? With this in mind, I turn to media reporting about the move to the detection of prodromal dementia.

Animating Public Interest in New Directions in Alzheimer Research

Between early 2010 and late 2011 the well-known science reporter Gina Kolata published 19 articles about Alzheimer disease in the *New York Times*. These articles were crafted around interviews she had conducted with leading scientists and delve into hot topics in the AD world. The essays quite often appeared on the front page of the *Times*, and some are challenging to digest due to the density of the technical language. Of the articles, 11 focus on the early detection of presymptomatic AD. Kolata notes in one essay that 2010 was a time "when news about Alzheimer disease seems to whipsaw between encouraging and disheartening."[37] Even so, her own reports between June and mid-December of that year struck a very positive note. In June her contribution to a series of articles titled "The Vanishing Mind" was about a discovery made by Daniel Skovronsky, whose small start-up company had been working on what she described as "a dye" and a "brain scan" that would show plaque building up in the brains of people with AD. Kolata reported on this same research a second time when it was presented at the major Alzheimer meeting in Honolulu in July 2010 and quoted a researcher who described these findings as among "the most highly anticipated data from the entire meeting." Scans of 35 people in hospices, whom it was judged would die in the following six months, some with AD and others not, had been assessed, and, according to Kolata, "the data showed that the scans were completely accurate in ruling out Alzheimer pathology" in those individuals who had not been diagnosed with the condition. Kolata noted, "[U]nlike doctors [making diagnoses in clinics], they [the scans] never said people had Alzheimer's pathology when they did not." Reisa Sperling, a Harvard neurologist whom Kolata interviewed, said she was convinced by the findings, and many experts readily agree that neuroimaging is proving useful in ruling out certain false positive AD diagnoses. But the bigger challenge, Sperling stated, will be to see if the imaging accurately *predicts* that people are developing AD before they exhibit symptoms.[38]

In August 2010 Kolata published another article in which the opening sentence stated, "Researchers report that a spinal fluid test can be 100 percent accurate in identifying patients with significant memory loss who are on their way to developing Alzheimer disease."[39] Kolata notes that such spinal fluid tests are already commercially available, in contrast to the brain scans described in her earlier articles. The day following this publication, Kolata did a nine-minute interview on the *Charlie Rose* show in which she essentially repeated the claim made in her article the previous day. This interview was not rated highly by viewers, receiving only three out of five possible stars, perhaps because people's minds went blank with talk of spinal taps; even so, its impact must have been great. A flurry of blogs appeared in response to the *Times* article, by far the majority written by people with statistical expertise, many of whom appended PhD to their name. One comment critical of Kolata's 100% claim appeared in the blog written by Arthur Brisbane, an editor who writes the Public Editor's Journal for the *Times*. He first made clear what the spinal fluid test had actually shown before hammering home his critique:

> The test showed . . . that 100% of the MCI patients who developed Alzheimer's over a period of five years had demonstrated the threshold protein levels and memory loss. What the test did not show, however, was that everyone with the threshold protein levels and memory loss actually went on to develop Alzheimer's. Some, in fact, did not. They might yet develop Alzheimer's, but they had not as of the study date. The key point here is that "100% accuracy" is only correct in the narrower sense that 100% of the Alzheimer's patients who had had MCI five years earlier were shown to have carried the threshold protein levels when they still had MCI. But "100% accuracy" is not correct in the sense that everyone who carried the threshold protein levels developed Alzheimer's. I could go on making further distinctions about the study, its structure and findings, but the risk of saying something inaccurate grows, so I will stop and ask the question: What went wrong here and what should the story have said instead? My take is that danger awaits stories that venture into the land of 100%—or any other absolute, for that matter. Stories that report on

something that is a "first," a "biggest," an "only"; stories that employ "never," and stories that predict with absolute certainty are often headed for trouble.[40]

The *Times*, acknowledging errors in Kolata's article five weeks after its appearance, eventually published a "correction," but there can be little doubt that many readers and TV viewers walked away with the idea that those among us destined for AD can now be picked out with certainty.

The most damning criticism of this article and of the *Times*'s publications on AD science in general came from the Alliance for Human Research Protection. The following comments were made by this organization in a September commentary:

> Kolata's article implied that a screening test to identify Alzheimer's was 100% accurate—there is no such test. Furthermore, it was tested in about 50 people. Furthermore, much like psychiatric screens for depression, schizophrenia, etc., the test is poor in specifically identifying who will develop Alzheimer's. Instead, it would misidentify far more people—"false positives"—who will never develop Alzheimer's, but they would be at risk of being exposed to experimental "prevention" treatments. . . .

> [A] series of articles by Gina Kolata (many on the front page) of *The New York Times*, under the heading The Vanishing Mind, purport to examine worldwide struggle to find answers about Alzheimer disease. However, the articles were written as if the *Times* were in the business of promoting dubious commercial screens and tests and hype about unproven therapies.

> The series has been a disservice to the public—and particularly to families struggling with the care of a loved one suffering from this devastating disease—or cluster of diseases—no one knows for sure, as there is no fail-safe test to confirm the presence of Alzheimer's.[41]

The Alliance points out that prior to this publication it had already criticized Kolata's earlier "promotional" articles in the *Times*, and suggested that readers would do well to read what Mark Smith had to say in *Forbes* magazine when interviewed by Robert Langreth (see chapter 2).

Prior to the appearance of the correction, Kolata went on to publish two more *Times* articles that July. The first was a summary of the new guidelines for the diagnosis of AD designed to enable the early detection of Alzheimer's, discussed at an international meeting in Hawaii in 2010. Kolata pointed out that, if adopted, this would constitute a change in the recognized criteria for AD for the first time in 25 years. And, she added, "[S]ome experts predict a two- to threefold increase in the number of people [diagnosed] with Alzheimer disease. Many more people would be told they probably are on their way to getting it." Kolata made it clear that although biomarker testing would enable an early diagnosis, to date none had been formally approved for routine use. She cited Paul Aisen, a researcher who works primarily on drug development, who said that he "foresees a day when people in their 50s routinely have biomarker tests for Alzheimer's and, if the tests indicate the disease is brewing, take drugs to halt it."[42] In this same article Ronald Petersen noted that such a move would result in some difficult conversations, but it would allow patients and their families to plan their lives. The psychiatrist P. Murali Doraiswamy added a cautionary note when he stated that there may well be unintended consequences associated with this testing: "We ought to be cautious that we don't stimulate all this testing before we can give people something to manage their disease. There is no point in just giving them a label."[43]

The second July article published by Kolata was titled "Drug Trials Test Bold Plan to Slow Alzheimer's." She noted that about 100 clinical trials were in progress designed to block amyloid accumulation (a number that continues to escalate to this day), and that companies are eager to apply them to combat the disease at its earliest stages, once the trials are successfully completed (Kolata did not note that all similar trials to date had failed). She observed that the FDA has "raised problems" because there is no satisfactory proof as yet that blocking amyloid makes any difference to AD onset or progression and, given how slow is the progression of the disease, it is extraordinarily difficult to obtain such proof. Kolata's article spells out the approach to this dilemma taken by the drug company Bristol-Myers. This company has designed trials in which patients with MCI and memory loss who have biomarkers compatible with AD (so-called prodromal dementia) are tracked using PET scans. The objective is to see if, when plaque formation is reduced in their brains on the administration of the drug being assessed, the patient's behavioral symptoms simultaneously improve. Paul Aisen argues that the moment a trial such as this shows symptom reduction, everyone will move into further trials designed to prevent AD. But Russell Katz of the FDA continues to be very cautious, as Kolata's article makes clear. Katz told her, "[T]he great fear is that maybe amyloid has nothing to do with the disease."[44]

Articles by Kolata continued to appear during the fall of 2010. Her December 13 publication was titled "Insights Give New Hope for New Attack on Alzheimer's."[45] Kolata attempts to summarize several of the numerous new approaches in molecular biology designed to tackle the Alzheimer puzzle in this article:

At this point, with so many threads of research pointing to so many ideas about Alzheimer's, everything is a target for treatments to prevent or slow the disease—enhancing the brain's beta amyloid disposal system, interfering with nerve cells' feedback loops, blocking tau, protecting the brain's default network by focusing on its unique metabolic properties.

But, she adds, researchers say that "the best hope for the immediate future is with experimental drugs, now being tested, that slow beta amyloid production." On December 17 yet another article appeared in the *Times* in which Kolata raised the matter of a moral dilemma—one of telling people that they may be on the road to Alzheimer disease years before the condition is likely to become manifest, especially when no medication is available. This is a dilemma that has been raised by virtually all of the AD researchers with whom I have talked over the years, the majority of whom believe that patients should not be given the results of tests designed to spot biomarkers, including neuroimaging findings. Mony J. de Leon, a neurologist whom Kolata interviewed, said that he tells people who come to his memory clinic that they are at "increased risk," "decreased risk," or "in the middle," based on results of tests, including spinal taps and neuroimaging. When pressed by patients, de Leon declines to give any further information and reminds them that they are in his clinic to do an experiment. He claims he is constrained by his ethics board, which does not support giving risk assessments to patients when no effective treatment exists.[46]

Another Kolata article, dated January 20, 2011, had the title "F.D.A. Sees Promise in Alzheimer's Imaging Drug." It was reported that the FDA had approved unanimously the use of brain scans to detect amyloid, provided that radiologists agree on what the scans "say" following training. Reisa Sperling told Kolata that this vote "has moved us a monumental step forward" so that "we will not just be guessing clinically." Murali Doraiswamy of Duke University was once again quick to provide a cautionary comment: "Some people have plaque without having Alzheimer's, so if a scan shows plaque, doctors will have to use their clinical judgment, taking into account a patient's symptoms, in deciding what the scan results mean." However, he added, if the scan shows no plaque then AD can be "safely" ruled out—a great advantage of this new imaging technology. Doraiswamy suggests that family doctors will soon be confidently using this type of neuroimaging, but he does not make clear how they would gain ready access to imaging machines, nor does he note the cost involved.[47]

The *New York Times* is not alone in such reporting, but this newspaper has been more prolific than any other publication written in English, and Gina Kolata is not the only reporter to write on this topic for the *Times* (and I have not discussed all of her publications). These articles did not appear in a vacuum, of course; on the contrary, virtually all of them are directly related to events happening in the AD research world, the most striking of which has been the move among experts to prevention of AD. Comments posted in response to the media included the following: "[A]n MRI or lumbar puncture as a requirement for running for office may avoid another Ronald Reagan, who most certainly had brain plaques during his Presidency," posted by "DP of New York City."

And it was noted on a Canadian website, "Governments are going to have to tax young people at 80% of their income to pay for the Baby Boomers."

Very few public responses to media articles about the new diagnostic criteria show concern about the uncertainties associated with predicting the future by means of biomarker testing; nor is the current limitation on their use to research settings noted—something that has been carefully reported in most media sources.

Revising the Definition of Alzheimer Disease

The Working group convened by Bruno Dubois and colleagues in 2005 (discussed at the end of chapter 2) continued to meet after its first publication in 2007. In July 2008 this group held a session at the annual meeting of ICAD (International Conference on Alzheimer's Disease, now known as AAIC, the Alzheimer's Association International Conference) that took place in Chicago, followed in 2009 by a research symposium held in Vienna during which time the number of people involved in this group had grown to 50 participants with both academic and pharmaceutical connections. In 2010, the group put out a second publication in *The Lancet Neurology* to which 24 authors contributed, 5 more than in the previous piece. This article was titled "Revising the Definition of Alzheimer's Disease: A New Lexicon." It is stated that one aim of the article is as follows:

> . . . to advance the new research criteria initiative by providing a companion lexicon wherein the different entities and concepts related to AD are defined and updated. This lexicon for AD is primarily intended to serve the research community by providing a framework of the disease that covers its full spectrum, and should be used for research protocols and clinical trials directed at early intercession in the pathogenic cascade of the disease. The potential to test disease-modifying interventions adds urgency to the need for a shared lexicon.[48]

A second objective is to provide clinicians with a clear view of this evolving field "in which use of biomarkers is advancing and might reach regulatory qualification and approval in the foreseeable future."[49]

The core intent of the article is to articulate a new definition of Alzheimer disease: "[W]e propose that the term 'Alzheimer disease' should refer only to the in-vivo clinicobiological expression of the disease and that it should encompass the whole spectrum of its clinical course." The intent is that, in the clinic, "episodic memory impairment" and "supportive biomarkers" will be regarded as a definitive diagnosis of AD; the qualifiers "probable" and "possible" that had until now to be affixed to a clinical diagnosis are no longer required, and postmortem evidence will no longer be the final arbitrator of the diagnosis. The article points out that further research will be needed before this proposal is validated, at which time the new diagnostic label will become "neuropathologically verified AD" determined in the clinic.

It is proposed that the early stages of this condition will be known as pro-dromal or predementia AD, and the term "AD dementia" (as distinct from other types of dementia) will be utilized when the cognitive symptoms are severe enough to interfere with daily functioning. The authors also propose two circumstances in which labels will be applied to individuals without any memory loss symptoms. These are "preclinical states of AD" including an asymptomatic "at-risk" state for AD to be applied when biomarkers alone can be detected without many memory loss. Second, "presymptomatic AD" will be used as a label for individuals without memory loss in the small number of families worldwide who are at risk for early-onset AD due to the presence of one of three autosomally dominant genes in their family known to cause AD (see chapter 5). Using this new lexicon, Alzheimer pathology refers to all the underlying neurobiological changes "responsible" for AD that span the range of pathogenic events that take place in the brain. No attempt is made to rec-oncile these new clinical diagnoses with Alzheimer neuropathology detected postmortem, even though it is acknowledged that there has been a great deal of discrepancy about this, due to the oft-repeated findings that so-called nor-mal individuals exhibit AD pathology at autopsy. This group also recognizes mixed AD as a valid diagnosis.

It is argued in this position piece that because MCI is heterogeneous and remains a syndromic classification, it has limited utility as a recognized med-ical entity. However, MCI will continue to be useful as a category for labeling individuals who do not meet the criteria for prodromal AD, notably because the biomarker evidence is lacking and uncertain, or testing simply has not been done.[50] The conclusion of Bruno Dubois et al. makes clear why a new "gold standard" for the diagnosis of AD is thought by them to be so important: "Uniformity of definitions will assist in constructing trial populations and comparing results across trials."[51] Kolata did not refer to this working group in her articles, perhaps because the group is based primarily in Europe and only 6 of the 24 researchers work in the United States. But this 2010 publication was extensively covered by Reuters and other major media sources and was cited over 60 times in its first year of publication.

It will be seen in the following chapter that several of the assertions made by this group are apparently set to one side by a yet larger, American-dominated group of experts who published a series of articles in 2011 about the preven-tion of AD. This chapter elaborates further on the reasons given for the aban-donment of the century-old dualistic clinical/neuropathological diagnostic approach, on the accounts by experts of what from now on will constitute em-bodied risk for AD, and on the ongoing efforts to standardize the "biological footprint" of AD as the principal signifier of molecular risk.

Chapter 4
Embodied Risk Made Visible

The relation between what we see and what we know is never settled.

—John Berger, *Ways of Seeing*[1]

At the 2010 Honolulu meeting of ICAD (International Conference on Alzheimer's Disease), a news briefing was held on July 13 in which information that had previously been embargoed from release was first made public. The briefing stated that by 2009 a broad consensus had become apparent among experts working on dementia that the criteria for an AD diagnosis needed to be revised on the basis of scientific advances made in the field over the past quarter century. With this in mind, the National Institute on Aging (NIA) and the Alzheimer's Association set up advisory meetings composed of individuals from both academia and industry, and it was decided that three working groups should be formed. The Honolulu briefing reported on the progress these groups had made over the preceding year.

The tasks that the groups had been assigned were to examine and make recommendations on three specific conditions: "preclinical Alzheimer disease" (to be detected through biomarker testing); "mild cognitive impairment (MCI) due to Alzheimer disease," and "Alzheimer disease." Two of the groups were composed of 11 members and the third had 12. It was striking that virtually every involved expert was associated with an American university and or drug company. Of the two exceptions, one was the neurologist Bruno Dubois of the Hôpital de la Salpêtriére in Paris, the driving force behind the European working group convened in 2005 to redefine what would count as Alzheimer disease. Despite the dominant American presence in these newer groups, it was explicitly asserted that a balance of expertise existed in and among the three groups, and that they had international representation. Following the Honolulu meeting, the recommendations for changes to be made to the AD diagnosis by these groups were posted on the Alzheimer's Association website and other sites, allowing for a period of public commentary. The plan was that the comments were to be assessed and incorporated, if appropriate, into the revised documents before publication in the journal *Alzheimer's & Dementia* in the spring of 2011.

Both critical and supportive commentary, scientific and otherwise, appeared online. One oft-repeated statement, to the effect that the groups were clearly not international in representation, resulted in a small modification to their final constitution. Immediately before the final round of discussion and revisions took place in preparation for drafting the prepublication documents, two experts from the United Kingdom and one each from Canada, Japan, Australia, and the Netherlands were distributed among the three groups. These researchers, together with Bruno Dubois, 7 in all, formed the international component among the more than 30 Americans. It is significant that particular emphasis was given toward the end of the prepublication discussion in the fall of 2010 to the necessity of striving for a "harmonization" of biomarkers.

Public Critique of the New Proposals

An op-ed appeared in the *New York Times* on July 19, 2010, written by Sanjay Pimplikar, professor in the department of neurosciences at a Cleveland research institute. Pimplikar had been present when the news briefing took place in Honolulu, and he had also read Gina Kolata's article of July 14 in the *New York Times* in which she reported on the release, citing researchers who claimed that these new guidelines would mean earlier diagnoses and that they represented a "major advance" in the Alzheimer field. Pimplikar's op-ed, titled "Alzheimer's Isn't Up to the Tests," takes strong issue with the notion that the new proposals represent a clear advance in the AD field; above all he expresses concern about the move to systematic use of biomarker detection in order to assist with early diagnoses. He writes, "Although these recommendations are well intentioned, evidence suggests that it would be a mistake to adopt them at this time."[2] Pimplikar points out with respect to the proposed use of PET scans to detect amyloid plaque in vivo, that patients would be systematically exposed to radiation. Furthermore, he reminds readers, plaque has repeatedly been found to be present in the brains of about one-third of normal elderly individuals. He adds for good measure that 11 drug trials had failed to show cognitive improvement in individuals after plaque had been removed from their brains.

Pimplikar goes on to consider routine use of spinal fluid analyses, a second biomarker recommended in the new guidelines. He notes that spinal taps apparently show promise, but the findings nevertheless remain uncertain and the cost is considerable. Moreover, the procedure can be painful, is liable to have side effects, and is not easy for practitioners to perform unless they are very skilled. One of the greatest difficulties in connection with biomarker detection emphasized by Pimplikar is the possibility for misdiagnosis and false positives. He makes an analogy to the routine use of the PSA test used to detect prostate cancer, in which it is now widely agreed that overdiagnosis and overtreatment have been common.[3] In conclusion, the op-ed reminds readers that no cure exists for Alzheimer's, with the result that early detection tests may well do more harm than good to public well-being.

Prior to and immediately following the appearance of this op-ed and associated articles, the following letters appeared in the *Times*:

To the Editor:

Re "Drug Trials Test Bold Plan to Slow Alzheimer's"

As many researchers have pointed out in the last 15 years, it is unlikely that amyloid is the causative agent of neurodegeneration, the real cause of dementia. Therefore, it is unlikely that removal of amyloid would stop the progression or reverse the disorder.

Unfortunately the amyloid hypothesis has acted too long as a red herring, robbing the Alzheimer disease field of precious financial resources needed to develop new avenues of investigation.

The persistent focus on amyloid may be best explained by both the lack of alternative scientific theories and the reluctance of the pharmaceutical industry to abandon years of sunk costs in amyloid-based research and development.

Nikolaos K. Robakis
New York, July 17, 2010

The writer is a professor of neuroscience and Alzheimer disease research at Mount Sinai School of Medicine.

To the Editor:

Sanjay W. Pimplikar is absolutely correct. Even assuming that the new diagnostic test for Alzheimer's is 100 percent accurate, what good does that knowledge do? There is no drug that cures the disease, only ones that mitigate the symptoms at an early stage.

My wife died four years ago from Alzheimer's at age 69. She and her family suffered with the disease for seven years after the initial diagnosis. Had we known earlier, everyone would have suffered even longer.

While early detection may have some merit, the real breakthrough will come when we know and understand the cause of this disease.

William Eisen
Philadelphia, July 20, 2010

To the Editor:

Over the last two months, your outstanding series of articles and an Op-Ed essay have emphasized the mysterious nature of Alzheimer disease. A particularly lively argument focuses on the role of amyloid, sticky clumps of material that build up between brain cells.

There are amyloid believers (like Dennis J. Selkoe and me, both of us quoted in your July 17 front-page article "Drug Trials Test Bold Plan to Slow Alzheimer's") and amyloid agnostics or naysayers (like Sanjay W. Pimplikar, whose Op-Ed article, "Alzheimer's Isn't Up to the Tests," appeared on July 20).

. . . The amyloid hypothesis may ultimately fail (although I suspect it is right), but it is backed by solid science and worthy of proper evaluation. We must not abandon the quest halfway, without performing prevention trials and getting a definitive answer to the amyloid question.

Sam Gandy
New York, July 21, 2010

The writer is a professor of neurology and psychiatry and an associate director at the Mount Sinai Alzheimer disease Research Center and the James J. Peters V.A. Medical Center.

To the Editor:

Scientists' understanding of Alzheimer disease may not be clear enough to develop tools for diagnosis and treatment. But Sanjay W. Pimplikar falls into the trap into which many of us physicians find ourselves: thinking that without a medical treatment, "many individuals would simply prefer to be spared the emotional trauma of a diagnosis."

This runs contrary to the spoken wishes of many people with memory loss. They are grateful to hear that their disorienting and frightening experiences have a name. And if the Alzheimer disease is diagnosed early . . . they can actively plan for their future.

Susan Czapiewski
St. Louis Park, Minn., July 21, 2010

The writer is a geriatric psychiatrist who treats Alzheimer's patients.

These opinions, strikingly at odds with one another, express common tensions at work in the AD world today that have been evident in earlier chapters. When I interviewed Sanjay Pimplikar in Cleveland two months after the publication of the op-ed, he stated ruefully that he had been wondering whether writing it had been "a smart thing to do." He had received negative feedback from several of his colleagues that disturbed him:

> [I]t really feels like being on a rubber dinghy out on the ocean in a storm—one moment you are high up, and at the next moment you are down there. I get so many emails and phone calls from people from all over. It's probably a ratio of 50:1—the 50 being people I don't know. But most of the messages make me feel good. There is a reward, but I haven't been able to get on with my science—that's a big price.

Pimplikar describes himself as someone who works on "membrane trafficking," and believes that he is in many ways an outsider to the AD field because initially he was interested in the APP (amyloid precursor protein) as a marker in a general way, but not in connection with AD per se. In his early work he studied a small cytoplasmic fragment that results from enzyme cleavage of APP that had not at that time been researched by anyone. His team found that this fragment has an effect on the expression of several genes, and the group eventually produced a transgenic mouse that expresses the cytoplasmic fragment alone; such animals show "many of the characteristics of AD—tau pathology, neurodegeneration and inflammation—but no plaques," says Pimplikar: "You have to ask a philosophical question; if they don't have plaques, do they have Alzheimer's?" The neurologist Peter Whitehouse was present at the interview, and he interrupted at this juncture to say, "I'm not sure that people have ever seen tangles in mice." Pimplikar responded, "The question is how do you define a tangle. By twelve to fourteen months, as the mice start to get old—they live up to two years [he inserted for my benefit], we start to see tangles with Bielschowsky's stain, but if you want them confirmed using electron microscopy—we haven't done that yet." This settled the matter for Whitehouse, exhibiting neither plaques nor having confirmation of tangles, he said emphatically, "no mouse gets Alzheimer disease."

When discussion turned away from mice to humans, it was clear that Pimplikar and Whitehouse have a fundamental difference of opinion that extends beyond the utility of animal models. Taking friendly issue with Whitehouse and the title of his recent book, *The Myth of Alzheimer's*, Pimplikar insisted, "In practical, real terms, Alzheimer's is a disease and by no means a myth . . . once we understand it, it will be preventable. Maybe not curable once it becomes fully a disease, but preventable . . . if I thought it was not preventable then I would probably be studying something else. . . . For me, Alzheimer disease is like a person who's not physically fit who finds him or herself on the expert ski slope. They go down, because they have no choice, but in a very bad way; out of control, and there's nothing you can do to stop it. . . . We all age, but some of us age in a graceful manner whereas others simply lose it." Pimplikar is in agreement with the position taken by his neuropathologist colleague, the late

Mark Smith, that inflammation may well be a crucial component of AD, and this is why he believes the condition may be preventable.

While talking with me, Pimplikar added that two principal concerns had made him write the op-ed: first, that the *Times* publications were, in his opinion, making the AD problem much too "black and white," whereas in real-life research "there is never a 'yes'–'no' answer—it's always, always a shade of gray." He added that Kolata's article "seemed to be trying to promise people something that is really not true." The second and yet more troubling concern for Pimplikar was the "enormous financial burden that would be placed on society" by moving to biomarker detection when the "outcome would be really miniscule."

Sanjay Pimplikar's op-ed had pleased Peter Whitehouse and brought about a restoration of their good relationship that, although it always remained congenial, had suffered somewhat following the publication of Whitehouse's book. In *The Myth of Alzheimer's* one of Whitehouse's major claims is that "so-called Alzheimer disease cannot be differentiated from normal aging . . . there is no one biological profile of Alzheimer's that is consistent from person to person, and all the biological hallmarks of AD are also hallmarks of the aging brain."[4] His argument is also one about inappropriate taxonomic classification in which he insists that numerous conditions, and not only several kinds of dementia, are routinely inappropriately sorted into the Alzheimer's category. His purpose in writing *The Myth* is to expound more successful ways of aging, based largely on lifestyle changes, family support, and community interactions, including cross-generational activities involving the young and the elderly.

As a basic scientist who works with mouse models, Sanjay Pimplikar's discussion of AD tends to be constrained by what can be produced and reproduced consistently in the laboratory. His approach and vision are very different from those of Whitehouse. He stated again and again during the interview that he is not a clinician and his task is not to tend to troubled patients; his work is to get the science right. But he added that he dislikes drug companies and, because they need payback, "we are stuck with amyloid because such a lot of money has been put into it . . . it takes a long time for science to be self-correcting. . . . And for those people who believe in the amyloid hypothesis, no amount of negative evidence is going to convince them to change their minds. It just doesn't . . . the human mind just doesn't work that way . . . it's going to be like the Copenhagen debate between Heisenberg and Niels Bohr. . . . Heisenberg, smart as he was, never got the probability part of quantum theory . . . but after a while, everyone moved on [it was actually fission theory that was at issue]." In contrast to Whitehouse, Pimplikar believes that we must struggle until we better understand the involved molecular biology; only then will we be able to tackle AD prevention. On the other hand, Whitehouse argues that because of the inherent, seemingly irresolvable entanglement of aging and dementia, we should focus on improving the quality of life of people as they age, a strategy that may well reduce the incidence of dementia. An incisive normal/pathological distinction cannot be made as far as he is concerned, and putting the bulk of our resources into drug development is not appropriate.

Many clinicians and researchers, regardless of their competing positions about AD ontology, have been quick to produce cautionary responses about the highly publicized move toward systematization of a presymptomatic diagnosis of AD. Even so, the AD juggernaut is proceeding inexorably toward identifying and attempting to standardize embodied biomarkers and associated estimates of AD risk. In order to carry out this feat, a new definition of Alzheimer disease is indispensible. The earlier formulated European publications discussed in the previous chapter were not sufficient; the NIA and the Alzheimer's Association in the United States had also to make their position very clear, with the result that their revised three-part report was published in April 2011, and a fourth part, on the neuropathologic assessment of AD, was published in 2012.[5]

Among the numerous authors of the companion articles that elaborate on the new recommendations, by far the majority had to report "conflict of interest disclosures," because they receive money from, consult for, and/or have stocks in drug companies. The exceptions are those individuals employed by the Alzheimer's Association or the government of the United States, and two university-based researchers.

Unhooking the Clinical-Pathological Entity of an Outdated AD

In previous chapters it has been pointed out repeatedly that over the years, and with increasing frequency in the past decade, research has shown that no tight correlation exists between clinical symptoms associated with AD and the neuropathology believed to underlie Alzheimer disease. The introduction in the journal *Alzheimer's & Dementia* that precedes the three-part companion articles not only is explicit about this lack of fit, but also makes it clear that this incongruence has been the wedge that has opened up a new approach to AD. With the purpose of moving AD research forward, the recommendations set out revised NIA–Alzheimer's Association criteria in which a "semantic and conceptual distinction is made between AD pathophysiological processes (AD-P) and the various clinically observable syndromes that result (AD-C)."[6]

The authors of the essay emphasize that research over the past quarter century has shown repeatedly that distinctive molecular changes in the form of biomarkers consistently occur in individuals *prior* to the onset of dementia due to AD. They add that these changes are often accompanied by mild clinical symptoms, changes that do not warrant a diagnosis of AD, but are characteristic of "an intermediate stage between normality and dementia."[7] The new recommendations state that from now on these intermediate changes—biomarkers and mild clinical symptoms—will be systematically incorporated into the diagnostic procedure.

The authors affirm that evidence of neuritic plaques containing beta-amyloid present in some parts of the neocortex, together with regional distribution of neurofibrillary tangles, will continue to be the defining features of AD-P.[8] An additional comment in the introduction about genetic findings makes the claim that "the available genetic risk data overwhelmingly point to the β-amyloid pathway as the initiating, or at least a very early pathophysiological event in the disease cascade."[9] Emphasis is also given to a time dimension associated with the development of AD pathology—individuals do not suddenly become demented, nor do they exhibit all the molecular changes at the same time. Notably, amyloid pathology and neurodegenerative pathology in the form of tangle formation and neuronal/synapse loss usually take place on different time scales.[10] β-amyloid pathology develops first, during a long presymptomatic phase of AD now well documented by neuroimaging. In contrast, neurofibrillary pathology and synapse loss begin to accelerate shortly before the appearance of the symptomatic phase of AD, during the "intermediate stage" between normality and dementia (emerging research suggests that this timeline is by no means as clear-cut as is claimed in these revised guidelines).[11]

If the diagnostic gold standard of plaques and tangles is to remain in place as the embodied representation of AD, has a significant change in AD conceptualization taken place or not? Given that a long presymptomatic time dimension of up to two decades or more is now acknowledged, during which supposedly significant molecular changes inherent to AD formation take place, AD is now undeniably understood as a process—a continuum—one that results in localized pathology. The current move to prevention, involving a temporal backtracking in order to bring about a nondemented future, is described in the medical literature as a "paradigm shift," and the emphasis on a continuum of long duration not fully appreciated until now by involved experts is indeed new. Biomarker technology has made this insight possible. However, the dominant model of AD causation survives unchanged, and amyloid deposition continues to be the prime event of interest—there is no new theory about AD causation—suggesting that Thomas Kuhn would almost certainly disagree with the neurologists now making a claim for a paradigm change.[12] Rather, the new approach to Alzheimer disease is perhaps an example of a shake-up in "normal science" in which the dominant model remains, although subject to considerable criticism and modifications.

However, given that it is freely acknowledged that persistent anomalies in research findings have been "the wedge" that has opened up a new approach to AD, it is possible that the field is entering a preparadigmatic stage prior to a major shift in orientation. Kuhn argued that "a proliferation of competing articulations, the willingness to try anything, the expression of explicit discontent, the recourse to philosophy and to debate over fundamentals" constitutes a crisis in a scientific field, and it is quite conceivable that out of such ferment new

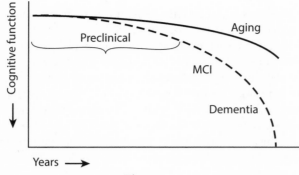

Figure 4.1.

The continuum of Alzheimer disease (AD). Model of the clinical
trajectory of AD. The stage of preclinical AD precedes mild cognitive
impairment. Note that this diagram represents a hypothetical model for
the pathological-clinical continuum of AD but does not imply that all
individuals with biomarker evidence of AD-pathophysiological process
will progress to the clinical phases of the illness. Reprinted from *Alzheimer's & Dementia*: 7, no. 3, Sperling et al., "Toward defining the preclinical stages of Alzheimer's disease: Recommendations from the National
Institute on Aging-Alzheimer's Association workgroups on diagnostic
guidelines for Alzheimer's disease", 280-292, © 2011, The Alzheimer's
Association, with permission from Elsevier.

ideas may well arise, leading to new methods, and finally to a new theory—this
would then constitute a paradigm change.[13] As Ian Hacking notes, for Thomas
Kuhn, novelty is the hallmark of science, "without revolution, science would
degenerate."[14] Kuhn was, of course, initially a physicist, and he had physics in
mind when he wrote *The Structure of Scientific Revolutions*. We will return to this
discussion in the concluding chapter with reference to the complexities of the
biology of Alzheimer disease, in which the law-like characteristics of physics
are nowhere apparent.

Striving for Radical Change

Following the introduction, the first of the three companion pieces in the report titled "The Diagnosis of Dementia Due to Alzheimer Disease" outlines diagnostic criteria that are flexible enough to be used by both general health care
practitioners and specialized investigators. It is argued that several features of
the original NINCDS-ADRDA criteria that have been in extensive use over the
past 27 years should be modified. In contrast to the European recommendations
noted in chapter 3, the categories of "possible" and "probable" AD are retained
in order to distinguish such patients from individuals with other kinds of dementia. The article concludes that a diagnosis of AD dementia should be made
by means of clinical assessment; in everyday clinical practice, such a diagnosis is

indispensible, and biomarker support, if used at all, should be secondary—this emphasis on clinical assessment differs from the European position.

The second of the companion articles deals with the diagnosis of MCI. This article stresses that it is particularly challenging for clinicians to identify transition points when patients progress from a presymptomatic to a symptomatic predementia phase because this progression is one of a slow steady change with no sudden events to define it. The core clinical criteria for determining MCI outlined in this article are expressly designed to be of use in every kind of clinical setting (again a position that differs from the European recommendations). Like the first of these companion pieces, it is emphasized that biomarker testing should be limited to research settings, but even so they can provide confirmation of a clinical diagnosis. The authors use the term "mild cognitive impairment (MCI) due to AD" to distinguish this state from cognitive impairment due to other forms of dementia. They stress that the degree of cognitive impairment diagnosed in this category is "not normal for age" (see Appendix 1 for a summary of clinical and cognitive evaluation for MCI due to AD).

The final companion article sets out by warning readers once again of the negative economic consequences looming on the horizon if nothing is done to lower the global incidence of dementia. Citing data from the Alzheimer's Association, it is noted that more than 13.5 million individuals in the United States will manifest AD by the year 2050 if the disease incidence continues at the present rate, with projected Medicare costs of $627 billion. A reduction in onset of the disease by five years would halve this bill.

This article makes an argument for moving to a preventive mode because of the dismal results to date in attempting to deal with AD, notably a lack of effective therapeutic interventions. Once again it is stressed that these recommendations are being made for use in research settings alone: "The extent to which biomarkers of AD-P predict a cognitively normal individual's subsequent clinical course [to AD-C] remains to be clarified, and we acknowledge that some of these individuals will never manifest clinical symptoms in their lifetime."[15] Subsequently it is claimed, "Recent advances in ante-mortem biomarkers now allow us to test the hypothesis that many individuals with laboratory evidence of AD-P are indeed in the preclinical stages of AD, and determine which biomarker and cognitive profiles are most predictive of subsequent clinical decline and emergence of AD-C."[16] It is suggested that once tested, individuals who are "cognitively normal" but biomarker positive be classified as "asymptomatic at risk for AD." Furthermore, it is pointed out,

> The difficulty in the field of AD is that we have not yet established a firm link between the appearance of any specific biomarker in asymptomatic individuals and the subsequent emergence of clinical symptomatology. If we can, however, definitively determine the risk of developing AD dementia and the temporal course of clinical progression associated with AD-P in individuals without dementia or

MCI, we will open a crucial window of opportunity to intervene with disease-modifying therapy. Although we hypothesize that the current earliest detectable pathological change will be in the form of Aβ accumulation, it is possible that Aβ accumulation is necessary but not sufficient to produce the clinical manifestations of AD.[17]

Following this comment, the authors write, "[W]e acknowledge that the etiology of AD remains uncertain, and some investigators have proposed that synaptic, mitochondrial, metabolic, inflammatory, neuronal, cytoskeletal, and other age-related alterations may play an even earlier, or more central, role than Aβ peptides in the pathogenesis of AD."[18] They also note that arguments persist as to whether the problem of Aβ lies with its "abnormal processing" or, alternatively, with its ineffective "clearance," and, furthermore, that it is now known that different forms of Aβ exist, some of which can have a protective mechanism on healthy brain functioning, although these forms may become faulty under circumstances as yet unknown. In conclusion, the authors remind their readers that genetics is implicated in AD etiology and, moreover, epidemiological data suggest that "there are significant modulating factors that may alter the pace of the clinic expression of AD-P."[19]

A list of caveats appears toward the end of this third companion article: First, many of the studies on which findings about biomarkers are based are likely to have "cohort biases" and are probably not representative of the general population, particularly when the "oldest-old" are implicated. Second, caution is noted because the biomarkers of interest are merely "proxies" for the disease, and may not fully reflect the processes taking place in the living brain. Last, but not least, it is reiterated that "the role of Aβ as the etiologic agent in . . . AD remains to be proven."[20]

The excerpt cited above together with the additional comments make it glaringly clear how many unknowns are involved in this so-called paradigm shift, and it takes little imagination to appreciate the many years of research that lie ahead to establish even a modicum of certainty as to what is involved in AD etiology and, above all, who among us is "at risk."

Even if biomarker evaluations are limited to research settings, individuals who become research subjects will be left in limbo, a heightened state of anxiety that can only intensify once they have tested positive for one or more biomarkers, and then must undergo repeat tests at regular intervals. Given that none of the biomarkers in use are "static" (in researcher language) and rates of change in each biomarker are not consistent over time and follow a nonlinear trajectory, it is unlikely that it will ever be possible to give individual research subjects, and eventually patients, information about their risk status or prognosis on which they might act. Whatever individuals are told could well be thoroughly misleading and, if imparted with appropriate caution, should be filled with equivocations. Even so, when I attended the 2011 meeting of ICAD in Paris four months after the new recommendations had been published, it came as no surprise to hear from one California neurologist as he delivered his research paper that his patients were asking for biomarker tests for AD and that, in his opinion, they have a right to pay for such tests and be given their results, if they so wish.

In February 2012 John Morris, long unwilling to accept the concept of MCI, published an article that clearly signals the extent of the difficulties, clinical, ethical, and social, that will arise as these new guidelines begin to be put into practice. Morris examined the records of over 17,000 individuals who had been evaluated at Alzheimer disease centers in the United States and whose files were submitted to the National Alzheimer's Coordinating Center. He concludes, under the revised guidelines for MCI (see above), "Almost all (98.8%) individuals currently diagnosed with very mild AD dementia and the large majority (92.7%) of those diagnosed with mild AD dementia could be reclassified as having MCI using the revised criteria. This would be based on their level of impairment in the Clinical Dementia Rating domains for performance of instrumental activities of daily living in the community and at home."[21] Morris concludes that the categorical distinction between MCI and milder stages of AD dementia that have been in place until recently are now compromised, particularly by a new expansive use of the idea of "independence in functional activities."[22] He is particularly concerned because the distinction will now depend on the individual judgment of clinicians that will "result in nonstandard and ultimately arbitrary diagnostic approaches to MCI" that will "confound clinical trials . . . complicate diagnostic decisions . . . and research comparisons."[23] Morris highlights the ambiguity already present in the MCI concept, and the repeated efforts over the years to revise and standardize the diagnosis. He insists that with the revised criteria in use

> the distinction of MCI from dementia now simply is a matter of an individual clinician's threshold for what represents one condition vs. the other. . . . Already many individuals with MCI are treated with pharmacological agents approved for symptomatic AD, indicating that clinicians often do not distinguish the two conditions when faced with issues of medical management. It is now time to advance AD patient care and research by accepting that "MCI due to AD" is more appropriately recognized as the earliest symptomatic stage of AD.[24]

Morris notes that in 2010, 1,220 publications appeared in which MCI was listed in the title, keywords, or abstract. It seems likely that publications will continue to appear at a great rate on this subject as experts work to sort out the ambiguities raised by the new criteria, and it is quite possible that diagnostic creep will take place in research settings due to the broadening of MCI diagnostic criteria. The question remains as to whether increased diagnoses of MCI will also take place in most clinical settings—a possibility that calls for systematic inquiry.

Future Prognostications

More than 15 biomarkers are already used in connection with AD research, subdivided into three primary divisions according to their molecular activity: (a) biomarkers of Aβ deposition, (b) biomarkers of neuronal injury (including changes in the protein tau associated with tangle formation, hippocampal volume, and evidence of brain atrophy, and (c) biomarkers of biochemical changes associated with AD (including inflammatory biomarkers, signs of oxidative stress, and/or cell death) (see Appendix 2). In addition many other posited biomarkers are being researched. It is assumed that if a plasma biomarker can be found in the near future, ensuring that a simple blood test is all that will be required to assess who is at risk for AD, it will then be possible to implement a public health strategy in which the prevention of AD takes center stage—but so far this is not in the cards. The two biomarkers that have been primarily examined to date are markers in cerebrospinal fluid (CSF), and the use of PET (positron emission tomography) neuroimaging to demonstrate deposits of plaque in vivo. In addition genotyping for the APOE gene is a third well-researched biomarker (see chapter 5).

A consensus report of the Working Group called together to consider "Molecular and Biochemical Markers of Alzheimer Disease" was published as early as 1998, sponsored by the Ronald and Nancy Reagan Research Institute, an affiliate of the Alzheimer's Association and an NIA working group. This report set out criteria for defining, developing, and assessing biomarkers for AD and stated that their detection should be, above all else, simple to perform, inexpensive, noninvasive, and precise, with high levels of sensitivity and specificity.[25] Genetic testing was included, but the group devoted most of its attention to CSF. In the late 1990s it was not yet possible to detect amyloid in the brain using imaging methods.

It is perhaps not surprising that attention was given early on to the possibility of CSF as a biomarker for AD; CSF has been described on occasion as a "biochemical window" into the brain. It is a clear liquid that surrounds the structures of the central nervous system (the brain and spinal cord), has more physical contact with the brain than any other type of fluid, and is not separated from the brain by the tightly regulated blood/brain barrier. Normal CSF contains Aβ in its usual soluble form, but it has been repeatedly shown that levels of this soluble peptide drop in the CSF when conversion into its insoluble form starts to take place in the brain, resulting in deposits of the plaque characteristic of AD and certain other conditions. It is assumed that a drop in soluble Aβ in the CSF is the first detectable biomarker associated with the onset of the AD continuum, and that it can be consistently detected prior to evidence of any clinical symptoms.

A second biomarker found in CSF associated with the onset of AD is an increase in the total amount of tau protein, and also of phosphorylated tau linked to tangle production. These changes become apparent later than the reduction in soluble beta-amyloid levels, and are associated with a stage in the continuum when MCI or early signs of AD become clinically apparent. A recent editorial in the *Archives of Neurology* titled "Sharpen That Needle" argues that what is measured in these CSF changes "reflects the biochemical composition of the anatomic lesions" of AD.[26] But the editors are quick to point out that other diseases bring about similar changes. It is suggested that these biomarker determinations should be used only as a "diagnostic aid." They add that the Reagan requirement for noninvasiveness is in the eye of the beholder, and that CSF taps are no more invasive than, for example, endoscopies. CSF tests have already been made use of in research settings for a couple of decades.

The writers of this editorial argue for patient education about the worth of spinal taps, and note that until relatively recently granting permission for autopsies was not regarded with favor, although this has changed in recent years—hence public attitudes about procurement of CSF can also be changed. Readers learn that physician training is needed, and it is suggested that specialist clinics might be set up in neurology departments because primary care physicians are neither "skilled nor experienced" in such procedures. These editorialists, "gazing into the future," as they put it, envision a time when CSF analysis will be implemented as a routine screening program to identify clinically healthy individuals at risk for MCI and AD, although the NIA–Alzheimer's Association guidelines specify that, for now, use of these biomarkers should continue to be limited to research settings.

In the same issue of *Archives of Neurology* an article appeared reporting results of a longitudinal study designed to identify biomarker patterns typical for AD without knowledge of information on the clinical diagnosis, using samples of "cognitively normal persons, patients with AD, and individuals with mild cognitive impairment." An AD "signature" was found in 90%, 72%, and 36% of individuals in the AD, MCI, and cognitively normal groups, respectively. The cognitively normal group proved to have a high proportion of people estimated to be at increased risk for AD due to their genetic makeup (see chapter 5). In a related data set of individuals diagnosed with MCI who exhibited these biomarker changes, 100% progressed to AD during the course of the following five years. The explicit assumption on the part of the researchers is that those normal controls who exhibited biomarker changes are without doubt on their way to MCI and then to AD. The authors argue, "AD pathology is active and detectable earlier than has heretofore been envisioned."[27] Emphasis is given to the value of being able to assess more than one marker at the same time, and it is concluded that "the AD signature appears to be naturally present in the data and is expressed as a homogenous group, consistent with a single pathological process underlying AD."[28] Their findings corroborated earlier findings that the first changes of the AD footprint appear to be a drop of soluble β-amyloid in the CSF, followed thereafter by changes in tau. The authors of this article argue that it is not surprising that their cognitively normal research subjects have both amyloid-containing plaques and tau-containing neurofibrillary tangles in their brains, and they cite two other earlier studies that have shown these findings.[29]

This particular article has been cited repeatedly during ongoing discussions in recent months about a preventive approach to AD. No caveats or cautionary notes appear in the article and no mention is made of recent findings indicating that β-amyloid is more complex than has usually been acknowledged to date, or that the changes in both β-amyloid and tau, assumed to be the AD signature, are regularly detected in patients with other neurological conditions.

In contrast, other researchers have voiced criticisms about the claims being made about the value of tracking CSF biomarkers, the most common being that there is, as yet, a lack of standardization, and no adjustment is made for the age of patients, even though the significance of β-amyloid levels in CSF may be quite different in 80-year-olds than in 50-year-olds. Those researchers who support the idea that there are many roads to dementia and do not recognize AD as a single disease entity are very uncomfortable with the assumption that CSF biomarkers are the undoubted predictors of future AD. One participant at the 2011 Paris ICAD meeting, David Brooks, a neuroimager based at Imperial College, London, argued that we should think of amyloid as equivalent to high blood pressure; not as a sign of a specific disease but, rather, as an indicator of risk for many possible conditions. He argued that the revised criteria should not become a new religion because the presence of β-amyloid deposition in the brain is not yet standardized among individuals or populations, nor is it known what its presence signifies.

Amyloid In Vivo

The second biomarker to receive a great deal of attention in the AD world, and frequently cited as exceptionally promising, is a recently developed application of functional neuroimaging with positron emission tomography (PET). PET is a technique where radioactive ligands (molecules) are injected intravenously into subjects, and then the distribution of the ligands in the brain is imaged using a highly sophisticated three-dimensional sensing machine. Among many other uses, PET imaging can be used to demonstrate amyloid plaque in the living brain. This is a technique that rules out AD if no plaques are demonstrated, but, if plaques are found, interpreting their meaning is challenging, as we will see below.

In 2002, University of Pittsburgh scientist Chet Mathis and geriatric psychiatry colleague William Klunk reported the first results obtained by using a radioactive PET tracer molecule known as PIB (Pittsburgh compound B) on humans. This report, demonstrating that PIB can bind to amyloid in the brain and then be imaged using PET scanning, was based on a single case from Uppsala University, Sweden. Researchers from Pittsburgh partnered with colleagues in Uppsala for this initial study. A second report by the team using the

PIB tracer molecule was published in 2004. These findings showed that when injected into 16 patients diagnosed with "mild" AD, PIB was retained in areas of the cortex known from autopsies to be associated with the accumulation of amyloid plaques, and also in the hippocampus (part of the limbic system associated with memory and emotions) where amyloid also accumulates. In 9 healthy controls little uptake of PIB was demonstrated.

The findings, described as "proof of concept," encouraged the authors to claim that further research was justified in connection with the relationship between β-amyloidosis and AD, and the following conclusions were drawn:

> As the technology of amyloid imaging moves forward, it will be important to avoid the circular reasoning inherent in the association of amyloid deposition with both the diagnosis and the cause of AD. Therefore, at the outset, it may be best not to equate amyloid deposition to clinical diagnosis. Rather than as a method of diagnosis, it might be best to first think of PIB retention more fundamentally as a method to detect and quantify brain β-amyloidosis.[30]

The authors note that the term β-amyloidosis was first used in 1983,[31] and they argue that plaque buildup can usefully be described this way. They point out that several basic, unbiased questions can be asked about amyloidosis: First, how does it correlate with a clinical diagnosis? Second, what is the natural history of β-amyloidosis and its onset relative to clinical symptoms of dementia? And third, can β-amyloidosis serve as a surrogate marker of efficacy for anti-amyloid therapeutics? In the future, the authors conclude, individuals diagnosed with MCI should be included in research protocols, as well as subjects at very high risk for AD due to familial AD genetic mutations. In addition, larger samples of individuals diagnosed with AD should be enrolled as research subjects as well as "normal" subjects for purposes of comparison.

In a 2007 article Klunk and colleagues, making use of a mixed sample of AD and MCI patients and healthy controls—nearly 100 subjects in all—using PIB scans, demonstrated that amyloid buildup did not change significantly over two years. The researchers argue that in nondemented individuals diagnosed with MCI, PIB binding and impaired episodic memory loss are clearly associated, but with progression to AD, the amyloid accumulation appears to reach a "plateau." They conclude that their findings support the proposal that Aβ imaging can detect the preclinical phase of AD, but that further longitudinal study is required.[32]

An article published in 2008 by Klunk's group, in which what was characterized as a community-based sample was made use of, reported results of PIB scans on 43 "nonimpaired elderly" volunteers aged 65 to 88 years. Once again amyloid is detected, but in healthy elderly subjects. The small sample size is acknowledged, but even so the researchers conclude that the findings show that elderly people with significant amyloid deposition can remain cognitively normal. It is noted that longitudinal follow-up of these subjects is needed, and that findings should be replicated using larger cohorts, with the objective of determining if amyloid deposition alone is insufficient to cause AD within some specified period of time.[33] Caveats such as this have not stopped the media from commenting on this research positively: "PET scans with PIB clearly distinguish people with Alzheimer's from healthy people. They may also help identify people with the progressive form of MCI."[34] Mathis and Klunk were honored in 2008 with the Potamkin Prize—the highest award for Alzheimer disease research.

A burgeoning number of research projects have drawn similar conclusions to the findings of Klunk and colleagues, and it is now commonly stated that between 20% and 40% of healthy individuals, also expressed as "up to a third of cognitively normal elderly subjects," carry plaque in their brains,[35] an anomaly that demands explanation. It is evident that these findings are dependent upon age and genotype (there is a greater likelihood of having amyloid plaques in the brain as an individual ages, or if that individual carries the allele of the APOE gene known to increase risk for AD) (see chapter 5), but this does not explain away the finding. It is also recognized that a nonconcordance exists between amyloid deposition, which takes place at a slow, constant rate but does not usually result in detectable cognitive decline, and neurodegeneration, which is commonly associated with the appearance of clinical symptoms and usually accelerates quite rapidly once it commences.

A 2011 summary article concludes that amyloid deposition can take place up to a decade or more before other changes take place. Furthermore, interindividual differences exist in the time lag before neurodegeneration starts, if indeed it commences before death. Such variation is "likely caused by differences in brain reserve, cognitive reserve, and coexisting pathologies."[36] The article points out that neurodegeneration usually both "precedes, somewhat, and parallels cognitive decline," and can be detected in the form of brain atrophy, notably atrophy of the hippocampus, using MRI. Furthermore, the "neurodegenerative element is the direct substrate of cognitive impairment and the rate of cognitive decline is driven by the rate of neurodegeneration."[37]

Luc Buee, from the University of Lille, in France, presenting research at the Paris ICAD conference, noted that tau, the protein associated with neurodegeneration, comes in six isoforms. He added that there is "good" tau and "bad" tau, the second kind being implicated in 20 neurological disorders. Buee used the yin/yang icon to depict how tau is in constant flux, and he explained that it has many more functions than had formerly been recognized, including bringing about changes in the cell nucleus—something that had been ruled out until recently. In short, there is a great deal more to be learned not only about amyloid but also about tau and, above all else, about their roles in neurodegeneration.

Neurodegeneration cannot be detected through PIB PET imaging, and researchers must couple detailed structural analysis of MRI images with PIB

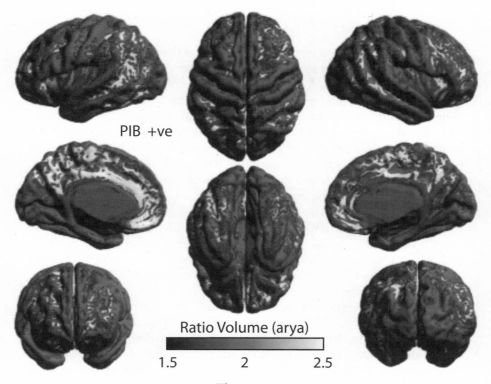

PIB +ve

Ratio Volume (arya)

1.5   2   2.5

Figure 4.2.

A neuroimage composed of multiple reconstructed images of one patient's MRI (magnetic resonance imaging) scan together with a PIB PET (positron emission tomography with Pittsburgh B compound) scan. The MRI images have been reconstructed with multiple surface renderings and appear as though gray. The PIB PET results are overlaid on the surface, with amyloid-rich areas appearing as whitish coloration. The PIB is distributed in regions of the brain typically known as the "association cortex," and primary sensory and motor areas are not affected. Reproduced with the permission of Howard Chertkow MD.

PET imaging (figure 4.2) to ascertain a more complete picture of what is taking place over time in the brains of individuals and/or in small population samples (by definition limited in size due to cost). Clifford Jack and colleagues at the Mayo Clinic argue that for the purposes of research trials both types of imaging are indispensible. PIB detection of amyloid is needed if removal of amyloid is the target of the trials. MRI detection of neurodegeneration is key if one is trying to track the relationship of neurodegenerative pathology to clinical manifestations of the disease.[38] One other application of PIB imaging, perhaps the most widely used at present, is to refine the AD diagnosis and rule out false positives—many researchers have told me that to date this is the only valid use for this technology in the clinic.

The New Phrenology

It was quickly recognized by involved researchers that the number and quality of plaques manifested in a living brain by means of PET imaging needed independent verification. A consortium of researchers decided that the best method to do this would be by comparing imaging findings with those obtained at postmortem—the AD gold standard. An ingenious, rather disturbing method of carrying out this comparison involved the voluntary recruitment of patients in hospice, long-term, and community health care facilities who were likely to die within six months. These individuals were invited to participate in a study in which they agreed to undergo PET imaging and then to donate their brains for autopsy once deceased. An article published in *JAMA* in 2011 was based on the first 35 individuals, average age of 79 years, to come to autopsy, all of whom had undergone a PET scan on average three months before their death.[39] Only 17 of these subjects had been diagnosed with AD. A control sample of 74 individuals aged 18 to 50 years of age (so-called young individuals) whose brains, it was assumed, would be free of amyloid also underwent scans.

This multicenter study, 23 sites in all, enrolled people from several parts of the United States and made use of a new PET imaging ligand. The compound, known as florbetapir, has a similar chemical mode of action to PIB in binding to amyloid plaque in the brain, but it has a distinct advantage over PIB because it has a longer half-life and can, therefore, be used in imaging centers that do not possess a cyclotron. The conclusions drawn from this study were that this first prospective study of florbetapir-PET imaging correlated well with the presence and density of β-amyloid at autopsy and therefore provided evidence of the effectiveness of a molecular imaging procedure that can identify β-amyloid in the brains of living individuals. It is noted in a final comment that individuals may have differing abilities to "tolerate aggregated amyloid" in their brains and that this depends upon "genetic risk factors, lifestyle choices, environmental factors, and neuropathological co-morbidities." The researchers acknowledge the small sample size of this study.[40]

This particular research project has received extensive criticism on several grounds, in addition to that of sample size, although none of the objections are based on ethical issues as far as I can ascertain. Writing an editorial in *JAMA*, the epidemiologist Monique Breteler points out that the participants in this study

> were not a random sample of individuals with cognitive problems who might be tested for presence and extent of amyloid pathology, nor were they a random selection of asymptomatic persons who might be evaluated to determine whether some of them already had amyloid pathology developing. Rather, they were a convenience sample of terminally ill persons who, for the most part, had advanced

dementia and non-dementia disorders, mostly cancer. The reported agreement of 97% between florbetapir-PET and the postmortem pathologic diagnosis is likely inflated because of the selection of participants in the study.[41]

Breteler, in striking contrast to the 20 authors of the original article, who have more than a column worth of conflict of interest disclosures, has none. She goes on to note that the absence of any scans rated amyloid-positive in the young control group was assumed to reflect a very high degree of specificity of the imaging. But she adds, "[J]ust as the observation of a flock of white swans does little to support the notion that no black swans exist, so the observation of amyloid-negative scans in young persons does little to establish the specificity of this imaging mode in persons aged in their 70s and 80s, who may or may not have varying amounts of coexisting brain atrophy and other pathological changes."[42] Breteler adds for good measure, "β-amyloid accumulation in the brain is likely not the culprit but rather an epiphenomenon of developing AD."[43]

A second major critique is to do with interrater reliability. In a letter to the editor published by *JAMA*,[44] two physicians, Carome and Wolfe, both involved with citizen awareness groups, pointed out that there was substantial variability in rating among extensively trained readers of the PET scans in this study and that the assessments of the three readers had been averaged out in the final publication. It was noted that should use of PET neuroimaging become more widespread, it would be very unlikely that readers would have had the extensive training given to these three individuals, with the result that "the variability and unreliability is likely to get worse," suggesting that false positives and false negatives would be the outcome. The authors of the study attempted to rebut the criticism of their publication and to justify at length the reasons for averaging out the reader findings in a letter published together with that of Carome and Wolfe in a later issue of *JAMA*.

When writing about the use of PET scanning in schizophrenia research, Joseph Dumit remarks that publications about illustrative cases show a clear differentiation between people labeled as schizophrenic and those without "even though there are many people diagnosed with schizophrenia whose brains look like those of people without, and people without schizophrenia whose brains look like those of people with it." The scans of volunteers are labeled as "normal controls" and the scans of those diagnosed with schizophrenia prior to scanning are labeled as schizophrenia. Dumit concludes, "[T]he image is thus labeled as showing the 'disease' itself, rather than a correlate symptom of someone found to have schizophrenia. Hence, the symptom has been collapsed into the referent."[45] Currently, considerable efforts are being invested in the AD world to replicate and standardize findings resulting from PET scanning of in vivo amyloid deposits in research settings. And other research is attempting to show the relationship between lowered β-amyloid in CSF fluid and an increase in plaques detected by means of this type of neuroimaging.[46]

The findings associated with schizophrenia are clearly problematic and much more elusive than amyloid. The "realness" of amyloid is not in dispute, particularly because it can be shown at autopsy, but what it signifies in vivo, especially for the future health of individuals, is debatable. A bias may well creep into the interpretation of images of amyloid in the living brain if it is categorically assumed that amyloid predicts an AD future, and such an assumption will in turn influence interpretations of the significance of biomarker findings that in effect come to assume the trappings of the disease itself.

Summarizing thus far, "slippage" between apparent neuropathology and cognitive well-being is overwhelmingly evident as a result of neuroimaging, but the interpretation of these findings continues to be disputed. Although the position taken in the new diagnostic guidelines calls for caution and argues explicitly that this anomaly cannot yet be explained, a good number of the individual authors who contributed to these guidelines, notably John Morris and colleagues, persist with their belief that every individual who exhibits preclinical biomarkers is on the way to Alzheimer's. The assumption is that only death can intervene in this unavoidable process. In contrast, other researchers focusing on elderly individuals who show few if any signs of cognitive impairment, even when neuropathology is evident in their imaged brains, assume that although death is clearly inevitable, AD is not; nor can it ever be shown to be so. Many among this latter group of researchers believe that greater efforts should be made to clearly establish, to the extent that it is possible, what it is that protects people from becoming demented well into old age. Yet other researchers equivocate (see below).

Rethinking Amyloid

One thing seems to be clear from the findings of epidemiologic community studies as well as PIB scanning thus far: those of us old enough to have trouble retrieving a word or two now and then can safely assume this is not a definitive sign of approaching Alzheimer's, although it is quite possible that β-amyloid may be beginning to clutter our brains and/or the effects of one or two small strokes may contribute to the situation. Plaques appear rather unsightly when neuroimaged, but these garbage bags (as one geneticist described them to me) may well simply accumulate unobtrusively little by little, and have never been proven to be definitive signs of a precarious future, even though they are almost universally present in the brains of patients diagnosed with AD. Furthermore, it is increasingly clear that, up to a point, the proteins that produce plaques are protective of brain function.

When I interviewed Mark Mintun, a neuroimaging specialist who was part of the well-known group researching Alzheimer's at Washington University in St. Louis (he has subsequently become an employee at the company that makes florbetapir), he made it clear that probably something "goes wrong in amyloid, or at least related to amyloid" at the beginning of the long decline into Alzheimer disease. But, he added,

> [S]omebody pointed out that the association between amyloid and AD may just be circular logic, although I don't think that's right. It seems now that first you get the plaques but they're not AD, so there has to be a second step. It seems that in some people, especially those who are homozygous for the APOEε4 gene [see chapter 5] the plaques buildup quickly and become more extensive and will just always push the next step to happen. But for a lot of people, I think that they can either develop some sort of equilibrium in the sense that plaques don't actually trigger the next step, or perhaps they don't ever get a high enough numbers of plaques. Plaques are there, but there are not enough of them. Or they're not of the right type. One of the things we are learning is that there are plaques that PIB binds to and there are plaques that it doesn't bind to. And you can actually have quite a lot of diffuse plaques and not any PIB binding. The neuritic plaques, dense extracellular plaques, have long been associated with AD, and have been the diagnostic gold standard since Alzheimer's day.

Mintun continued,

> [I]t's true you almost never have AD if you don't have neuritic plaques. But if you lived in a moderately toxic environment this might affect diffuse plaques, at this stage no one can say that those diffuse plaques are safe; it's just that they aren't, at the point that we look at them, leading clearly to AD. But they could transition quickly under certain circumstances. So we may be missing a lot of the iceberg that's really close to us.

> For all we know if everyone lived to 118 we'd all get AD. But some people get there much earlier, and your genetic load and the environment can accelerate it. Diabetes and heart disease and not doing crossword puzzles can accelerate things. But the person who exercises and is mentally active may live with amyloid in his brain for 20 years because he has sort of, you know, he's growing more neurons and synapses as fast as the amyloid is killing them, or something like that. And his immune system is properly balanced, and he doesn't have a run-away inflammatory process. But what we have now is a gap between the two anchors involved with AD. We now know much more about the first anchor—the neuritic plaques, but we don't know what sets off the

second anchor. Right now I'm looking at the relationship between plaque and free radicals. (May 2009)

Discussion then turned back to the PIB findings, and Mintun reminded me that PIB is negative in 60% to 70% of the people who become experimental controls in their research. But he added that this does not necessarily mean that they do not have any plaques at all: "[T]hey may have some that I simply cannot see. Or they may just have diffuse plaques." When conversation turned to the 30% of "normal non-demented controls" who test PIB-positive, Mintun made it very clear that he believes they are indeed "drifting" toward AD, as opposed to individuals who exhibit no plaques. He argued that cognitive testing carried out regularly over 10 or more years begins to show a drift downward in the plaque positive group, whereas those without plaques retain their cognitive status. Mintun is quick to add that it is not possible to see this with only a 5-year follow-up; a longer time frame is essential, and the sample followed must be large enough—several hundred individuals—such as they are currently studying at Washington University. In an email exchange several months later, Mintun made the following comment: "But, this is not a slam-dunk. There is very much the possibility that people who are in the normal range for cognition but have amyloid in their brains may not all end up with Alzheimer disease symptoms. [An analogy is that not all people with atherosclerosis and coronary artery disease end up with heart attacks]. However, it is reasonable to hypothesize that the fraction that go on to be impaired may be very high." Mintun supported this final comment on the basis of research in progress, but noted the shortcomings of the studies to which he refers.[47]

Peter Nestor, who does neuroimaging at Cambridge University, also believes that those people who are healthy but positive for PIB scans are on their way to becoming demented, but have not, as yet, reached a "threshold" for symptoms: "[I]t's an insidious process, and you can see from biomarkers that it's a pathology years before that threshold is reached." He adds,

[O]f course, a positive PIB scan doesn't tell you when its going to start, and so if you're completely healthy and you're 75 years old and you get a positive scan, how useful is it to know that? I find that I can often cheer people up when they worry if they are going to get Alzheimer's because their mother and grandmother had it. I remind them that their relatives were both in their 90s when they died, but I also point out that no one in my own family has been diagnosed with AD, and that's because no one in my family has ever survived beyond 80—my family all die relatively early of something else! For some reason this seems to cheer some people up—it makes them think that maybe they are lucky in a way! I think it's really helpful to remind people gently that they have to die of something and, yes, their chances of getting AD may be a bit higher than someone else because of their family history, but it's not necessarily destiny. We still have so much more research to do before we can make big claims about AD. (September 2010)

Howard Chertkow had the following to say when asked what he thinks is the significance of findings indicating that approximately 30% of healthy elderly people have tested PIB-positive:

I think it can mean one of several things: that some of these people are indeed in a presymptomatic stage of dementia; alternatively, it may be that some, or many, have sufficient cognitive reserve for one reason or another that they will eventually die of something else without crossing the threshold that precipitates dementia. And, most important, we simply don't know enough as yet. It's only six or seven years since we have been doing PIB scanning. Insisting that everyone with amyloid in their brains is going on to get AD is an act of faith—we have no proof of this. You would have to follow a cohort to age 100 and if any of them were not demented then you'd have to follow them to age 120 to be sure! You see the problem! So, it's a bit of a religious assertion to insist that "every older person with plaques will eventually go on to get AD." You can never prove this kind of statement. Marsel Mesulam [a Chicago-based neurologist] and others say the opposite: "the brains of old people have a lot of abnormalities in them and since most of these individuals never get demented, it is obvious that the mere presence of some plaques or tangles, or even infarcts, is not enough to guarantee eventual dementia." This is actually a more scientific viewpoint. (October 2010)

Chertkow went on to remind me that one has to ask, "What is normal?" (He might have said, "Normal for what? Normal for whom?" as did Redlich 30 years ago.)[48] And he made it clear that this question cannot be considered in abstraction from aging: "All 20-year-olds are PIB-negative, and 50% or more of people aged 90 and over are PIB-positive. If one chooses only 75-year-olds as normals for research, and their reaction times to cognitive testing are average, then about 95% of this population will, by definition, be excluded, because they have slowed down a bit and no longer count as 'normal.' And should people with diabetes and obesity be excluded? These conditions are now well recognized as risk factors for AD." Chertkow pointed out that researchers such as himself who make a rigorous selection of "normals" for PET scanning by drawing only on volunteers from the population at large who have no subjective memory complaints and who cope well with cognitive testing find that only 10% are PIB-positive. This is in striking contrast to research where normals are chosen from among people who have voluntarily presented themselves to a memory clinic or similar facility for testing, in which case the number of PIB-positives can be as high as 45% or 50%. This is the situation with the ADNI (Alzheimer's Disease Neuroimaging Initiative) study, to which we now turn.

Toward Neuroimaging for All

The idea of a national initiative to prevent Alzheimer disease has been on the table since 1987, but despite ongoing effects to reduce its incidence and standardize diagnoses, relatively little progress was made, as we have seen. By 2009 a renewed effort, focused on early detection and prevention, was set in motion. The promotional banner this time around is the "Campaign to Prevent Alzheimer Disease by 2020." This latest national strategic goal, argued Zaven Khachaturian, the editor-in-chief of *Alzheimer's & Dementia*, is "mandated by the looming financial catastrophe facing the U.S. national health care system."[49] In an editorial the previous year, Khachaturian, whose mother died of probable Alzheimer disease, wrote the following passage:

> The array of challenges for the mission to prevent AD within a decade are no less daunting than those faced by similar national endeavors such as the Apollo space program, the Manhattan Project, or the Human Genome Project. Ultimately, the execution of this national enterprise will require decisive actions by both public and private entities, as well as bold public policies that foster radical changes in: 1) the governance and organization of research, 2) mechanisms or programs of research funding, 3) deployment of resources and infrastructure, and 4) paradigms for developing interventions/treatments. . . . One of the most critical challenges for the national mission to prevent AD is the need for centralized control and coordination of all AD-related activities.[50]

Khachaturian went on to list the programs and consortia that have been put into place over the past three decades—11 of them in all, many managed by different governmental agencies. And yet other assemblages exist that are globally distributed.

ADNI, a public–private partnership, was created in 2004 and is central to the "goal of the proposed National Strategic Plan . . . to create a new paradigm for planning and supporting the organization of worldwide cooperative research networks to develop new technologies for the early detection and treatment of various forms of memory impairments."[51] The objective of the ADNI partnership is to research changes in cognition, brain structure and function, and biomarkers in elderly individuals selected as research controls, subjects with MCI, and subjects with Alzheimer disease. The technologies made use of by ADNI include MRI and PET scanning, as well as the systematic examination of CSF and blood biomarkers. The data generated by this research are available to researchers worldwide without embargo and may be neither patented nor owned. The governance and administration of ADNI are exceptionally complex, and will not be elaborated on here.[52] William Thies of the Alzheimer's Association made the following observation about the onus that will fall onto ADNI research subjects:

> I think in the Alzheimer world we are blessed with having people with the disease who are remarkably willing to give back in the hopes of helping the next generation. You know, they're not only going to get lumbar punctures but some of them will get three kinds of scans every six months . . . that's not trivial. You sit in this machine and it's claustrophobic and it's banging away and noisy. With the MRI machines you can hear everything bumping and crashing and sometimes people spend a total of almost 60 minutes in the three scanners and they will be exposed to radiation, of course . . . this is really an amazing gift to give back to society. (April 2010)

John Morris made it clear just how difficult it is to recruit volunteers from among "normal" people who will become controls for AD research:

> We were inundated with phone calls when we put out the call for research subjects, but these came from offspring of individuals who had Alzheimer's. We had a much harder time with people who don't have Alzheimer's in their family. Why would they come in? So we tried to get people from Alzheimer families to bring their spouse in as a control, or a friend. We used the families to help recruit the controls. Now we have the largest group anywhere of cognitively normal people who have undergone PIB studies. (May 2009)

A 2010 article in the journal *Alzheimer's & Dementia* explicitly states that the ADNI project is not built around, nor does it depend upon, the amyloid hypothesis, despite evidence "in favor" of this hypothesis. It is also made clear that to date there is "limited information concerning the pathophysiological sequence of events of AD in human beings from autopsy studies and studies measuring only cognition."[53] The overarching goals of ADNI are to "determine the relationships among the clinical, cognitive, imaging, genetic, and biochemical biomarker characteristics of the entire spectrum of AD as the pathology evolves from normal aging through very mild symptoms, to MCI, to dementia, and to establish standardized methods for imaging/ biomarker collection and analysis for ultimate use in clinical trials."[54] Of the 58 involved sites, 53 are in the United States and 5 are in Canada. Globally, sites in Europe, Australia, and Japan are affiliated with the project, but have their own neuroimaging initiatives, and Chinese and Korean sites are about to be added. The number of subjects involved, over 1,000, is large for neuroimaging studies.

The limitations of ADNI have been clearly set out. First and foremost, the population studied is not an epidemiological sample, but represents what is termed "a clinical trial population" in which individuals who have had a cortical stroke, cancer, heart failure, substance abuse, and so on are excluded. Second, the age range of subjects is from 55 to 90 years, and hence, given what is now known about amyloid deposition in the brain, the lower age limit should, ideally, be younger by a decade. A third limitation is that no lifestyle information is being collected from subjects, and a fourth limitation is that not all of the measurements taken thus far have been made on all subjects, limiting the possibilities for comparison. The principal objective is that of standardization to provide "an infrastructure for world-wide clinical trials by the pharmaceutical industry."[55]

To date, amyloid imaging showed 89% PIB positivity in patients diagnosed with AD in the ADNI study. Other PIB studies have shown that 10% to 20% of AD patients are PIB-negative—a finding that is not yet explained, and may or may not be due to a doubtful clinical diagnosis. Among the "normals," a very large number—47%—tested PIB-positive. This number is considerably higher than in other studies to date, but is not regarded as an anomaly because it correlates well with the findings of CSF β-amyloid measures in the same subjects. The plan is to follow these individuals over the coming years to determine whether they begin to evince changes in cognitive function.

Gina Kolata reported on ADNI in an August 2010 article in the *New York Times*. When asked by Kolata to comment on the ADNI approach to sharing of data, John Trojanowski, a specialist in the pathobiology of neurodegenerative disorders with a prodigious publication list, replied, "It's not science the way most of us have practiced it in our careers. But we all realized that we would never get biomarkers unless all of us parked our egos and intellectual property noses outside the door and agreed that all of our data would be public immediately."[56] Kolata reported that companies as well as academic researchers are using the data, and that there have been more than 3,200 downloads of the entire massive data set and almost a million downloads of the data sets showing brain scans. This consortium, between January 2010 and September 2011, published 135 articles.

ADNI is a work in progress—a prescient example of the biosciences of the future organized to defeat AD—a complex amoeba-like public–private consortium designed principally to bring about success with drug development for this baffling condition. Readers of the 2010 *New York Times* article by Kolata learn that in the interests of cost containment the entanglement of aging, lifestyles, sociopolitical conditions, and dementia have been set to one side. It was also pointed out that biomarker measures "have not yet been demonstrated to have high value as either predictors or outcomes," and, furthermore, the "subject burden" in the form of repeated tests is high, and additional tests will impair enrollment and increased dropout rates.[57]

Brain Identity

The sociologist Nikolas Rose has hypothesized that the very idea of what constitutes "self" is being transformed in light of developments in molecular neurobiology. He has documented evidence for what he describes as "neurochemical selves" in which "[m]ind is simply what the brain does. And mental pathology is simply the behavioral consequence of an identifiable, and potentially correctable, error or anomaly in some of those elements now identified as aspects of that organic brain."[58] With respect to Alzheimer nosology, mind effectively disappeared from view from the latter part of the 19th century, displaced by a powerful discourse about localized neuropathology, although in certain eras efforts were made over the years to challenge or dislodge this dominant discourse. A focus on neuropathology permitted both the medicalization and gradual destigmatization of dementia and, at the same time, bolstered the common belief that once individuals become demented they are no longer their former "selves" and that a radical, irreversible rupture has taken place. Initially, early findings about the genetics of AD reinforced this situation, as we will see in the next chapter, with the result that a fundamental materiality has pervaded Alzheimer discourse over the past two decades. This position is consolidated further because cognitive functioning is today comprehended as the direct product of neurological activity, although a good number of experts believe that this situation will hold only until such time as the neural networks and pathways that activate consciousness and individual awareness are spelled out.

The findings of Simon Cohn, a medical anthropologist who interviewed psychiatric patients who had undergone brain scans and were given copies of the scans to take home, alert us to what may be in store with increasing routinization of neuroimaging. The patients in this study told Cohn in no uncertain terms that a scan is, for example, a "picture of who you really are. On the inside. I tell people it is my self-portrait." Further investigation by Cohn made it clear that the brain scan in effect verifies the diagnosis for patients, thus making them into living cases of the label they have been given.[59] It is not yet known if a similar transformation is likely to take place when people undergo repeated PIB scanning, individuals who, on the basis of their selection for trials, will already be confronting the possibility of dementia in their lives. It appears from the literature and the interviews I have conducted that perhaps the majority of neurologists believe that evidence of amyloid in the brains of healthy individuals is a definitive sign of incipient AD. They take this position despite repeated findings that very large numbers of old people with amyloid deposits are cognitively intact when they die, but simply assume that patients would have died of AD if they had not previously been picked off by some other condition. It is quite possible that neurologists and healthy research subjects together may well interpret findings of a positive PIB scan as a definitive sign of approaching AD with, no doubt, significant effects on trial subjects and their families.

A neurologist who comes from a family where AD is present told me that if he found amyloid in his own brain he would be very worried indeed (although he has no intention of checking), although finding out about his genotype would not concern him much, whatever the result. The question arises as to whether or not people who are informed about amyloid deposits in their brains will have the fortitude to situate the risk estimates handed out to them in the broader social context of their life experiences. Or will they, in effect, be transformed into Alzheimer patients years before symptoms are detectable, if indeed symptoms ever do manifest themselves—living out a slow, unrelenting death that in reality may not be taking place at all? Alternatively, will some of these research subjects, whatever their own test results, be able to sustain a conviction that the trial may eventually be successful, in which case they will have contributed to research that has potential benefit to millions, should an effective drug be developed? Such individuals will be stoic corporeal citizens— those who choose to make their bodies available for the betterment of society at large. Given the enormity of what is involved, it will be important to watch carefully for dropout rates from drug trials that carry out repeated biomarker testing. And presumably some subjects monitored over the months and years in connection with incipient Alzheimer's will be vulnerable to serious depression.

Pasting Over Incommensurabilities

It is just possible that AD research is at the beginning of the long road to a paradigm shift, because acknowledgment of stubborn anomalies is often a first step to a major change of orientation. The work of Julie Schneider and Carol Brayne, together with their respective colleagues, among other recent research findings, highlights incommensurabilities that can no longer be brushed to one side.[60] And as early as 1997 Zaven Khachaturian noted,

> The life of a neuron hangs on many interdependent systems, particularly the systems for neural intercommunication, metabolism and repair. Those three systems ordinarily work in sync, like highly trained acrobats. But internal factors such as changes in a person's nutrition, immune response or neuroendocrine status, and external factors such as toxins, trauma or infection can shake one of the systems and upset the entire delicate balance.[61]

Despite commentary such as this for more than a decade, a dramatic transformation in the order of a conceptual revolution is not in sight. On the contrary, the primary hope associated with the application of technologies that allow assessment of biomarkers is, it seems, that the findings will confirm and consolidate the causal model centered on the amyloid cascade hypothesis. This model is the one that continues to attract drug companies as they plan the direction of future research. And, as William Thies notes,

> People have the idea that the biomarker puzzle and the therapeutic puzzle are sort of independent. But they're not. They go hand in hand. They'll be solved at about the same time because one really needs the other to really come to closure . . . ultimately therapies will be the key tools because they'll give us the ability to then confirm our biomarker or imaging markers of particular subsets. At the moment you can hypothesize all you want about a certain biomarker indicating a certain kind of change but until you can change that biomarker [by "treating" it] and see the disease change you really don't have the ability to confirm what biomarkers are doing. (April 2010)

Strategic use of the concept "paradigm shift" is very frequent in the AD world today. This seductive hype, above all else, is a means of raising money and of sustaining hope among the public for a cure for AD and, ultimately, for its prevention. Formerly researchers often made claims that AD would be defeated before too long. One researcher broadcast in 2001 that "within 10 years drugs will be developed to prevent Alzheimer's."[62] In recent years, although such hype has continued to appear in the media at times, it is striking among researchers that this kind of talk has become relatively unusual, and caution and even skepticism have been quite often evident as people confront the slew of unknowns that riddle the field.

The next chapter introduces the complexities associated with the genetics of Alzheimer disease and discusses genotyping as yet another type of biomarker. In contrast to the biomarkers discussed above, in a small number of families worldwide, genetic knowledge can predict AD occurrence with a high degree of certainty, but, for by far the majority of aging individuals, genotyping for AD provides no definitive answers.

Appendix 1

Summary of clinical and cognitive evaluation for MCI due to AD

Establish clinical and cognitive criteria

Cognitive concern reflecting a change in cognition reported by patient or informant or clinician (i.e., historical or observed evidence of decline over time)

Objective evidence of impairment in one or more cognitive domains, typically including memory (i.e., formal or bedside testing to establish level of cognitive function in multiple domains)

Preservation of independence in functional abilities

Not demented

Examine etiology of MCI consistent with AD pathophysiological process

Rule out vascular, traumatic, medical causes of cognitive decline, where possible

Provide evidence of longitudinal decline in cognition, when feasible

Report history consistent with AD genetic factors, where relevant

---

Abbreviations: AD, Alzheimer's disease; MCI, mild cognitive impairment.

Reprinted from *Alzheimer's & Dementia*: 7, no. 3, Albert M.S. et al., "The diagnosis of mild cognitive impairment due to Alzheimer's disease: Recommendations from the National Institute on Aging and Alzheimer's Association workgroup", 270-279, © 2011, The Alzheimer's Association, with permission from Elsevier.

Appendix 2

Biomarkers under examination for AD

Biomarkers of Aβ dispositionCSFAβ$_{42}$
PET amyloid imaging
Biomarkers of neuronal injury
CSF tau/phosphorylated-tau
Hippocampal volume or medial temporal atrophy by volumetric measures or visual rating
Rate of brain atrophy
FDG-PET imaging
SPECT perfusion imaging
Less well-validated biomarkers: fMRI activation studies, resting BOLD functional connectivity, MRI perfusion, MR spectroscopy, diffusion tensor imaging, voxel-based and multivariate measures
Associated biochemical change
Inflammatory biomarkers (cytokines)
Oxidative stress (isoprostanes)
Other markers of synaptic damage and neurodegeneration such as cell death

---

Abbreviations: Aβ, beta-amyloid protein; AD, Alzheimer's disease; BOLD, blood oxygen level-dependent; CSF, cerebrospinal fluid; FDG, fluorodeoxyglucose; fMRI, functional magnetic resonance imaging; MR, magnetic resonance; MRI, magnetic resonance imaging; PET, positron emission tomography; SPECT, single photon emission tomography.

Reprinted from *Alzheimer's & Dementia*: 7, no. 3, Albert M.S. et al., "The diagnosis of mild cognitive impairment due to Alzheimer's disease: Recommendations from the National Institute on Aging and Alzheimer's Association workgroup", 270-279, © 2011, The Alzheimer's Association, with permission from Elsevier.

Chapter 5
Alzheimer Genes: Biomarkers of Prediction and Prevention

Every morning she wakes up a little further away from us. Yet she still manages to convey something, if only with a glance, of the world she is entering. Beyond the fear and the loss, she seems to say, there is life of a sort here, at the dark edge where everything is crumbling.

—Michael Ignatieff, *Scar Tissue*[1]

In a 1997 article titled "Plundered Memories," Zaven Khachaturian, then the director of the Ronald and Nancy Reagan Research Institute of the Alzheimer's Association, and later director of the Office of Alzheimer Disease Research Center at NIH, wrote,

Some critics ask whether genetic research is worth the resources it consumes and the anguish it will bring to those who test positive for a harmful gene—when a cure still seems so far away. In my view, however, the genetic approach is on the right track, and I think the continuing research on Alzheimer disease may soon confirm that belief. Those of us in the front lines of the fight against Alzheimer's have never been closer to unmasking this mysterious thief, the robber of the very thing that makes human beings unique.[2]

In this article, Khachaturian is scornful of the evolutionary biologist Richard Lewontin who had recently suggested that genetic research into Alzheimer's is like going down a blind alley as far as cures are concerned (a comment that to date has proven to be the case). Khachaturian, passionate about the need to combat Alzheimer disease, has devoted much of his life to this cause, and raised millions of dollars for research. His mother died of probable AD, and over a decade ago Khachaturian started to think that he himself was beginning to suffer from it; in retrospect it seems that this is not so.

Given the findings about AD genetics that had appeared just prior to the publication of "Plundered Memories," the apparently extravagant claims made by Khachaturian do not seem so wildly out of line, Lewontin's sagacious comments notwithstanding. It is of note that the early articles about the discovery of genes "causal" of AD raised many of the questions and uncertainties that continue to plague the AD world today. Moreover, their discovery precipitated the formulation of the amyloid cascade hypothesis that continues to be the dominant model of AD causation.

Dominantly Inherited Alzheimer's

In a 1996 article, the neurologists Ephrat Levy-Lahad and Thomas Bird pointed out that the high prevalence and late age of onset of the common form of Alzheimer disease are features not usually associated with "genetic disease." This was an acceptable statement in the days before molecular genetics had become firmly entrenched, but the authors then went on to self-consciously rebut their own statement, presaging some of the dramatic changes about to take place in the world of genetics as a whole. The body of their article was a discussion of three autosomal dominant genes[3] that, using linkage studies followed by positional cloning had, in the 1990s, been shown to be causal of what is today termed "familial Alzheimer disease" or "dominantly inherited Alzheimer's disease." The first of these genetic loci to be uncovered, the amyloid precursor protein (APP) encodes the β-amyloid peptide associated with the buildup of amyloid plaque (see chapter 2). Originally four mutations of this gene were found, but more than 30 are now recognized, at least 25 of which cause dominantly inherited AD. Individuals are usually affected between ages 40 and 60.[4] It was relatively recently recognized that the APP gene has what may well be a critical role in learning and memory function during development, and that maintaining an appropriate level of APP expression is important but, if disrupted, then excess β-amyloid may result, leading to plaque formation.[5] It is currently estimated that APP mutations cause approximately 15% of all cases of familial AD.

Two related genes known as presenilin-1 and presenilin-2 are also associated with dominantly inherited Alzheimer disease. Presenilin-1, first mapped fifteen years ago, is involved with brain and spinal cord development and with the processing of APP into smaller fragments. Over 185 mutations of presenilin-1 have been shown to affect more than 30 large extended families worldwide, with a very early age of onset, on average between 35 and 55 years. These mutations are associated with plaque and tangle pathology, and account for approximately 70% to 75% of the cases of familiar AD.

The presenilin-2 gene was discovered when research was carried out among Volga German families living in the United States—families who emigrated from Germany to Russia in the 1760s, many of whom eventually migrated to the United States. A second large extended family that carries this mutation is Italian.[6] Presenilin-2 is involved with cell growth and maturation as well as the processing of APP into peptide fragments. Presenilin-2 is also associated with plaque and tangle pathology. At present, 13 known mutations are associated with this gene, accounting for approximately 5% to 10% of dominantly inherited Alzheimer cases.

Presenilin-2 mutations are not strictly speaking autosomal dominant because a few individuals who carry these genes do not develop AD, even at a great age. Furthermore, although the mean age of onset of presenilin-2-initiated AD approaches 55 years, it is striking that the age range can vary from 40 to 75 years, even among families who carry identical mutations. For familial AD in general it has been shown that the age of onset for identical twins can vary by as much as a decade, strongly suggesting that environmental factors influence to some degree the "penetrance" (phenotypic expression) of these genes.[7] Research indicates that these 230 or so presently known mutations associated with the APP gene, presenilin-1, and presenilin-2 account for between 5% and 10% of all AD cases (this figure is a rough estimate given that, globally, cases of Alzheimer's, even cases of early onset Alzheimer's, are not inevitably diagnosed). Yet more mutations are regularly confirmed.

Using new sequencing technology, it is now apparent that rare mutations of all three of the genes associated with dominantly inherited AD are at times present in individuals not diagnosed with dementia until later in life. This research was carried out with individuals who come from families where four or more members are affected with AD. This finding strongly suggests that an unknown factor or factors must function to defer the age of onset of dementia in these instances.[8] Increasingly, as research into AD genetics advances, it is clear that boundary making among what are thought to be different forms of AD on the basis of age of onset, as has usually been the case, may not be highly informative.

It is usually stated that it was a dominantly inherited form of AD that caused dementia in Alois Alzheimer's patient Auguste D, but there is no conclusive evidence that this was so. It is known that a good number of cases of so-called sporadic or late-onset AD can affect individuals at a relatively young age, but it will not be possible to establish definitively what "type" of AD so tragically ended Auguste D's life.

In an editorial published in 1996 written for the *Annals of Neurology*, Bradley Hyman pointed out that during the course of the 1980s research had shown that the neuropathological features—the plaques and tangles—of early-onset dominantly inherited AD (known as "presenile dementia" throughout most of the last century) were essentially identical to those in the more common form of late-onset Alzheimer's. With the discovery of the genes associated with familial Alzheimer disease, Hyman suggested that a reevaluation was required as to whether or not early-onset AD and late-onset AD are indeed a single process happening on different time scales or, alternatively, "do these various genetic factors produce a group of ADs that share some phenotypical commonality but are distinct processes?"[9] As we will see in later chapters, this question continues to be of great significance, particularly in connection with drug development.

Hyman recognizes that his concern is, in effect, a "philosophical question"—a matter of ontology—but he uses findings from molecular epidemiology to conclude that genetic factors associated with familial AD, while they may result in what appears to be a worse disease, do not apparently culminate in a qualitatively different pattern of brain involvement. In other words, β-amyloid deposition seems to be the neuropathological factor most strongly influenced by the known genetic factors, and ultimately this leads to "'Alzheimer disease' rather than 'Alzheimer diseases.'" Hyman concludes, "In this way, Alzheimer's is probably similar to atherosclerosis and other complex trait diseases in that there are multiple possible influences, some genetic, some 'familial' and only defined in a vague way, and potentially, even some environmental. These all, presumably feed into a common pathophysiological process that leads to a very similar clinical and neuropathological phenotype."[10]

John Hardy and Gerald Higgins published their hypothesis about the amyloid cascade in *Science* in 1992, and it was the discovery of the APP gene, details about which were published in *Nature* in 1991, that made their hypothesis so appealing to numerous researchers. Discovery of the genes associated with dominantly inherited Alzheimer disease also ensured that those families in which the genes were present would be regarded as desirable research subjects, particularly in connection with drug development designed to stop amyloid deposition. Interest in recruiting members of these families as research subjects has come to a head in the past few years, particularly with the move to bring about AD prevention.

The Paisa Mutation

Little by little, studying the infinite possibilities of a loss of memory, he realized that the day might come when things would be recognized by their inscriptions but that no one would remember their use.

—Gabriel García Márquez, *One Hundred Years of Solitude*[11]

A large group of 25 extended families, originally of Basque origin, about 5,000 in all, live in both urban and rural areas of the province of Antioguia, Colombia, scattered mostly around Medellín. These families compose the most concentrated group of individuals known to date who carry a specific mutation of the gene presenilin-1. Close to one-third of the population—approximately 1,500 people—harbor what is known locally as *paisa*—a word that simply refers to the people of the region. Typically, memory loss becomes manifest around 45 years of age, and most affected individuals progress to full-blown dementia by their early 50s—a very rapid progression to *La Bobera* (the foolishness). Francisco Lopera, a neurologist associated with Antioquia University in Medellín, was the first scientist to pay serious attention to the devastating condition that affected so many Antioguian families. Once Lopera understood that during the course of three generations, half of the children were affected, he realized that he had stumbled across a form of dementia that must surely be due to a single gene mutation. Only a decade later, in the 1990s, was the gene isolated, at which time the finding began to attract international attention. Despite the serious risks posed by drug traffickers and Farc rebels, who were very active in the 1980s and '90s, Dr. Lopera and his colleagues managed to collect DNA samples and assemble a genetic pedigree extending back nearly 300 years, to around the time that these families first arrived as part of the ongoing colonization of Colombia by Spaniards, searching for gold.

In late 2010, the executive director of the Banner Alzheimer's Institute in Phoenix, Arizona, Eric Reiman, a newly recognized rock star of science, appeared on Fox News following an appearance in *GQ* magazine together with the reality star Bret Michaels in a Los Angeles studio. This event took place in order to raise money for Alzheimer's research, some of which would be directed toward a planned trial with the Colombian families in collaboration with Francisco Lopera in Medellín. A 2011 BBC news commentary noted that one scientist (unnamed) described these extended families as a "natural laboratory."[12] Researchers are excited, listeners were told, because the interrelated families provide them with an "enriched sample." Earlier, Neil Buckholtz of the NIA informed a reporter from the *New York Times* that these individuals will have "cleaner brains that can give a better picture" than a sample made up of older subjects.[13] In other words, because an autosomal dominant mutation is involved, it is virtually certain that at least one-third of the subject population will become demented and, significantly from the point of view of the researchers, at a young age.

This same newscast reported that the Banner Alzheimer's Institute planned to use subjects selected from among the Colombian families to test drugs designed to attack plaque in the brain as part of a newly formulated Alzheimer's Prevention Initiative (API). It is acknowledged in the newscast that it is not known if amyloid plaque is the cause or an effect of Alzheimer's, but, as Joseph Arboleda, a Harvard-based researcher interviewed by the BBC who works with Dr. Lopera, stated, "the trial puts this hypothesis to the test. . . . It is possible that drugs will inhibit the brain plaque and yet the family will still get dementia. Such results would prove devastating for current research." However, "If in the extended family the onset of Alzheimer's is delayed, or stopped, then the researchers will have hit the mother lode—a potential cure for sufferers worldwide. That remains a big if."[14]

Eric Reiman has also been very explicit in suggesting that, should "the disease be halted," this "could generate treatments to protect millions worldwide from common Alzheimer's."[15] Commenting on the proposed trial in 2011 Reiman said,

> [I]f there is no effect on the biomarkers after 2 years in the right direction [we will] declare futility and give these people at the highest

imminent risk access to the next promising treatment. If, however, they do budge in the right direction [we will] continue to follow them a little bit longer to see if it slows even subtle memory decline . . . if it does, I have a feeling that may be enough evidence for regulatory agencies to consider using biomarkers under their accelerated approval mechanism in other [AD] populations.[16]

Reiman does not make clear what would be the "next promising treatment," should the trial drug prove ineffective. In 2012 a Reuters report was emphatic that "[t]he trial in Colombia could offer the most definitive test yet of the amyloid theory of Alzheimer's. . . . This trial [will] be different because it will be tested on people before the disease has done much damage to the brain cells."[17] The excitement associated with this research is remarkable, and I was told informally in the summer of 2012 that Francisco Lopera was receiving so many email communications from interested parties of one kind and another that the Banner Alzheimer's Institute scientists had taken it upon themselves to protect him as best they could from this onslaught.

Very moving stories accompanied by striking images about elderly parents caring for their affected adult children accompanied both the BBC news program and the *New York Times* report. These accounts are disturbing, not only because of the pathos involved, but also because the trial raises many troubling ethical questions. Eric Reiman is reported to have said that his team is "trying to manage expectations within the Colombian family, making clear the drug is experimental and risks and benefits are uncertain,"[18] something he reiterated with me on the phone. When I spoke in July 2012 with Pierre Tariot, an experienced geriatric neurologist and an associate director of the Banner Alzheimer's Institute, also involved with the trial, he told me that discussions have been taking place since 2006 about whether or not the trial was really feasible. Matters such as how best to "operationalize" this complex, potentially exploitative project were of great concern to involved researchers from the outset. Among the many questions posed were these: How can the trial be carried out in a way that would be "respectful and humble"? Should the biological materials obtained from these peoples be regarded as, in effect, a "sacred resource"?

A registry was set up in Medellín in 2012 enabling potential participants to make a connection with Dr. Lopera, and signatories were asked to attend seminars where the ultimate objectives of the trial; what would be required of research subjects, and the process of randomization were explained at length. Many of the trial subjects have at least 10 years of formal education, but others have only three years or so of schooling. Dr. Tariot noted that it was readily apparent that the information was attentively received and mutually discussed. Not everyone has proved eager to participate in the trial, but the community as a whole has accepted the plan very favorably, and many have expressed willingness to participate, even among those who have no signs of dementia.

These families are practicing Catholics, and, furthermore, young women are raised to believe that they should have children—that this is part of God's will. Given this situation, it was important to involve local priests in discussions from the outset. One requirement of the study is that women should not become pregnant during the trial, and therefore should be using contraceptives. The priests proved to be cooperative, and apparently had no difficulty with contraceptive use for this purpose. The media reported that at least one woman of reproductive age had obtained a hysterectomy in order to avoid passing the disease on to the next generation—confirmation of the inescapable fears that people in this community confront, even when childbearing is considered so important.[19] Kenneth Kosik, a neurologist from UC Santa Barbara who has been involved with the community since the 1990s, encountered a young woman who very much wanted to have children, but was terrified at the prospect of passing down the mutated gene.[20] Kosik further noted, when commenting on the possibility of genetic testing among this population, "It's very dangerous knowledge . . . I saw a 23-year-old man who said that if he found out he had the mutation, he would commit suicide."

In late 2011, members of a few of the affected Medellín-area families were brought to Phoenix where they were given PET scans, among other preliminary tests, that will provide a baseline for amyloid deposition in their brains before the trial commences. As of mid-2013 ethics clearance of the trial had not been completed; the intent had been agreed upon in writing by the FDA, but the trial still awaited clearance from both the Colombian Ministry of Health and the FDA. Should this trial go forward, it will be a novel approach to testing drugs for AD; prior to this time subjects have been identified as having dementia before entering trials.

Trial Preparation

The randomized trial, with an estimated cost of $100 million, will be coordinated out of the Banner Alzheimer's Institute and is now slated to begin in late 2013 with the hope of producing interim results by 2017. Collection of data will be carried out entirely in Colombia, and a cyclotron has been set up for the first time at great expense in Medellín in order that neuroimaging can be carried out on-site. In 2012, the NIH announced that it will provide $16 million toward the cost—part of the U.S. government initiative taken by President Obama to provide funds for AD research. Another $15 million comes from private sources, and the drug company Genentech, the U.S.-based biotechnology unit of Roche based in Switzerland, will provide the biggest amount—$65 million. Selected study participants must be at least 30 years of age, and within 15 years of the age when their parents' symptoms began. A total of 100 healthy individuals who carry the paisa mutation will be enrolled in the trial and be administered the drug Crenezumab, an experimental antibody that targets amyloid buildup (it is claimed that this drug has fewer side effects than other anti-amyloids used to date, and was selected from among "25 rivals").[21] The researchers point out that because the drug is less toxic, it can be used at higher doses, and will attack pathological forms of both soluble and insoluble forms of amyloid in the brain. The drug will be administered subcutaneously every two weeks.

Among the 200 controls, half will be individuals who carry the mutation and the other half who do not. All trial participants will be administered lumbar punctures, brain scans, and other tests at regular intervals over the course of 24 to 60 months to track biomarkers and other possible clinical changes unless, of course, the trial is stopped prematurely. Among the tests, several are designed to detect subtle signs of memory loss—described as the "primary outcome measure" of the trial. Noninvolved neurologists have made passing comments to me about the duress that these subjects will be under by submitting themselves repeatedly to these procedures, all the while receiving little or no communication about data findings. The expressed hope is to get an early indication that the drug is indeed removing amyloid. Participants will learn ahead of time that they will not be told whether or not they are carrying the mutation, even after the trial is completed. And it appears that the research subjects have no wish to learn this information, a position that causes the researchers to feel at ease, because no genetic counselors are available in Medellín. The three-armed trial will mean that the research subjects are "blinded," that is, they will be unable to tell who among them carries the mutation.

It remains an open question as to what extent the research subjects fully comprehend the uncertainties associated with the trial in both its implementation and its outcome, despite the best efforts of the research team to explain matters. Given the tragedy that stalks the lives of all these families, it surely must be the case that many of these individuals would be willing to undergo extensive personal discomfort and risk for even a small possibility that a medication may be discovered that will lift the monstrous burden that their families must bear.

Pierre Tariot is convinced that this research will not result in harm and that, whatever the outcome, it will be informative. When he talked with me, he emphasized that millions of dollars had already been spent in preparatory work. Tariot noted that if benefit clearly results, then, of course, that will be the best possible outcome, but this may well not be the case. He is convinced that the dose level that will be administered of the selected drug is correct but that quite possibly it should be started at an even earlier age than 30 to be effective. He emphasized, in contrast to much media reporting, that should the drug prove effective, it could be used only for kindred of the Colombian families with the same mutation. Paisa is a founder mutation, one that was present among the families who made up the original Basque immigrants, or else arose shortly after their arrival in Colombia, thus substantiating Tariot's opinion, that much more research would be required before it was known whether or not the particular drug being used in the trial could be applied to other populations. Eric Reiman, in common with many other AD experts, has made it clear to the media, in contrast to his colleague, that in his opinion dominantly inherited, early-onset AD is essentially the same condition as is late-onset AD in its final stages. Hence any drug that proves effective in stopping the former condition is likely work in a similar manner with the latter. However, the position taken by John Hardy, the originator of the amyloid cascade hypothesis (see chapter 2), and increasingly other commentators, is that this may well not be the case. Preliminary findings from exploratory research with Colombia subjects suggest that the latter position may well be correct, as we will see shortly.

A Question of Beneficial Distribution

Concerns that will arise in the minds of social scientists in connection with this trial are reminiscent of problems associated with the Human Genome Diversity Project, and also with the expanding universe of drug trials in which "naïve" subjects (individuals who do not routinely imbibe medication) often living in economically deprived conditions are systematically recruited for trials. In recent years the "importation of subjects" to carry out research in "developed countries" has started to take place.[22] This method is usually adopted when small, novel research strategies are implicated, such as gene transfer, or isolated populations of interest are involved, as is the case with the Medellín subjects for pretrial testing. After extended scrutiny in connection with both its science and the involved ethics, the Human Genome Diversity Project was, in the event, never funded. The strongest criticism of this project, had it taken place, and also of the exponential increase in the "offshoring" of clinical trials, is that the final "products" of these endeavors, if successful, will primarily benefit the developed world, and not the subjects under study.[23] The Colombian trial has been set up in such a way as to avoid this type of blatant exploitation. The arrangement with Genentech is such that, should the trial prove successful, medication will be made available to all affected families on an ongoing basis. Presumably too, local health care facilities will have to be expanded. The anthropologist Nikola Bagic, who has worked in the region, stresses that by far the majority of Colombians live in relative or extreme poverty and lack access to anything more than very basic primary health care. His ethnographic research among families who harbor the paisa mutation has made it clear to Bagic that their travails must first and foremost be recognized as a political matter. Pensions and other forms of social support do not exist, resulting in extraordinary stress for affected families, who inevitably become desperately impoverished particularly because the disease strikes young and middle-aged adults. However, Bagic is encouraged because the local physician/researchers based in Medellín have had considerable resources made available to them in preparation for the trial, resources that they plan to use not only to improve their own decrepit computers but also to improve family well-being (personal communication). It may well be that this potential influx of social support provides the greatest incentive of all for these families to participate in the trial.

Despite Pierre Tariot's cautionary comment above that trial findings would not apply to populations other than that of the Colombian families without extensive further research, media reporting of comments made by other researchers, some of which were cited above, make it very clear that many think a great deal more is at stake. Benefit to the local population does not appear to be uppermost in the minds of many researchers, and the Colombians are indeed thought of by many noninvolved researchers as a "treasure trove." This means, of course, that Francisco Lopera is poised in a very delicate position; on the one hand he must be a protector of his compatriots, among whom he has worked for decades, and on the other hand, as a participating member of the Banner Alzheimer's Institute research team he has obligations to the team. He is not the first scientist living in a poor country to find himself in this position, but it is an extraordinarily delicate matter to negotiate.

An experienced Montréal gerontologist with whom I talked told me that, in his opinion, the proposed trial must be done. As far as he is concerned the findings will either make or break the amyloid cascade hypothesis once and for all. Not everyone agrees with this position, and many assume that even if the trial is a failure the argument will be that it should be started at a younger age or a different dosage should be used. But preliminary findings published in late 2012 (see chapter 8) suggest that this trial may indeed bring about quite a shake-up in the AD world.

By 2010 it had been shown that amyloid plaques may begin to accumulate in the brains of Colombian individuals as early as 28 years of age, and that plaque deposits increase steadily until about age 38, after which they tail off. Other research has suggested that CSF changes may be present in individuals who have a mutation for dominantly inherited Alzheimer's as early as 18 years of age. As noted above, it has been assumed that among the Colombian families, cognitive impairment is usually readily detectable by age 45, and dementia sets in, on average, by age 51. However, a study published in *The Lancet Neurology* in 2011 that presented findings from the largest and longest retrospective study of predementia clinical stages in familial Alzheimer's disease to date forced reconsideration of the age of onset of clinical symptoms in the Colombian families. Based on a sample of more than 1,700 individuals in Medellín, that included 449 carriers of the paisa mutation, it was shown that measurable clinical deterioration could be detected using cognitive testing in individuals who carry the mutation two decades before dementia onset, that is, at a median age of 35.[24] It is evident that changes associated with dominantly inherited Alzheimer's, both molecular and clinical, commence earlier than had been appreciated prior to this research and, furthermore, as anticipated, the molecular changes precede cognitive changes by a good number of years.

Research that complements that taking place in Colombia is being carried out under the umbrella of the multisited consortium DIAN (Dominantly Inherited Alzheimer Network). In this consortium, cross-sectional, longitudinal studies are under way with over 260 research subjects in the United States, Great Britain, and Australia, all of whom carry one of the three genetic mutations associated with familial AD. Preliminary findings from this research, based on the tracking of biomarkers, suggest that pathological changes associated with AD may be detectable up to 20 years before the appearance of clinical symptoms of dementia. John Morris, director of the Alzheimer research group at Washington University, heads up this project. When he talked with me, he made his position clear: biomarkers "very objectively predict decline." And he added, "People [with altered biomarkers] should be considered the real treatment target"; furthermore, this long "prodromal" phase will be very helpful in assisting drug research and the development of potential treatments.

Genetic Testing for Early-Onset AD

To date, no research has been published in which people from families that carry one of the mutations associated with early-onset AD have been asked to comment on their experiences as research subjects. Genetic testing will be a key part of the research process, but Morris insists that most people from families with familial AD with whom he has close contact, similar to the Colombian families, do not wish to know their genetic status, and they will not be informed about the results of their biomarker testing nor if they are carrying a genetic mutation for AD in the project he is conducting.

In Montréal I was able to interview just two people, brother and sister, who come from a family that carries a presenilin gene. Brenda was eager to talk; she had already been tested when I met her, and her result had been positive, although she is not clear about the name of the gene she carries. She had decided to be tested she said because, as a single woman, she wanted to have time to set her affairs in order should she test positive and arrange for appropriate help long before the AD symptoms leave her helpless. Brenda had been forced to retire at age 45 from her teaching position that she loved because her memory difficulties made it impossible to manage large classes of students studying music. As we talked it became clear that, at 52 years of age, she was resigned to a bleak future. Her father and two of her three uncles had died of the disease in their 60s, but Brenda had not been involved with caregiving because she lived in a different part of Canada at the time—that task had fallen to her mother. When I met her, she had recently undergone neuropsychological testing at a memory clinic, and said that she had really looked forward to it because she "loves tests" and had done well on the same test on previous occasions. But this time round, Brenda found that she simply could not answer the questions. As an excuse for her difficulties, she said that the kind of questions she had been asked again and again when doing these tests repeatedly in recent years now bored her. Later on in our conversation, Brenda said that she started taking the tests because her doctor had told her it was good to have a "baseline" from which to assess her condition. She finally added, hesitantly, "In the last year or so I've noticed that I'm really slipping." As we talked, it became clear to me that Brenda was struggling to answer the questions I asked, and she dealt with this difficulty by taking over the conversation. I was feeling deeply saddened as we parted after talking for over an hour, and was worried about the effects our meeting might have on Brenda, although she said spontaneously that she had enjoyed our talk, and had gone out of her way to meet with me.

Brenda's brother, Alan, lives in a different city than his sister, and they rarely communicate. Alan has not yet undergone genetic testing, having pushed the matter to one side during younger years. He has three children from his first marriage, but only after he divorced, remarried, and found himself thinking about having another child had he stopped to reflect on his family history. This time round he wanted genetic testing so that he and his present wife could "plan things well," as he put it. When asked about his children, Alan did not elaborate on their lives other than to say that they are now adults, and it would be up to them to get tested if they so wish.

It is no doubt the case that for Colombians of Basque origin who carry the paisa mutation, termination of pregnancy should a fetus test positive for a familial AD gene would not be acceptable (I assumed, perhaps wrongly, that Alan would take a different position, given that he talked about planning "things" well). For many patients who attend the clinic run by John Morris in St. Louis, it is quite possible that pregnancy termination is also not acceptable, and this might in part account for the apparent lack of willingness to hear about the results of genetic testing that these people profess to. These are matters worthy of sensitive ethnographic investigation, particularly because examples exist from other parts of the world where genetic screening prior to marriage has proven to be an option that many individuals choose to take when families must face up to the devastating effects of single gene disorders.[25]

An Elusive Susceptibility Gene

The APOE gene, located in humans on chromosome 19, is essential for metabolism, transportation, and processing of lipids and LDL cholesterol. This gene comes in three common, universally distributed allelic variations, APOEε2, APOEε3, and APOEε4. Each variation produces one of three different proteins, and in human populations the ε3 variant is the most common.

In 1993, a research group headed up by the geneticist Allen Roses at Duke University (who later became vice-president of genetics at GlaxoSmithKline Inc.) published articles that were the first to made an explicit association between the APOEε4 allele and increased risk for the common, late-onset form of AD.[26] It was shown that this allele is implicated in both sporadic cases of AD (in which there is no apparent evidence of AD in other family members, although this is often impossible to determine due to death from other causes), and in late-onset familial AD (in which other family members have clearly been affected). In the short space of time of two or three years, this association was confirmed in over 100 laboratories worldwide, on the basis of results obtained from both clinical and population samples.[27] These findings forced second thoughts about the received wisdom of the day—namely that all instances of Alzheimer disease in older people are "sporadic." They also created a great deal of excitement because it was assumed by some researchers that the Alzheimer puzzle would now soon be solved.

From the outset, it was clear that the APOEε4 allele is a susceptibility gene; that is, it is neither necessary nor sufficient to cause AD. It is estimated on the basis of population studies that at least half the individuals homozygous for ε4 never get AD, and it has also been shown that between 30% and 60% of those who develop AD do not carry the ε4 allele. It was evident right away that other genes and also no doubt environmental and possibly social variables are in all likelihood implicated in disease causation. In European and North American populations its distribution is between 14% and 16%, but it is higher among Pygmies, the Khoisan, the indigenous peoples of Malaysia and Australia, Papuans, and some Native American groups—a finding to which we will return below.[28]

Research quickly showed that when the ε4 allele is implicated in AD, the same "final common pathway" resulting in plaques and tangles appears to be exactly the same as that associated with the autosomal dominant genes responsible for the early-onset form of the disease. However, the neuropathology associated with APOEε4 usually becomes manifest later in life, between the ages of the late 60s and the mid-70s,[29] although increasingly research has shown that these age differences are not hard and fast. It has also been shown repeatedly that the "gene dose" of APOEε4 is significant, and that individuals who are homozygous for the allele will be affected at an earlier age than those who are heterozygous, and that these individuals usually, but not always, exhibit a greater amyloid burden.[30] Given that somewhere between one-third and over one-half of patients diagnosed with late-onset AD do not carry the APOEε4 allele, it was apparent early on that there must be at least one other pathway, and probably several, that results in what is recognized as Alzheimer neuropathology.[31] Today it is assumed virtually unanimously that mutually interactive genes constitute these pathways, in addition to which DNA regions with functions other than protein coding, in conjunction with environmental factors internal and/or external to the body, are all involved. These latter pathways to AD relatively rarely result in clinically diagnosable symptoms until late in life, usually after age 70 to 75, or even later, but it continues to be assumed, by perhaps the majority of researchers, that the same final common pathway is involved as that for both early-onset and APOEε4 linked AD. However, as we have seen throughout the book, these hardened assumptions are increasingly open to question as more findings are made.

Furthermore, a satisfactory explanation does not yet exist as to what sets off this train of molecular events, and basic science research increasingly makes it clear that the final common pathway is extraordinarily complex and subject to micro-molecular variation. Even so, John Hardy's argument in a 1994 editorial in *Science* that genetic findings (in connection with APOE) give support to the amyloid cascade hypothesis as being causal of AD continues to be taken seriously,[32] in large part because APOEε4 is associated with amyloid burden.

From the time of the discovery of the APOEε4 allele, first by means of linkage studies, followed by its mapping, it was assumed that as yet "undiscovered" genes must be implicated in AD, and gene hunting continues to be an important activity to this day.[33] Using the citation index PubMed, two neurogeneticists, Bertram and Tanzi, showed that in 2003 alone a total of 1,037 projects were carried out in which 55 genetic loci were examined on 20 different chromosomes for association with risk for AD. These authors then summarized the situation with respect to AD genetics as follows: "First, and most importantly, the heritability of AD is high.... [T]his had been demonstrated in various studies ... over the past decades." But, they go on to note: "most of the research currently being done has faulty methodology, lacks replication, and is inattentive to haplotype structure."[34]

The article concludes with a caveat: "while the genetic association *per se* [of APOEε4 with AD] has been extremely well established ... there is no consensus as to *how* this association translates pathophysiologically," nor, they add, how it functions "in conjunction with the other numerous candidate genes"[35]—a statement that holds to this day.

Readers will recall that the diagnosis of AD in the majority of clinical settings is not reliably replicable and, furthermore, that very many instances of AD, probably the majority,[36] are never seen in clinical settings. The result is that although claims about the relationship of APOEε4 and increased risk for AD have the appearance of being very robust, considerable variation is seen from sample to sample as to just how predictive is this relationship. For example, estimates of the number of individuals with AD who carry the ε4 allele range from 30% to 90%,[37] and many studies do not specify whether these numbers refer to those who are hetero- or homozygous.[38] In addition, researchers report that between 23% and 68% of AD patients do not have the APOEε4 allele, serving to highlight the complex and elusive nature of the association between susceptibility genes and the pathology of AD.[39]

In addition to retrospective studies of individuals who already have AD, other studies attempt to predict the number of people with APOEε4 alleles who will eventually develop AD. Considerable discrepancies exist between the estimates made in these prospective studies. Depending on the study consulted, the number of individuals who are heterozygous for the APOEε4 allele and who are expected to develop AD range from 7.6% to 47%. The range for homozygous individuals is between 21.4% and 91%.[40] The literature suggests that a person with one ε4 allele has 3 times the odds and a person with two ε4 alleles has between 8 and 30 times the chance of developing AD compared to individuals described as ε-negative.[41] However, the baseline on which these probabilities are estimated is rarely provided, making such estimates questionable.

For those individuals with two APOEε3 alleles (about 60% of the population in Europe and North America), risk for AD is estimated as "average," and it is calculated that about a quarter will develop AD once over the age of 80. Relatively few people carry APOEε2, although it is higher in some populations than others. Those who inherit two copies of this gene are thought to be at low risk of contracting AD, and ε2 appears to be protective, although two of the individuals I know who chose to have themselves privately tested because AD is present in their families found that they are carrying ε2 alleles that apparently do not protect their relatives.

One of the principal causes of confusion about genetic risk for AD is inherent to the design of much of the research. One problem is that research is often based on clinical and not population samples, thus introducing bias.[42] When general population samples are used, the relationship between APOEε4 and AD appears to be significantly weaker than is commonly suggested.[43] Not surprisingly, given the complexity involved, it has been repeatedly confirmed from the latter part of the 1990s that the ε4 allele does not determine progression to dementia, and community studies have shown that up to 25% of "normal" individuals in their 80s carry an ε4 allele.[44] This matter is made yet more complex because the incidence of APOEε4 and its relationship to AD varies significantly according to the human groups under study and their geographical/ environmental locations (see below). In chapter 7 ethnographic findings will be presented based on interviews conducted with individuals who have been tested for their APOE status and informed of the results of their tests.

Setting aside for one moment the question of the accuracy of reporting and of risk estimates, several researchers have argued for years that too much weight has been given to the contribution of the APOEε4 allele to AD. More than a decade ago, the biological anthropologist Alan Templeton was critical of the conclusions drawn by many researchers in connection with the significance of APOE to AD incidence. He pointed out that genomes are "commonly organized into clusters of functionally related genes," and that APOE is part of one such large cluster. Templeton argued that when this type of gene is associated by linkage with a specific phenotype, great caution is called for, because the gene may simply be a marker for another gene or genes located nearby on the same, clustered segment of DNA.[45] It is of note that APOEε4 was first associated with increased risk for heart disease and hypoglycemia, and is implicated in yet other medical conditions (in technical jargon it exhibits pleiotropy), but its association with risk for AD has attracted the most attention in both the research world and in the media.

In 1996 in an article that was part of a three-volume issue devoted to the APOE gene published by the *Annals of the New York Academy of Sciences*, Allen Roses and colleagues concluded their contribution with a provocative statement. They argued that, if genotyping for APOE were to be linked with the use of other biomarkers such as neuroimaging, particularly in individuals with mild cognitive complaints, it would soon become possible to administer "anti-dementia treatments before extensive brain damage develops."[46] This statement makes clear that a move to the prevention of AD was already being considered as a realistic option in the mid-1990s. However, in a second article published in the same volume of *Annals*, Roses modified his previous statement significantly, now making it clear that predictive genetic testing of "cognitively intact individuals" should not be recommended, and that APOE genotyping should be used only to confirm a diagnosis of probable AD in patients with symptoms of dementia.[47] Guidelines similar to this latter position have recently been loosened in the United States but not in the United Kingdom, Canada, or France. Even so, the new NIA–Alzheimer's Association criteria for diagnosis discussed in chapter 4 exhibit caution and state several times that their recommendations apply to research settings alone and are not as yet for use in the clinic. In a third article published in 1998, once again in the *Annals*,[48] Roses, by then affiliated with Glaxo Wellcome Research and Development Unit in North Carolina, stated that a second locus for a susceptibility AD gene on chromosome 12 would shortly be announced but, despite further research, this locus was never verified.

A summary article published in 2000 by Ann Saunders, Allen Roses's spouse and a prominent member of his research team, argued that clearly APOE is a multifunctional molecule with potential roles in amyloid deposition and clearance, microtubule stability, intracellular signaling, immune modulation, glucose metabolism, oxidative stress, inflammation, and other cellular processes. She added that the presence of an ε4 allele had been associated with poor recovery from brain trauma. Saunders also stated that the alleles of APOE differ from each other by only one amino acid substitution (an inaccurate claim; it has been known for many years that two amino acid substitutions are involved).[49] She pointed out that the differential effects on CNS and brain function are very dramatic, given these small differences among the alleles (very evident, even though two substitutions are involved).

The APOE gene has clearly presented a challenge to both researchers and clinicians from the time that its confounding role in the AD conundrum was first made apparent. But there are more findings, created largely by population geneticists and biological anthropologists, making the picture yet more convoluted. However, such findings have rarely been noted in the majority of articles written about APOE by neurologists and other clinicians, with the notable exception of a recent issue of *Alzheimer's & Dementia*.[50]

Human APOE—Which Allele Came First?

Since the late 1990s there has been debate among geneticists about the evolution of the APOE gene. This gene is well "conserved," meaning that, in terms of an evolutionary time scale, it has been in existence for millions of years and is widely present in mammals, including nonhuman primates. However, other than in humans, APOE exhibits only one variant that resembles the human ε4 allele, but is not exactly like it.[51] The association of this gene with deposition of amyloid plaques, neuron atrophy, and other changes in the brains of apes and certain other mammals in the latter part of their lives has been known for some time,[52] and behavioral signs of dementia have been reported in older animals who are pets or in captivity.[53] It was proposed in 1988 that the ε4 allele must be the ancestral mammalian form of APOE, including that of *Homo sapiens*.[54]

The well-known primatologist Robert Sapolsky coauthored an article in 1999 with Caleb Finch, whose specialty is the neurobiology of aging, in which they postulated a connection between the so-called grandmothering hypothesis and selection for sporadic mutations of the APOEε3 allele during the early history of *Homo sapiens*. The grandmothering hypothesis postulates that, as a result of the extremely long dependence of human infants after birth, it became adaptive for grandmothers during their postreproductive life to contribute extensively to child care, leaving younger women free to forage for food. Recently this hypothesis has been called into question by certain researchers, but its use to bolster a thesis about the evolution of the ε3 allele is nevertheless interesting. Finch and Sapolsky claim that if the grandmother role is important, and there is evidence from existing hunting and gathering societies that this is indeed the case, then a predisposition for dementia and cardiac disease would be selected against. This selection process may account for why the ε3 allele is by far the most common among human populations today (up to 77% among so-called Caucasians). At least one well-known biological anthropologist, Ken Weiss (personal communication), remains entirely unsatisfied with this kind of argument. It is equally possible that genetic drift rather than adaptation accounts for the increase in the ε3 allele and, given that it is estimated that currently 95% of humans carry at least one copy of ε3, it is quite possible that ε4 and ε2 will eventually in effect disappear. What we may be seeing today is probably an "evolutionary snapshot" of a process that will eventually drive these alleles to "extinction."[55]

However, Finch and Sapolsky, good scientists that they are, also ask if the apparently deleterious ε4 allele may confer some advantage on young adults in connection with neurodevelopment. In attempting to explain why the ε4 is more prevalent in parts of Africa and other isolated tropical regions, they note that it appears at times to confer resistance to infectious disease. The question of which APOE allele came first remains unsettled, but these ongoing arguments make it clear that, without reference to a context larger than clinical settings, knowledge about the functioning of susceptibility genes such as APOE is impoverished—genes have a deep history encompassing two time dimensions simultaneously, evolutionary and historical, and as the following chapters will make clear, the question of how exactly they are "expressed" in any one individual depends upon local environments, macro and micro, and on individual behavior.

With this in mind, the Italian geneticists Rosa Maria Corbo and Renato Scacchi carried out an extensive study in the late 1990s to examine the distribution of the APOE gene. They found that among Pygmies, the Khoisan, indigenous peoples of Malaysia and Australia, Papuans, some indigenous North American peoples, and the Lapps, the ε4 allele was proportionately the highest. Their argument is that these are groups of people whose subsistence economy was until relatively recently predominantly that of hunting and gathering and who have lived in situations were food shortages are common. Corbo and Scacchi postulate that ε4 can best be understood as a "thrifty gene," similar to the genes associated with diabetes, in that it assists with a higher absorption of cholesterol that would have been protective in times of food scarcity. They cite James Neel, who first set out the thrifty gene hypothesis, an argument that has taken some battering lately, largely because many readers have assumed that Neel was making causal arguments about the genes associated with diabetes, which was not the case. Corbo and Scacchi, like Neel and other thoughtful geneticists,[56] make no such claims about causality, but emphasize that gene/environment interactions are at work. They suggest that the ε4 allele may have become disadvantageous only once the human life span was routinely extended due to technological and environmental changes accompanied very frequently by the adoption of a sedentary lifestyle and, particularly in recent years, the adoption of diets rich in carbohydrates and fat and low in fiber. These authors conclude that to carry APOEε4 in a "Westernized environment" may well have become disadvantageous.[57]

But this is not yet the whole story. It is unusual for genes to exhibit only three common alleles, and evolutionary biologists and population geneticists believe that this apparent lack of variation should be accounted for. Several researchers have reported on an APOEε5 allele found at very low frequency in the Horn of Africa. And, in contrast to clinician researchers, biologists make it clear that many other APOE alleles in addition to the three ε alleles exist, several of which are in the regulatory area of the APOE gene and may well contribute to APOE "effects." These alleles are relatively recent discoveries and are in effect absent in discussions about APOE and AD.

One group of molecular biologists carried out comparative research in which they looked for variation within the alleles ε2, ε3, and ε4. By examining haplotype variation (sets of single nucleotide polymorphisms [SNPs] that are transmitted together) they concluded that prior to the past 200,000 years ε4 was the only allele in existence, and therefore must be the ancestral one (even so, they take issue with the grandmother hypothesis). They also found that when they "looked beneath the surface" of these three APOE protein polymorphisms they could readily detect substantial variation (heterogeneity) in the sequencing of each of them—in other words, each allele, whether ε2, ε3, or ε4, exhibits variation depending in part on the population in which it is found and the geographical location.[58] This could account for a great many of the contradictory results that have always plagued population genetic research in connection with the APOE gene.

The Indianapolis–Ibadan Dementia project headed up by the psychiatrist Hugh Hendrie for over 17 years working together with Nigerian collaborators has shown that the incidence of AD and dementia among Yoruba are less than half that found among African Americans, even when controlling for age. However, the frequency of the ε4 allele is not significantly different in the two cohorts. Furthermore, the Yoruba have a lower incidence of both vascular disease and vascular risk factors, including hypertension, than do African Americans, and, significantly, cholesterol and lipid levels are much lower among the Yoruba. Hendrie's group concludes, reasonably, that genetic and environmental factors may well both be responsible for these differences. They also note that considerable variation is likely to exist among the genomes of the studied Yorubans, as has been well established for African populations in general. As yet, they do not know how well the African American admixed population corresponds to the Yoruban population that was studied, calling for further investigation.[59] Research findings based on 338 African American subjects showed that the ε4 allele confers about the same increased risk as it does for white populations living in the United States;[60] however, the probable extent of vascular dementia among these subjects may have been a confounding factor. Clearly, examining haplotype variation, particularly in Nigeria, would be advantageous.

The methodology of a good deal of the comparative epidemiological research into APOE and AD has been criticized, almost exclusively by epidemiologists, and it appears to be virtually impossible to eliminate bias in this type of research;[61] even so, the data appear sufficiently robust to draw the conclusion that both risk-reducing factors (in Africa) and risk-enhancing factors (in North America) must surely be implicated, among them other genes, their protein products, diet, environment, and possibly yet other factors.

APOEε4 and Neurodegeneration

Yadong Huang is a neuroscientist who works at the Gladstone Institute, University of California, San Francisco. He exudes excitement about his research on APOE, a topic that has occupied him for over 20 years since his time as a graduate student in China. We talked for well over an hour when I met him.

He started out by saying, "I know the APOE proteins quite well." And his concluding comment was, "I always feel good when I talk about APOE!"

Huang's position is that, without doubt, the ε4 allele puts people at increased risk for AD due to its interaction with β-amyloid—two decades of research have demonstrated this repeatedly, he insists. But, he adds, "this is not the full picture—APOEε4 clearly has its own effect, or effects, independent of amyloid that are associated with cognitive decline and, furthermore, whatever effects the ε4 allele has, some of these changes could well take place early in life." As an example, Huang cites a body of research showing that ε4 carriers exhibit a decrease in glucose metabolism in certain parts of their brains during their 20s. This decrease, associated with the mitochondria—the powerhouses in cells—suggests a vulnerability to AD, years before similar changes are seen in other individuals. Then Huang adds, "and there is a difference in activity that has been detected in the 'default mode network' [that includes the hippocampus] between people who are ε4 positive and ε4 negative.[62] He goes on to elaborate on ε4 and its properties in association with Aβ, tau, and other molecules, and then cites animal studies that his team has carried out on the relationship of APOE and the gene known as GABA.

When I mentioned that I had been told by a number of experts that genetically modified mouse models are simply not adequate as an AD substitute for humans, Huang replies,

> The mouse is a good model but we have to keep in mind that a mouse is just a mouse—they are not humans. For one thing, human APOE and mouse APOE, although they are similar, they are not the same. Of course all mice are "ε4-like," but also they clear β-amyloid differently from humans. However, sometimes the mouse APOE behaves functionally a bit like APOEε3. This has enormous significance for clinical trials when Stages 1 and 2 are conducted in mice. And it also indicates that we shouldn't assume the amyloid cascade is wrong on the basis of what we learn from mice. But this goes back to the point I want to emphasize: we shouldn't put all our eggs into one basket. Right? Above all, we shouldn't mix our efforts at drug development— where most of the money goes—with research designed to work out the disease mechanism. It's time to think seriously about more than Aβ; we need to diversify the research.

After a pause Huang adds,

> Alzheimer disease is a kind of syndrome—it's a mixture of many subclasses. We should probably call it Alzheimer diseases. It's like hypertension—many paths can lead to increased blood pressure. In the end there will be more than one drug for AD. . . . It's too limited, just to play with the Aβ pathway. We've put all the money there, but we

haven't had any good luck so far. The theory, the amyloid theory, may be OK, but it's only one of several pathways. Because of the genetics John Hardy emphasized that early onset AD and late-onset AD are caused by the same mechanism. This idea could be true, but perhaps it's not. From accumulating evidence I'm thinking more and more that it's not true, that it may be only part of the story. We've been too limited in our thinking. And, of course, we need to know why so many people who carry ε4 don't get AD—this is a really important question that interests me greatly.

Huang has coauthored articles with his colleagues at the Gladstone Institute, one of whom is R. W. Mahley, cited above. Like Mahley, Huang is convinced that the ε4 allele is the ancestral form of APOE, and, moreover, Huang very much likes the "grandmother hypothesis" to account for selection for the ε3 allele during the course of human history. A summary of the complexity of the thinking currently made use of by Huang and his colleagues in connection with APOEε4 is made clear in a recent article; their fundamental position is that this allele is "much more than a contributing factor to neurodegeneration" due to the remarkable number of its functions critically relevant to the brain.[63] One striking thing to emerge from Huang's comments is that, for him, the so-called final common pathway may not be as straightforward as has always been assumed. Although the core of the amyloid cascade hypothesis cannot be denied, a great deal of research is still called for. A position apparently also supported by John Hardy (see chapter 2) and Rudolph Tanzi, a Harvard neurogeneticist, has the following to say about amyloid:

> [A]myloid is not just junk; it has a role in protecting the brain, particularly against the effects of infection. Amyloid is, in effect, part of the brain's immune system. I'm a Baptist, for sure. We have had a knee-jerk reaction to the failure of the drug trials, but the first drugs in any trials always fail. It's been like a 5th grader shooting a ball from the midline on a soccer field. Now we are starting to work down the field. So far we can't agree as to whether the problem is with an increase in Aβ levels, or whether Aβ starts to aggregate, or whether it's to do with the ratio of Aβ42 to Aβ40—most people go for the ratio these days [Aβ42 is associated with amyloid deposition and Aβ40 appears to be protective against amyloid deposition]. (March 2010)

Tanzi added that he had published an article 20 years ago with the neurogeneticist Peter Hyslop showing how, following injury, Aβ "gums up" to protect the immune system. He insists that "we should learn from genetics," by which he means that more effort should be spent on asking what it is that "starts the disease off," rather than trying to "hit the clinical phenotype." He adds, "Don't go too far downstream; it's better to start with the genes." Tanzi made these comments to me as a prelude to talking about the latest technology being made use of in the world of molecular genetics—genome-wide association studies (GWAS)—to be elaborated on in the next chapter.

In 2009 when I talked with John Morris he explained that one of the projects his group was working on was to try to better understand the relationship between the APOE gene and deposition of amyloid in the brain. Using a sample of 241 individuals, the largest group to date anywhere to be repeatedly PIB scanned, resulted in the following findings:

> . . . 75% of people who are . . . ε4 positive in our oldest age group and still cognitively normal are PIB-positive. Now why?! And will they dement? These are the questions that we are asking. . . . We call our people "participants," not patients, not subjects, okay? We try to make them partners in our research and every year we hold a quote, "meeting," unquote, where our researchers give back to our participants the results of the research in which they've been engaged. We only enroll people who indicate from the start that they understand and are willing to participate in all our studies, including the spinal tap. So, you know, the people who come are in a way pre-selected and quite committed. But we don't divulge to them their PIB results, nor their APOE results, nor CSF results. But this is something we may need to revise in the future.

The Morris group published an article in 2010 in which findings derived from this sample of "cognitively normal individuals" age 45 to 88 years old were set out; all individuals had been genotyped and PIB scanned, and 168 had undergone CSF taps. They concluded, "[A]ge and APOE genotype interact to increase the frequency of cerebral Aβ deposition in cognitively normal older adults."[64] The study found that Aβ deposition begins in middle age and increases in frequency with age, so that 50.0% of the individuals in the study aged 80 to 89 exhibited significant CSF changes, and in 30.3% PIB scans detected amyloid deposits in their brains. These results are comparable to an age-related frequency of neuropathological AD found by the same research group in "cognitively normal older adults" at autopsy.[65] The 2010 publication confirms that APOEε4 has a powerful dose-dependent effect on cerebral Aβ deposition (that is, two copies of the ε4 allele have a greater effect than one), but individuals who do not have ε4 alleles also exhibit age-dependent increased changes in CSF and amyloid deposition, although not to the same extent as carriers of APOEε4. This study also demonstrated, perhaps convincingly for the first time, that APOEε2 is protective against amyloid deposition; however, what is not noted in the article is that the APOEε2 allele places individuals at increased risk for hyperlipidemia.

The authors acknowledge certain weaknesses in the research protocol, but, even so, on the basis of their findings, they concluded, "Aβ is central to the initial detectable pathological changes in preclinical AD, with changes in tau likely occurring later."[66] But, they go on, "[T]he concept of preclinical AD must remain speculative" until there is enough evidence that "cognitively normal older adults" with reduced CSF Aβ42 who are also PIB-positive, and who have other clinical indicators of preclinical AD, are at a disproportionally greater risk for developing AD than individuals without evidence of these biomarker changes. This publication concludes, "Alzheimer disease is a complex disorder and its pathogenesis almost certainly cannot be explained simply by abnormal metabolism of Aβ. However, we find powerful evidence that cerebral Aβ deposition with age is the pathobiological phenotype of APOEε4, the strongest genetic risk factor for late-onset AD. . . . We also find evidence that Aβ abnormalities . . . initiate the pathological cascade of preclinical AD." The final comment argues that "presumptive preclinical AD" has been found in "a substantial number of cognitively normal adults"—these individuals should be tracked using longitudinal studies to "determine their risk" for Alzheimer disease.[67]

Entanglements of aging and dementia, normal and pathological, continue to grip the AD world, as does profound uncertainty about the role of amyloid-β in AD causality. A 2011 summary article about the amyloid cascade hypothesis as causative of AD makes the following points, among others: It is still not clear how much amyloid-β production should be lowered, or to what extent amyloid-β clearance should be facilitated to modify the disease. Nor is it clear at what stage in the disease process modification of amyloid-β is likely to have clinical efficacy. It is also noted in this article that there is considerable heterogeneity among human brains with AD, and that it is impossible to assess accurately the quantity of amyloid deposition in the brain, especially because different parts of the brain are implicated. The authors insist that it has proved to be "very difficult to derive predictive data on the dynamics of plaque deposition," and, furthermore, "The data . . . suggest that absolute levels of amyloid-β are not a key determinant in the age of onset of AD."[68] What is more, the deposition of amyloid-β does not correlate with the presence of neurofibrillary tangles, cell loss, or dementia. These researchers nevertheless continue to support the amyloid cascade hypothesis, but ask for minor modifications so that amyloid would be understood as a "trigger." This small change captures much better, the authors insist, the temporal separation of the complex processes that appear to take place at the molecular level over years. It is concluded that large "natural history" studies are called for in which young "normal participants" should be consistently evaluated over years. And, furthermore,

> to test the amyloid-β trigger scenario is beyond the resources of a single pharmaceutical company, but if this truly reflects the AD process, then a radically different model of pharmaceutical development will need to be established to prevent the public health disaster that awaits us.[69]

When two scientists who are not involved with AD research reacted to a query I raised as to why the amyloid cascade hypothesis has such staying power, their answers were revealing. One likened the situation to a horse race in which the only horse left running is lame, but even so you hang on to your bet. Another said that amyloid deposition is clearly like the outcome of a car crash, but the investigator ends up looking at the debris instead of what happened to bring about the crash—he then likened this to clearing up a car crash and assuming that will prevent future injuries.

Thomas Kuhn argued long ago that until such time as a new apparently robust hypothesis is put firmly on the table, researchers will not give up on an earlier hypothesis because of concerns about a situation bordering on chaotic, with undeniable funding implications. And clearly it cannot be flatly denied that amyloid is indeed in some way a major actor in lifelong neurological activities. Research has shown repeatedly that amyloid *must* in some way be implicated directly or indirectly in the causes or effects of dementia-like conditions, notably those usually diagnosed as AD. And amyloid is also undeniably involved in other related neuropathological conditions. Hence, to date, the prevailing paradigm has been tweaked several times, funding has been raised, and research proceeds apace with the cascade hypothesis as its anchor.

Chapter 6
Genome-Wide Association Studies: Back to the Future

Each cell expresses hundreds or thousands of genes, and most genes are expressed in many different types of cells or in varying cellular contexts. Complex traits are spatio-temporal collectives of multiple interacting gene products, changing over time (from conception onward) in any target organ or tissue, and sculpted by increasingly well-documented phenomena including largely unknown environmental influences.

—Anne Buchanan, Samuel Sholtis, Joan Richtsmeier, and Kenneth Weiss, "What Are Genes 'for' or Where Are Traits 'From'?"[1]

Polymorphisms—variations at a single DNA site or locus—are common in nature, and much more frequent in the human genome than had been anticipated prior to its mapping. They constitute the genetic basis for human variation and diversity.[2] The APOE polymorphic variation, the alleles ε2, ε3, and ε4, are clinally distributed, that is, they occur universally, but they do not everywhere have the same prevalence. And, as we have seen, it is well established that individuals with an 4 allele are at increased risk for AD under circumstances that are as yet poorly understood. Given that APOEε4 is neither necessary nor sufficient to cause AD, it has long been assumed that other genes must be involved in the incidence of this condition. It is also widely accepted that gene/gene and gene/environment interactions without doubt contribute to the initiation of AD pathology. Furthermore, given that intensive research over the years has not revealed any other gene that, similar to APOEε4, clearly has a moderately large effect on AD risk, current understanding has it that it is very unlikely that more low-hanging fruit will be found, although other genes that contribute in one way or another to AD causation are probably numerous. These genes are presumed to have small effects that may be cumulative in circumstances yet to be spelled out. Until relatively recently it has not been possible to carry out research to investigate this knowledge lacuna effectively.

The development of high-speed technologies over the past eight years or so (and hence not available when the human genome was first mapped) permits so-called deep sequencing of genomes, a technology that has enabled the investigation of AD genetics to advance in new directions. One such sequencing technique is known as GWAS (genome-wide association studies); this is an approach in which a great number of DNA sequences are randomly assayed (examined) for comparative purposes. The usual objective is to detect genetic variants known as SNPs (single nucleotide polymorphisms) in individual subjects.[3] SNPs are understood as biological markers that identify a region of the human genome believed to increase risk for a disease under investigation, but SNPs do not *specify* genetic causation. However, when SNPs occur within a gene, or in a regulatory region near a gene, they may have a significant role in disease occurrence by affecting the function of the gene in question.

Over 1,200 GWA studies had been carried out by the end of 2010, designed to uncover differences among SNPs that may be implicated in over 200 diseases. In contrast to other methods of DNA investigation that specifically test one or a few genetic regions, GWA studies examine the entire genome. GWAS has been put to use most frequently to determine the genetic contribution to several common multicausal diseases including heart disease, cancer, and Alzheimer disease. If one type of variant is more frequent in people who have the condition under consideration, that SNP is said to be "associated" with the disease. The purpose when investigating AD, for example, is to systematically detect SNPs that appear to be associated with AD traits, such as memory loss or amyloid deposition, across the population under examination.

Today, SNP assays allow researchers to sample and identify 500,000 or more SNP sites in the genome of individual human subjects, that can then readily be compared with the SNPs present in the genomes of very many, often thousands, of other subjects. This process facilitates an examination of all or most of the genes of any given organism on a massive scale. The cost of such sequencing has been greatly reduced over the years, and the time to conduct a SNP assay has also been shortened, making use of GWAS increasingly attractive. Research subjects donate a sample of cells, usually procured using mouth swabs, or at times from hair; DNA is then extracted from the cells and spread on SNP chips, from which millions of DNA sequences can then be read. It must be reiterated that GWA studies identify chromosomal sites of interest and not genes per se, from which deductions are then made about which genes, their protein products, and noncoding parts of the DNA are likely to be implicated. Such research was originally based on the assumption that the genetics of complex, "common" diseases such as heart disease, cancer, and Alzheimer's, among others, would be best explained using the "common disease-common variant hypothesis," in which it is assumed that one or else several common genetic variants account for disease causation.[4] In practice, once Alzheimer GWA studies commenced it quickly became evident that this hypothesis did not appear to apply, as a good number of involved experts had already suspected, and that many genes with small effects appeared to be of more significance. However, further investigation has shown that the matter is yet more complex, as we will see below.

Most GWAS are described as hypothesis-free (a fishing expedition as some biologists describe it), although there is nevertheless an underlying assumption that the trait under examination has an identifiable genetic component. GWA studies usually compare the DNA of two groups of participants: identified cases of the disease or trait under study and unaffected controls. A problem immediately arises with such an approach in connection with AD research because one cannot be certain that healthy controls are not harboring prodromal AD. As one researcher noted when talking with me, "[T]here is always the fear that you have people in your so-called healthy control samples who are already harboring AD but are not symptomatic. I read somewhere that something like 10% of the controls turn out not to be controls at a later date, or when autopsied. Of course, this effect can be diluted a bit if you use really large samples." A further problem raised in previous chapters compounds this matter, because heated debate persists as to whether or not biomarkers do indeed reliably detect so-called prodromal AD, and, furthermore, the replicability of the clinical AD phenotype is questionable. The situation is one in which both an evasive genotype and an elusive phenotype are under investigation, making it exceptionally difficult, perhaps impossible, to establish robust associations between the two.

GWAS follow-up studies often limit the assay to SNPs located in known or predicted regions of the genome, as established by a first run. This practice results in an "enriched" sample of potentially functionally relevant areas of the genome, at the expense of aiming for genome-wide coverage—a practice that raises yet more problems when interpreting findings.

Prior to 2011, findings from eight Alzheimer GWAS had been published or were in press, all of which confirmed that the APOEε4 gene puts certain individuals at risk for AD, especially in its homozygous form. The neurogeneticists Bertram and Tanzi carried out one of these studies and published a review article in 2009 summarizing what they believe had resulted to date from GWAS research. They commenced their review by reiterating the cautionary note they first made in 2004, cited in the previous chapter, about the conflicting results that have emerged over the past 30 years in connection with AD genetics; then they set out GWAS findings of the day:

> The most common perception is that late-onset AD is likely to be governed by an array of low-penetrance common risk alleles across a number of different, currently only ill-defined loci. These genes likely affect a variety of pathways, many of which are believed to be involved in the production, aggregation and removal of Aβ.

Their summary position is as follows:

> As additional GWAS are carried out on larger datasets and higher-resolution arrays, we can expect the list of novel AD gene candidates to keep growing over the coming years. For all of these putative

associations, replication attempts and meta-analyses across multiple independent samples will be essential to determine the identity of *bona fide* AD susceptibility genes. Despite the rapid progress being made in these still early days of the GWAS era, it should be emphasized that for none of the novel AD candidate genes that have thus far emerged from genome-wide screening, do we have conclusive functional genetic evidence that would allow us to unequivocally establish any of these loci as genuine AD risk genes.[5]

Bertram and Tanzi note that because of the exceedingly large number of studies taking place, it has become virtually impossible to systematically follow and evaluate them all. With this in mind they set up an "AlzGene" database, easily accessible on the Internet. As of 2009, drawing on the findings of various meta-analyses, AlzGene was listing 32 loci that contain at least one genetic variant that shows a "nominally significant association" with AD causation. However, an assumption has to be made that, statistically, all the studies used in the meta-analysis are comparable, which is often not the case.[6] The findings of 10 of the completed GWA studies are assessed at length by Bertram and Tanzi in their review article noted above, and they conclude that over two dozen novel potential AD-associated loci have been uncovered that demand careful, repeated replication.

One of the first efforts to carry out a "cross platform comparison" on four of the Alzheimer GWA studies reached the following conclusion: "The number of replicating association signals we observed is no higher than would be expected due to chance." The researchers suggest that increasing the power by using additional data from larger studies may result in more encouraging findings.[7]

On September 7, 2009, AlzGene issued a "Paper Alert" in which they reported on findings made by the two largest GWA studies to date, findings that had first appeared online. This research was published as a pair of articles in *Nature Genetics*, a company-run journal devoted almost entirely to the distribution of GWAS findings. AlzGene comments, "Netting three new risk genes between them, these two studies stand out not only for their collaboration and data pooling—together they engaged several dozen institutions from 10 countries—but also for fingering three new genes robustly, without the lingering doubt of small sample size and missing replication that has accompanied most prior gene association results except APOE."[8]

The chief scientific director of Alzheimer's Research UK, Julie Williams, was lead author of one of the articles. The research conducted by Williams's group had been funded by the Wellcome Trust, the Medical Research Council, the Alzheimer's Research Trust, and the Welsh Assembly Government, among other funders. The project involved scientists at universities in Cardiff, London, Cambridge, Nottingham, Southampton, Manchester, Oxford, Bristol, and Belfast who collaborated with researchers at Irish, German,

Greek, Belgian, and American institutions. Julie Williams notes that in the study "over half a million differences in the DNA of each of 4,000 people with AD were compared with 8,000 people without the disease." In addition to APOE, two other genes, CLU (clusterin, also known as apolipoprotein J) and PICALM, showed "overwhelming evidence for a relationship with AD." These findings were then replicated using another sample of over 2,000 diagnosed AD cases and 2,000 controls. Williams claims, "The findings are significant and conclusive."[9] The chief executive of the Alzheimer's Research Trust provided a nationalistic footnote: "These findings are a leap forward for dementia research. . . . The work of Professor Williams and colleagues shows how British researchers lead the world in the struggle to understand and defeat dementia."[10]

The second GWA study was led by Philippe Amouyel, the director general of the Institut Pasteur in Lille, an MD with a PhD in cellular and molecular biology. This study made use of 2,032 cases and 5,328 controls, and the findings were replicated in a second study involving 3,978 AD cases and 3,297 controls—samples that originated from centers located in Belgium, Finland, Italy, and Spain. The findings of the Williams and Amouyel teams together allowed researchers to conclude that three genes in addition to APOE had now been shown to be definitively associated with risk for late-onset Alzheimer disease. Clusterin, produced by the CLU gene, is associated with the clearance of cellular debris and with apoptosis (cell death); it had already been found prior to this study in high levels in the blood of patients with Alzheimer's, and is associated with cognitive decline. PICALM is involved in the transport of molecules into and inside nerve cells, and is associated with memory formation in addition to other brain functions. CRI (complement receptor I gene) is the third gene that proved to be significant, initially only in the French study, but after combining the findings from the French and British projects, the finding was regarded as solid. CRI appears to play a key role in connection with the functioning of the immune system, is associated with inflammation, and, if overproduced, can be very damaging to tissue, including that of the central nervous system. Prior to the GWA studies, CLU and CRI had already been shown to be involved with clearance of Aβ. In the following years a small number of other GWA studies replicated these findings.

The researchers involved with this research made it clear at the time that, in their opinion, yet larger samples were needed if the "promise" of GWAS was to be fulfilled. Accordingly, a massive intercontinental project was undertaken, the results of which were published in May 2011 as a pair of articles, once again in *Nature Genetics*. This time the input from the United States was substantially increased. One article lists 155 authors and the other lists 172 authors whose specific contributions to the article are set out under the following headings, among yet others: "Study Management and Coordination," "Statistical

Methods and Analysis," "Study Design," "Manuscript Writing Group." Several authors contributed to both of these articles. The following excerpt gives readers a glimpse into the complexity of these undertakings:

> To identify genetic variants associated with risk for Alzheimer disease, the [Alzheimer's Disease Genetics Consortium] ADGC assembled a discovery data set [over 8,000 cases and over 7,000 controls] using data from eight cohorts and a ninth newly assembled cohort from the 29 National Institute on Aging (NIA)-funded Alzheimer's Disease Centers (ADC's) with data coordinated by the National Alzheimer's Coordinating Center (NACC) and samples coordinated by the National Cell Depository for Alzheimer's Disease (NCRAD) For the stage 2 replication we used four additional datasets and additional samples. . . . The stage 3 replication used the results of association analyses provided by three other consortia [the largely European-based samples discussed above] . . . because the cohorts were genotyped using different platforms, we used imputation to generate a common set of 2,324,889 SNPs. We applied uniform stringent quality control measures to all data sets.[11]

Findings reported on the basis of this research state that, in addition to APOE, nine other genes had now been shown to be associated with risk for AD. Although each of these genes individually is attributed with a very small risk, the "cumulative population attributable fraction" of these new loci is estimated to be 35%, although it is acknowledged that these estimates will probably vary widely among studies as further research findings accrue. The authors conclude that although their study was "well powered" in order to reveal risk alleles of "small effect sizes," finding additional loci, that they are certain exist, will require yet larger studies with "increased depth of genotyping to test for the effects of both common and rare variants."[12]

Clearly, numerous questions are raised by these findings, above all, under what circumstances are these genes of low effect expressed or inhibited in vivo? Are their respective effects cumulative, and what exactly are the biological pathways that they instigate, overactivate, or block, and under what circumstances? It is understandable that the involved scientists are excited about what they are doing; the technology alone and the speed with which an enormous number of samples can be screened are mind-boggling, and no doubt highly stimulating to manipulate and manage. But I am reminded of the much greater hype that surrounded the mapping of the human genome, and the cautionary note sounded over 10 years ago in connection with the limits of what such mapping could tell us, a caution that was disregarded for many years by a large number of experts.[13] The majority of involved GWAS researchers are well aware that, if the AD world is to evolve toward prevention as its primary

approach to management of this condition, then it is essential to move beyond association studies in order to understand how these genes and their
products function throughout the life course of individuals—in other words,
to document in detail the pathways that are involved through time. This is a
daunting task, one that GWAS cannot begin to address; on the contrary, the
findings make it abundantly clear that decontextualized information about
genes is merely a first, hesitant step toward confronting the complexity associated with AD genetics—a situation fully appreciated by the GWAS scientists
to whom I talked. Gina Kolata published a predictably upbeat article about
GWAS findings in the *New York Times*,[14] but the comments she cites by knowl-
edgeable scientists are carefully measured: "This is a big, solid step," states Dr.
Nelson Freimer, who directs the Center for Neurobehavioral Genetics at the
University of California, Los Angeles.

Questions Triggered by GWAS

I traveled to Cardiff, Wales, to talk with Julie Williams and two of her colleagues at the Center for Neuropsychiatric Genetics and Genomics in November 2009, shortly after the publication of the findings of the first two GWA
studies. Williams made it clear at once that she does not believe that "we're
anywhere near the stage of being able to predict from the genes we've found
if someone is going to get Alzheimer's or not. What we have are at least four
susceptibility genes in addition to APOE [details about one were unpublished
when we talked], and each one possibly accounts for around 2% of actual risk."
When I asked if one can assume that these effects are additive, Williams agreed
that this is probably the case, but even so such knowledge does not improve
individual risk prediction. She went on to stress that simply asking whether or
not these genes have an effect on the clearance of β-amyloid is not adequate:

> [A] number of things contribute—genetic risk factors and probably
> environmental risk factors, and once you are over a certain burden
> you cascade into the disease . . . but there are going to be several pro-
> cesses involved, and some of the genes will code for molecules that
> have a number of roles, I suspect, so it's not going to be straightfor-
> ward. But what finding these genes will do is that it will help to fo-
> cus research in the right place—we'll be dealing with the right causal
> pathways.

Williams is particularly excited by findings that strongly suggest that the immune system and inflammation are implicated in AD causation. She reminded
me that clusterin has a role in dampening down inflammation in the brain and
went on to emphasize that usually inflammation, long noted as associated
with risk for AD, has been viewed as a secondary effect of the disease, whereas

amyloid as primary. She states that the GWAS results suggest that researchers may have been thinking about things "the wrong way round," and that inflammation may precede amyloid deposition. Williams adds,

> I think the real motivator for us, rather than just identifying a list of genes that can be used to predict individual risk or possibly develop drugs, is to understand the complexity of the pathways involved so as we can understand causation much better. My view is that plaques and tangles may be correlational but I think there are new mechanisms that we are now beginning to see that we haven't been measuring well. What we haven't looked at much is absolute cell loss, but this actually correlates better with cognitive changes and this fits with the complement story we are now beginning to unravel—recent work suggests that complement proteins have functions we hadn't paid attention to, like synaptic pruning, and this is something I'm excited about. GWAS really makes it clear that we have to think about much more than the production of APP [amyloid precursor protein]—we are beginning to see new patterns, and that's exciting. And because so many "normals" with plaques don't have AD, it seems clear that plaques are a risk factor but not the whole story. But at the moment I think some neuropathologists feel threatened by this.

At one point I asked if any attempt was being made in GWA studies to control for ethnicity in the sampling procedures. The reply from Williams's colleague sitting in on the discussion was that in effect the research team had deliberately controlled their sample to be as Caucasian as possible. This researcher added that the team wants to investigate different populations based on ethnicity, but that it is virtually impossible to obtain the large number of cases that are needed to carry out such a study. Williams added, "[R]ace and ethnicity are confounding factors—they are a genetic conundrum, so we have to try and simplify the question for the time being." In effect, the demands of the technology and the urgent need for continued funding constrain the research methods.

In May 2010 I met with Gerard Schellenberg, who heads up the major Alzheimer GWAS project taking place in the United States. I had previously met Schellenberg in Montréal right after he had agreed to become the director of this project, at which time he had said that he was having trouble sleeping at nights, wondering what he had let himself in for. Sitting in his office at the University of Pennsylvania School of Medicine at 8:30 in the morning a year later, he seemed more relaxed, but even so emphasized the incredible pressures being experienced by his team, several of whom regularly sleep in the lab. Schellenberg has had years of experience working on the genetics of AD and, in 1995, along with Rudolph Tanzi and other colleagues, identified presenilin-2, which is causal of familial AD present in the families of Volga Germans.

Like virtually every other researcher in molecular genetics I have talked to, Schellenberg stressed that carrying out genetic tests on individuals to predict their risk for late-onset AD makes no sense at all. We talked at some length about sampling for GWAS, and he emphasized that despite a great deal of care with sample collection "there will never be a perfect AD diagnosis and so samples cannot be perfect." Schellenberg added that difficulties exist with "harmonizing" the huge samples that his team works with because they have been procured from different "genotyping platforms."

When asked about ethnic representation, Schellenberg replied that about 1,000 African Americans are in their GWA sample, and lower numbers of Hispanics and Asians. He added, somewhat ruefully, that in order to say anything significant about a minority population, one would need at least 10,000 cases.

Schellenberg is skeptical of claims beginning to be made about AD pathways based directly on GWAS findings:

> Pathway analysis is in its infancy. Researchers claim that here's a pathway that shows up statistically significant, but this pathway has not been assembled or vetted by an expert in the field. And then you look at this pathway and you say this is . . . not real. If you are somewhat familiar with the pathway you know that it just doesn't make sense what is being claimed directly, just from GWAS findings.
>
> You know, the experience with diabetes was when they got up to 30,000 cases [using GWAS] they'd got about 30 genes, or something like that. It's not clear when there would be an end to finding more genes of small effects, so working out pathways is quite a problem. They're probably additive; so the more of them you know, the more risk you'll be able to predict, but it won't be useful to give this kind of information to patients.

When I asked Schellenberg whether he thinks it is possible to make a clear distinction between normal and pathological aging, he responded immediately by saying that such an argument seemed to make sense in the 1960s but today things are not so clear:

> What's interesting is that when we started to understand something about APP and presenilin-1 and -2, you could argue that the onset is earlier for this kind of AD, and the course is a little bit faster (although not always). But, if you hand a neuropathologist two brains, they cannot distinguish between a brain from an early-onset family and a late-onset brain. And when my group showed, years ago, that APOE interacts with presenilin-1 to affect the age of onset, and APOE also acts in the same way with late onset populations, then this all strongly suggested that we are looking at one single disease. But recently there's

been a lot of research with tau and ubiquitin and other molecules that suggest we should be able to separate this condition into two or more diseases, based on pathways—but we haven't got there yet, and in the end it may turn out that they are just variations that influence the eventual expression of Alzheimer's.

Schellenberg concluded our hour-long discussion by stating firmly that he is "highly skeptical" that we will ever be able to set up trials for preventive medications: "You'd have to give somebody who doesn't have Alzheimer's, or even any symptoms, trial drugs for ten years at least, and then at the next stage you'd have to give these drugs to 100,000 people for ten years, and you're going to have to follow them all this time—I'm not optimistic about this!"[15] I met with Schellenberg before the trial using Basque Colombians as subjects was being discussed in public. From what Schellenberg told me in 2010, it appears that he would be ill at ease with this trial in terms of both sample size (too small) and duration (too short).

Born in Kenya, Peter St George-Hyslop, like Gerard Schellenberg, Rudolph Tanzi, and other neurogeneticists I talked to, is someone who has been deeply involved in Alzheimer research for many years, and whose team identified several of the key genes and a key protein implicated in nerve cell degeneration. Hyslop's primary interest for many years has been in dominantly inherited Alzheimer disease. I first talked to him at the University of Toronto where he has worked since 1991, becoming director of the Center for Research in Neurodegenerative Diseases in 1995. Then in 2009, I talked with him again, in Cambridge, where he also holds a position as the Wellcome Trust Principal Research Fellow at the Cambridge Institute for Medical Research. At our second meeting, immediately following the initial publication in *Nature Genetics* about the GWAS findings of three new AD genes in addition to APOE, Hyslop expressed considerable concern about GWAS as a method:

> A major problem with GWAS is that when you get a statistical result and then someone else gets the same result, particularly when they use an overlapping population, you always have to be very cautious about what this really means . . . does it mean that there is a genetic signal there, or is it a statistical explanation that is not due to genetics? Second, I think you have to be very cautious about the magnitude of the results. People [researchers] are very attracted to the *p* value and less attracted to the biological impact the genes may have.[16] So, you know, if you test 1000 people and you get a *p* value of 0.003, for example, and then you test 10,000 people and you get a higher *p* value, you believe the higher *p* value, but the biological information is exactly the same. The real information is in the odds ratio,[17] and the odds ratios they are getting with GWAS are not high. So what are you getting out of these GWA studies? What exactly have you got? Not a lot.

Especially because at least two of the genes being reported are ones that we already know about—there are already literally hundreds of studies out about complement and APOEJ.

Hyslop then went on to stress that no doubt many different pathways are likely to be involved, but added that he believes the most "robust" way to proceed at this point is to work out several of the pathways in detail, prior to making claims about risk associated with specific genes. One cannot infer pathways from genes alone, he insisted, and although his name is attached to one of the GWAS projects, his primary interest is now in proteomics, with the hope of obtaining a deeper understanding of involved pathways. Hyslop is a strong supporter of the need to better recognize gene/environment diversity among populations, and in this respect he also finds GWA studies to be wanting. But proteomic findings will not radically change the dominant, reductionistic direction of AD research, although it moves to the cellular level, rather than focusing exclusively on genes.

Two years after he talked with me, Hyslop did an interview with CBC radio in April 2011, in which he discussed the contribution made to current understanding of AD genetics by the Canadian team he leads that contributes to the NIA-funded Alzheimer's Disease Genetics Consortium (ADGC) that conducts GWAS research. Hyslop had the following to say about the 5 genes recently added to the list of genes believed to increase risk for AD, making 10 in all:

> This represents new information about the pathway that causes Alzheimer's. . . . Several of the genes were genes we didn't know about and they are going to quite richly tell us more about the disease. I think they are going to be very valuable in the next few years and might even lead us to diagnostic or treatment markers with potential to slow down the disease. . . . Now we will need to go back and look at a whole new range of possibilities. . . . Things that we thought were end stage of life events will have to be looked at again. We have to do a full scale re-think.[18]

It seems that Peter Hyslop was taken by surprise with the latest round of GWAS findings, and that he now believes it is quite possible that this technology will result in significant genetic findings that he had not previously anticipated.

Raising the GWAS Stakes

In November 2009 Rudolph Tanzi, the Harvard neuroscientist, participated in an event at Capitol Hill together with Francis Collins, the director of the National Institutes of Health. These scientists, both billed on the Internet as rock stars of science, joined up with Aerosmith's Joe Perry and together they played a rendition of "The Times They Are a-Changin'" in which the second verse was liberally adapted to direct the attention of Congress to the urgent need for more funding for Alzheimer disease. Joe Perry performed "unplugged," Collins was the lead guitar, and Tanzi played the harmonica, an ensemble strikingly portrayed in a poster to commemorate the event. A line appears on the video informing viewers that the sponsors of the event were the pharmaceutical companies Wyeth and Elan. In an accompanying interview conducted on the steps of Capitol Hill, Tanzi, who is an excellent guitar player, informs everyone that he has been an Aerosmith fan since he was 16 years old and adds that Joe Perry and he are in agreement that what is needed in both science and music is "novel thinking; thinking outside the black box." Tanzi then puts in a serious plug for Alzheimer funding (which is why he takes time out from his lab to participate in a performance such as this) noting that at present only $400 million a year is spent on AD—a drop in the bucket, as he describes it.[19] A 2012 request sent to the U.S. government is for funding to a level of $2 billion per year.[20]

I met Rudy Tanzi in his Harvard office attached to his laboratory facilities in early 2010. Tanzi, grinning, started out by stating,

> [W]e are sadomasochists here, we don't believe in using core facilities and outsourcing, we do everything ourselves because then you can really trust it. Our GWAS is family-based and we used four different samples of over 1,300 families.
>
> GWAS is telling us we have—if you take it at face value—just looking at the SNPs, many AD candidate genes that are common variants exerting tiny effects on risk. Is that what AD is? Well, it fits very nicely with a model I helped propagate in '99: common variant-common disease, rare mutation-rare disease. But that's in question now. Here's the thing: penetrance and prevalence affect what determines the effect of a gene—right? Penetrance means the chance of getting the disease when you inherit a mutation or a variant. Prevalence is the percentage of the population or patients carrying it [the mutation or variant]. And for early onset AD it looks like this: heavy weighting on penetrance and low on prevalence—rare mutations, full penetrance, in effect. APOE looks like heavy prevalence with moderate penetrance, and with GWAS what we are getting is huge prevalence, tiny penetrance—if you take them at face value.

When I asked if the GWAS genes found thus far work in consort, Tanzi said,

> They probably all work in consort with each other, but it's difficult to do gene, to gene, to gene, to gene, to gene, to gene interaction studies. You get Bonferronied [a technique that makes use of multiple testing to avoid false positives] to death, as I say, we can't get a reliable signal

to noise ratio with all these interactions. You can imagine, we might have a risk factor in the same biological pathway as a protective factor sometimes, and at other times other risk and protective factors are in different pathways. So, depending on pathways they may be simply additive, or synergistic, or antagonistic.

So dozens of novel AD candidate genes contain common variants of SNPs that appear to exert tiny effects on the risk for AD. But are they actually pathogenic? Or, and this is the most important thing I want to say to you, do these common variants tag ancestral haplotypes [a set of alleles at closely linked loci on a chromosome that tend to be inherited together over generations] that harbor and cluster rare mutations with large effects on AD risk in a small portion of the population?

At this juncture, Tanzi referred to the recent work of several molecular geneticists who have produced widely cited articles in which attempts have been made to account for what is known as the "missing" heritability of complex diseases.[21] One such article published in *Nature* in 2009 sets out with this statement:

> Genome-wide association studies have identified hundreds of genetic variants associated with complex human diseases and traits, and have provided valuable insights into their genetic architecture . . . most variants identified so far confer relatively small increments in risk, and explain only a small proportion of familial clustering, leading many to question how the remaining, "missing" heritability can be explained.[22]

These authors then make an interesting observation, namely that the vast majority of disease-associated variants (greater than 80%) fall outside coding regions on the genome. They then list the usual suggestions made to account for "missing" heritability, including the following: that there are many more variants of small effect to be found, and this demands further GWA studies with larger and larger sample sizes. Second, there may be variants that are very rare, possibly with larger effects that are unlikely to be detected using the present GWAS approach, because it focuses only on those variants that occur in at least 5% or more of the population under study. Third, detecting gene-gene interactions is not well "powered" in GWAS, and, last, there is inadequate accounting of shared environment among the families that are studied. It is noted, "Consensus is lacking . . . on approaches and priorities for research to examine what has been termed 'dark matter' of genome-wide association— dark matter in the sense that one is sure it exists, can detect its influence, but simply cannot 'see' it yet."

Among the proposed reasons for "missing heritability," the one that most attracts Rudolph Tanzi and certain other AD geneticists is the theory about rare variants with large effect size. The article in *Nature* elaborates:

> A probable contributor to small genetic effect sizes observed so far [by GWAS] is that current investigations have incompletely surveyed the potential causal variants within each gene. . . . Notably, 11 out of 30 genes implicated as carrying common variants associated with lipid levels also carry known rare alleles of large effect . . . suggesting that genes containing common variants with modest effects on complex traits may also contain rare variants with larger effects.
>
> An important consideration is that the overwhelming majority of GWAS and other genetic studies have been limited to European ancestry populations, whereas genetic variation is greatest in populations of recent African ancestry and studies in non-Europeans have yielded intriguing new variants.[23]

These authors add that even when numerous rare variants can be detected in a gene or region, they may have disparate effects on the phenotype, thus greatly complicating matters. Nevertheless, their conclusion is that GWAS continues to be an efficient means of examining unexplained heritability, and that sampling should be extended to non-European populations.

A 2010 article published by the group headed up by David Goldstein at Duke University extends the above discussion along similar lines. The concept of "synthetic association" is introduced in which it is suggested that many of the signals being detected in GWA studies could very well result from rare variants with large effect sizes capable of acting over large genomic distances. These variants, because they are rare, are not themselves detected by GWAS as yet but form crucial associations with the much more common variants being detected in GWAS. Furthermore, if this is indeed the situation, examination of single haplotypes is unlikely to be sufficient to account for the associations being observed; years of further empirical work is called for.[24] Related research introduces a concept of "phantom inheritability" in which it is argued that insufficient attention has been paid to "epistasis"—the interaction of genes one with another—leading to overestimations of heritability.[25]

The recent review on the genetics of Alzheimer disease published by Lars Bertram, Christina Lill, and Rudolph Tanzi in *Neuron* in 2010 points out that virtually all of the recent AD GWA studies have detected loci that are linked to Aβ metabolism in one or more ways, notably the aggregation or clearance of Aβ in the brain. They argue that with improved high-sequencing technologies "missing heritability" will increasingly be assessed in "unprecedented" detail, and that rare sequence variants that are no doubt implicated will come to light.[26] They imply in this article that the amyloid cascade hypothesis stands firm as a result of genetic research, but, as Tanzi emphasized when I talked with him, he believes that researchers must now move upstream:

> [W]hat people are saying about AD is that lots of SNPs with tiny little effects all work together somehow to finally give you the disease—they hit a threshold. No! What is *really* happening is that all these little effects of these SNPs is just an echo, the wake of a large boat that they are following. GWAS is the very first step in the journey, a necessary first step. GWAS tells you which haplotypes to look at to find rare mutations in clusters. We have rare mutations associated with early onset that we know about and we have late onset mutations also with strong effects, but they don't guarantee the disease until 100 or 120 years old. And most people die before that! So, it only looks like partial penetrance because people don't live long enough.
>
> Well, guess what, when people were dying around 40 years old APP and presenilin mutations [associated with dominantly inherited Alzheimer disease] looked as though they had partial penetrance too.

It is of note that Evelyn Fox Keller, in her book *The Mirage of a Space between Nature and Nurture*, argues that because interactions of genes and environment are inseparably entangled throughout the lives of individuals, their relative contribution to phenotypes cannot be assessed independently of each other, thus making the very concept of "missing heritability" questionable, see chapter 9.

Tanzi and his colleagues have recently been doing research on a gene called ADAM10 using samples taken from their "enriched" GWA study composed of families they have been following for years. Out of 1,004 families screened, they found 7 families that all appeared to be at an increased risk for AD on the basis of common SNPs clustered in each of these families. The GWA research with much larger cohorts had shown that these particular SNPs increase risk for AD by a small amount. Tanzi's group found that these 7 families all had very rare mutations of the well-conserved gene ADAM10. These mutations were present in 70% of the affected family members and were found in only one of about 1,000 subjects unaffected by AD. What was striking about this finding was that these mutations apparently did not bring about dementia until the age of 70 or thereabouts. "We got lucky with this one," said Tanzi, "imagine how we are going to document the ones that don't strike until 80 or 90 years old." He added, "[T]he odds ratio we're finding for ADAM10 for increased risk of AD is actually a little stronger than APOE. But it's rare, and its effect is huge, and it is associated with a whole lot of common variants with tiny effects. Here you have an example of what Goldstein's group calls 'synthetic associations.' I was really happy to see that Goldstein who was trashing GWAS has turned right around and now he's telling us how we can do GWAS."

Tanzi believes that similar findings will be made once researchers start to "dig deeper" in connection with GWAS research. He thinks, for example, that this kind of approach will show how rare variants are associated with the innate immune system, inflammation, and in turn AD. However, when I checked in July 2012 on the AlzGene list of genes, ADAM10 was ranked as only 39. The fortunes of genes come and go on this ranking list; two others, TOMM40 and SORL1, have both been reported in recent years as very promising candidate genes, but they are currently ranked at 22 and 24, respectively, and neither came up in the large GWA studies. Just about every GWA investigation reveals new candidate genes, simultaneously heightening excitement and uncertainties, but these findings are rarely repeated by other studies, and how to interpret them remains unanswered. Clearly some deep digging is in order, or even "some other kind of digging altogether"![27]

It seems likely that this recent insight into the possible contribution made by rare variants to AD causation may account for Peter Hyslop's apparent change of heart when he commented in an interview that recent GWAS findings indicate that a "rethink" is called for. It appears that at times rare variants can bring about complex effects as a result of "synthetic associations" in their genetic milieu and also that common variants with small, gradually accumulating effects are indeed implicated as part of this extraordinary story. The biologist Ken Weiss insists that greater understanding will come about only once it is better recognized by researchers that "a mix of allelic variation in all likelihood contributes to various levels of risk that differ according to context. No one hypothesis about risk will be appropriate for all situations." Weiss insists that scientists have to learn to deal with the fact of the mix and the epistemological challenge it presents (personal communication).

In February 2012 an article appeared with findings about the genetics of 439 individuals selected from families where at least four relatives had been diagnosed with late-onset Alzheimer disease (see chapter 5). The study showed for the first time how rare variants and pathogenic mutations of the APP, PSEN1, and PSEN2—the genes associated with dominantly inherited Alzheimer's—are at times implicated in cases of late-onset AD. The researchers conclude, "Clearly factors other than the mutation can impact the age of onset and penetrance of at least some variants causing AD." And, furthermore, "*familial aggregation is more important than age at onset* in determining the likelihood of an individual carrying a disease-causing variant."[28] It is noted by the authors of this article, as other researchers in the above discussion have noted, that GWA studies are thus far able to identify only common variants associated with disease, and that rare variants go undetected in such studies. In this recent study it is concluded, in common with Tanzi, that rare variants could indeed explain a significant proportion of genetic heritability of AD. But the epigraph at the head of this chapter makes clear that the contribution of variants, common or rare, to phenotypic traits cannot be assessed decontextualized from the environments, ranging from cellular to social, in which they function.

Clearly much remains to be uncovered in connection with AD genetics, and it is striking that research has barely begun to touch on gene/gene interactions, life course exposures to toxins and other types of environmental effects, and the social milieus in which people are raised, all of which can modify the expression of genes and influence risk for AD.

Thanks largely to technological transformations that make it increasingly possible to manage ever larger data sets at a lower cost than previously, and at an increasingly rapid rate, many geneticists, "hungry" for more data, are strongly advocating whole genome sequencing. I have been told informally that several researchers, changing their former positions, now firmly embrace GWAS, and that this "crossing of the floor" is almost certainly in part due to political expediency as far as funding goes, in addition to the lure of manipulating such huge data sets. No doubt the image of scientists united in a worldwide humanitarian effort to defeat "the robber of the very thing that makes human beings unique,"[29] also captivates a good number of researchers.

The senior researchers cited above have known each other for more than 20 years and meet regularly at conferences and workshops. My impression is that, on the whole, their relationships are of reasonably friendly competition. John Hardy told me in 2010 that he and Peter Hyslop have breakfast about once a month; this is time well spent, no doubt, when future directions are taking shape for research involving massive amounts of money and hundreds of labs worldwide. But it is important to mark what is so often forgotten in the world of AD research. In a 2011 issue of the journal *Alzheimer's & Dementia* a long detailed article prepared by the Alzheimer's Association sets out the "facts and figures" of Alzheimer disease. The closing sentences run as follows:

> Although available data do not permit definitive estimates of how many individuals have undiagnosed dementia, the convergence of evidence from numerous sources indicates that as many as half of [the] people satisfying diagnostic criteria for dementia have never received a diagnosis. Some lines of evidence suggest that as many as 80% or more of affected individuals are never diagnosed.[30]

This article is focused entirely on the American situation, and a large portion of it falls under the heading of "Caregiving," in which the plight of the nearly 15 million Americans who provide unpaid care for demented relatives is discussed at length. Such care amounted to 17 million hours of work in 2010, a contribution to the nation of approximately $202 billion. The political and economic pressures to defeat AD in the United States, and virtually everywhere else these days, are enormous, but because a cure, or better still prevention, is always just around the corner, or so it seems, the incredible burden that falls on caregivers, especially those who are not well-off, continually drops out of sight. In 2011 the National Alzheimer's Project Act was signed into law by the Obama administration, and in 2012 the war on Alzheimer's was upgraded when a plan was announced that targets 2025 as the goal for preventing or treating Alzheimer's disease.

The following chapter is devoted largely to a consideration of the responses of healthy individuals who come from families where AD has been diagnosed and who volunteered to have their APOE genotype tested as part of a randomized controlled trial. What was striking while interviewing these people was that very many of them, above all else, wanted to tell my research team about the demands that they have to shoulder daily in connection with caregiving in their families. Pondering the implications of the results of their own genotyping, as we asked them to do, was very often relegated to a distant second place in our discussions with these individuals, many of whom were exhausted when we met them.

Chapter 7
Living with Embodied Omens

[A] man's future health and happiness depends on conditions that are already in existence and can be exposed by the oracles and altered. The future depends on the disposition of mystical forces that can be tackled here and now. Moreover, when the oracles announce that a man will fall sick . . . his "condition" is therefore already bad, his future is already part of him.

—E. E. Evans-Pritchard, *Witchcraft, Oracles and Magic among the Azande*[1]

From time immemorial divination has been an integral part of the everyday lives of humankind. Whether its practice involves an examination of the entrails of sacrificed birds and animals, consultation with oracles in trance-like states, or sessions with one of any number of kinds of fortune-tellers, divination produces knowledge not readily available to ordinary people—knowledge that has the potential to incite action. Historical and anthropological research suggests that one objective of divinatory proceedings is to provide explanations for what has already taken place, for it is in the reconstruction of past events that causes of misfortune are uncovered and moral responsibility is assigned, thus indicating what action should be taken.[2] However, divination does not simply link the past with the present; uncovering omens about the future is also integral to these practices.

With the rapid accumulation and dissemination of findings from molecular genetics, genomics, and epigenomics,[3] a new form of divinatory space has arisen with the possibility of creating a highly potent zone of anxiety about what the future may have in store for the health of individuals and families. This future depends, of course, on past human activity—reproductive activity. Moreover, an accelerating ability to bring "potential futures into the present"[4] means that in theory every one of us is "presymptomatically ill" because we are liable to be susceptible to many conditions as a result of our genetic heritage.[5] It is tempting to think that having one's genes named by an expert will bring about greater insight into future probabilities than do fables told by fortune-tellers, but, paradoxically, the more we learn about the universe of molecular biology, including interactions among genes, interactions between genes and their environments, the activity of genomes as a whole, and developmental biology, it is increasingly clear that for virtually all common conditions, on their own, genes are not powerful agents of disease causation.

Even though in the case of so-called single gene disorders, genetic technologies permit us to speculate with much greater precision than was formerly the case about who may be struck with misfortune, a characteristic feature of all forms of divination nevertheless persists: in seeking to take control of the future, new ambiguities and uncertainties come to the fore, as anthropological research has repeatedly shown,[6] and that will be readily apparent in the ethnographic findings set out in this chapter. Significantly, knowledge of individual genetic embodiment raises questions about shared bodily substance across generations—knowledge that may well raise moral concerns within families. Another troublesome matter arises when it comes to the allocation of responsibility for ill health. The location of agency for the causes of conditions thought of inherited has changed over time—in the 18th and 19th centuries human behavior was implicated, but from the beginning of the 20th century scientists began to conceptualize genes as entirely autonomous material entities, and hence devoid of moral import. Over the intervening years, increasingly, such an understanding has become common knowledge, although in practice moral approbation continues to be attached to carriers of certain genes, as will become apparent below.

Today, it is understood that genes are implicated in all biological processes and pathways; it is no longer appropriate to think of certain rare illnesses such as Huntington or Tay–Sachs disease, where a single genetic mutation is causal, as the only type of "genetic" disease. In recent years, now that gene/environment interactions are gaining increasing attention, and the conditions under which genes are expressed have taken center stage, the idea of societal and individual responsibility for illness causation is gaining ground once again. What is more, it is recognized, as a result of exposure to environmental variables, that genes are modified during the course of individual development—a phenomenon known as "epigenetic inheritance." And it is postulated, but by no means well established as yet, that certain of these changes are passed on directly to succeeding generations, raising further questions about embodiment and the allocation of responsibility for illness causation (see chapter 8).

Exposing the Genetic Body

Over the past two decades social scientists have been documenting certain of the societal changes that have taken place as a result of accumulating knowledge about human genetics. They have also made projections about what is in store as genetic testing and screening become increasingly routinized. Edward Yoxen suggested in the 1980s that newfound abilities to detect "presymptomatically ill" individuals would ensure that virtually all of us would be subject to increased medical scrutiny in the near future.[7] In the early 1990s the epidemiologist Abby Lippman coined the term "geneticization" to gloss this new form of medical surveillance that she characterized as a process "in which differences between individuals are reduced to their DNA codes."[8] Above all, she was concerned about the possibilities for a reinforcement of racism, social inequalities, and discrimination against those with disabilities, the result of a rekindled conflation between social realities and an essentialized biology grounded in small differences in DNA sequences, a sentiment shared by other social scientists.[9]

Nikolas Rose, drawing on Foucauldian biopolitics, suggested that in advanced liberal democracies where life is "construed as a project," values such as autonomy, self-actualization, prudence, responsibility, and choice are integral to "work on the self." He argued that genetic forms of thought have become "intertwined" in this project, and the merged language of genetics and risk "increasingly supplies a grid of perception that informs decisions on how to conduct one's life, have children, get married, or pursue a career." Rose is quick to add that there is little evidence that people labeled as genetically at risk are inevitably dominated by medical expertise.[10] And Sarah Franklin cautions that social scientists and ethicists may be accused at times of "over-sensationalizing" the potential of genetic technologies to transform society at large.[11]

Nuanced investigation of the way that individuals and families respond to and are affected by the introduction of molecular genetics into the clinic and public health screening programs have increasingly been carried out.[12] Not surprisingly, to date virtually all of this research has dwelt upon the subjective experience of individuals who have undergone testing for single gene disorders. Such research has also shown that in families where lethal childhood diseases are common, parents frequently become politicized in order to advance research that may contribute to vanquishing the particular disease that affects their family.[13] Certain of these findings have documented the way in which individuals actively interpret the knowledge imparted to them about molecular genetics, and how they frequently exhibit resistance to using genetic explanations alone to account for the illnesses that "run" in their families.[14] Furthermore, when knowledge about genes is incorporated into accounts about illness causation, such information may supplement, but rarely displaces, previously held ideas about kinship, heredity, and health. For example, Cox and McKellin, writing about Huntington disease, have shown that lay understandings of heredity can conflict with theories of Mendelian genetics because scientific explanations prove to be inadequate for families dealing with the lived experience of genetic risk. On the basis of empirical findings, they argue that "theories of Mendelian inheritance frame risk in static, objective terms. They abstract risk from the messiness of human contingency and biography."[15] Their findings show graphically how at certain junctures in an individual's life cycle, notably when making decisions about reproduction, knowledge obtained from genetic testing is recognized as useful, but, in general, factors such as social proximity to an affected family member and the family's inter-subjective construction of risk are the most influential variables that inform everyday life.

The sociologists Kerr, Cunningham-Burley, and Amos suggest that it is rea-sonable to assume that lay people are their own authority when it comes to appreciating and understanding how exactly genetics may shape their lives,[16] and responses to the uptake of genetic testing when it is offered add support to this claim. It is estimated that only between 5% and 25% of adults designated as at risk for a named genetic disease, or for carrying a fetus believed to be at risk for a genetic disease, have been willing thus far to undergo genetic testing, a finding that has held for over 10 years (although these numbers vary from country to country and differ according to the disease in question).[17] It has also been shown that a good number of people when tested ignore or challenge the results.[18] No doubt this situation exists because uncertainty, disbelief, doubt, and kinship concerns color people's responses to genetic test results. In addi-tion, worries about social discrimination, including stigma, insurance cover-age, and possible employment difficulties, also contribute to the reluctance of many people to consider testing.[19]

Further concerns arise because the bodily effects of specific genetic muta-tions vary greatly, and are dependent upon the "penetrance" of the involved gene(s). Clearly the impact of learning that one has an autosomally inherited disorder such as a presenilin mutation will be very different from learning which of the APOE variants one carries. Moreover, when considering genetic testing, individual reactions are inevitably affected by the age of onset of the disease in question, and whether or not reproductive decision making is im-plicated (as it may well be when dominantly inherited AD is involved). In ad-dition, testing cannot always predict the severity of many conditions or, for some conditions, whether or not symptoms will manifest themselves at all during one's lifetime. And for by far the majority of single gene disorders, few if any preventive measures can be taken, and no treatments are available as yet. In other words, interpreting and acting on the results of genetic testing de-pend to a great extent on the bodily effects of the condition in question and whether treatment is available.

Monica Konrad, an anthropologist, drawing on ethnographic findings, describes "the making of the pre-symptomatic person." She uses Hunting-ton disease as an illustrative example and documents the situation in families where some people choose to be tested and others refuse. Inspired in part by anthropological research into divination, Konrad explores the "prophetic re-alities" unfolding in contemporary society as a result of genetic technologies. Konrad is concerned with "moral decision-making," and her emphasis is on how, when bodies are made into oracles, "moral systems of foreknowledge" thus produced are enacted both within and across generations. Her work, in common with other social science research on genetic testing, makes it abundantly clear that the accepted precept in bioethics of a "right to know" and an assumption of individual autonomy and agency with respect to decision making in connection with genetics are both extremely problematic. Konrad discusses at length the "pragmatics of uncertainty" that infuse the everyday lives of people living with genetic foreknowledge and, furthermore, the new forms of "relational identity" that testing brings about, including how and when to inform one's children of one's own test results, whether to be entirely "truthful" or not, and deciding whether children should be tested and, if so, when. The very ties of kinship are medicalized, she argues, as a result of genetic testing, or even its prospect.[20]

Alice Wexler's book *Mapping Fate* draws her readers with heartbreaking force into the life of a family where Huntington disease is present. Alice, a historian, and her sister Nancy, a geneticist who first mapped the Huntington gene, come from a family where their mother and three brothers all died of the disease. The book weaves together family life with its struggles, secrets, and interludes of calm, with accounts about the activist work of the Wexlers' physician father in forming various committees and foundations to combat Huntington disease and promote research into it. Also recounted is the fascinating history of Nancy Wexler's work with a Venezuelan community where Huntington disease is highly prevalent, and where a great deal of significant research has been carried out in connection with the genetics of this disease that led to the mapping of the Huntington gene. Alice and Nancy both struggle with the question of whether or not to participate in the game Alice describes as Russian roulette and have their genetically allotted "fate" tested. Fragments from Alice's diary reveal the real torment that the possibility of having the test precipitates:

> The immensity of it scares me shitless. The idea of really knowing—and what if it is positive? Or if Nancy is? Once we do know, there's no going back. . . .

> Alternately exhilarated, depressed, elated, overwhelmingly anxious, and frantic. I find it hard to think about the test at all. My mind tries to escape to other things. I only want to take this test to find out that I don't have the illness. . . .

> Now the plan is for us all to get together to talk about the test—get everything clear before we do it. . . . I'm not sure if Dad is really prepared for this.[21]

While struggling to reach a decision, Alice talked with people who had taken the test and found, not surprisingly, a wide range of responses. One individual recounted the way her life had changed radically when she tested negative—she was now married and had changed her job, but she could not escape

Huntington disease because her two brothers had tested positive. Another person stated that he had lost his "creative terror" when he found he had tested negative for the disease, and then later had been arrested for writing bad checks. Several people whose results were positive, contrary to the published findings of psychologists, argued that their uncertainty was not reduced. They now had the enormous worry of when the disease would strike, and many found this harder to bear than a generalized uncertainty about whether or not they would get the disease. Alice Wexler opted not to be tested and writes that she perhaps actually enjoys the ambiguity she settled for when she resisted confronting "sharp categories and binary definitions."[22]

The research of the sociologist Nina Hallowell makes it abundantly clear that the gene in question has an enormous influence on how people are likely to respond to being offered genetic testing. Hallowell interviewed women in the United Kingdom who come from families where cancer is very common and who were undergoing testing at a specialty clinic for the BRCA genes associated with increased risk for breast cancer. Almost without exception, she found that these women believed that it was their duty to themselves and to their children to undergo testing. Moreover, many women who had already borne children believed themselves to be responsible for having unknowingly put their children at risk. On the basis of these findings, Hallowell argued that women, more so than men, are likely to develop feelings of "genetic responsibility"—that is—to experience an obligation to undergo testing and reveal the results to kin.[23]

In summary, knowledge about genes can initiate or inhibit action, and increase, reduce, or transform anxiety about genetic embodiment. But does such knowledge produce what might be called an embodied "genetic subjectivity"? And is there a tight looping effect, to use Ian Hacking's term,[24] between genetic disclosure and a transformation in subjectivity? Are people consumed by the idea that they are their genes? Or is genetic knowledge most commonly absorbed into preexisting beliefs about the risk posed by one's genetic endowment to self and family, the busy realties of daily life, and the broader social, cultural, and sometimes moral and religious dimensions that inform people's ideas about life, sickness, and death, as the majority of social science research findings suggest?

In the mid-1990s the anthropologist Paul Rabinow coined the term "biosociality" to gloss what he predicted would be the likely formation of new group and individual identities arising from emerging genetic knowledge.[25] Rabinow was careful to note that support groups formed on the basis of shared disease experiences were already in existence well before genotyping became available; such groups, he noted, would continue to function with respect to "pastoral and political activities." But, he suggested, new congeries of people would emerge as a result of individualized knowledge about molecular genetics that would assist them in coming to terms with, intervening in, and understanding their common fate.

Recently reassessing the situation, Rabinow concedes that today, an era when genes have been demoted as the determining forces in by far the majority of illness situations, the idea of biosociality must be modified.[26] The very idea of coming together as a group on the basis of possession of an APOEε4 allele, for example, makes no sense at all, given that this gene is neither necessary nor sufficient to cause AD. Alzheimer support groups have been in existence since the 1970s, and families dealing with this condition are brought together in such gatherings designed, above all else, to give support to exhausted care-givers and to people feeling sad and often guilty for having "abandoned" their relatives in nursing homes. Other groups are formed so that individuals in the early stages of AD can meet, but as the condition progresses, such group participation is no longer realistic, except when therapeutic activities including art and theatre projects are the reason to gather together. Taking part in groups with other people like oneself who are destined to lose their memories and indeed their very selves does not make for convivial discussion on an ongoing basis. AD society support is highly valued by many affected families, but group discussion based on similar genotypes is not part of the agenda (see below).

It must be reiterated that the conditions most commonly researched to date by social scientists have been the rare single gene disorders that affect approximately 2% of the human population. This by no means diminishes the significance of these findings, especially because they stimulate the production of generalizations that extend beyond each disorder investigated. Furthermore, one distinct shortcoming has been that virtually all research in connection with genetic testing and screening to date has been carried out in North America, Europe, and Israel, one notable exception being that of Duana Fullwiley working on sickle cell anemia in Senegal.[27]

Testing for the APOE Gene

When individuals undergo testing for single gene disorders, they are informed whether or not they carry the gene associated with the disease in question, and on the basis of this information are expected to make relevant decisions about reproduction, pregnancy termination, health insurance, family support, and so on. With the escalation over the past decade of research findings conducted globally in connection with molecular genetics and genomics, it has become clear that assessing future risk for disease on the basis of segments of DNA alone is more problematic than has usually been assumed, even for many of the approximately 4,000 single gene disorders.[28] It has been shown that a large number of the mutations involved in disorders such as cystic fibrosis and Tay–Sachs disease among many others have numerous allelic variations, sometimes in their hundreds, and the situation is less predictable than was at first envisioned, with the result that risk predictions are by no means easy to make.[29] Discussion in the previous chapter about dominantly inherited Alzheimer disease made it clear that the situation is not as straightforward as was formerly thought to be, and new mutations are being found on a regular basis.

Uncertainty is magnified enormously when dealing with susceptibility genes, as we have seen is the case for late-onset AD. With the move to a preventive approach to Alzheimer's, the APOE gene has attained the status not merely of one of embodied risk, but of a biomarker to be assessed in conjunction with other biomarkers believed to be prodromal signifiers of dementia. Under the circumstances, there is little doubt that in the not too distant future increased routinization of APOE testing will take place, especially as costs of genotyping continue to come down steadily.

Universal agreement has existed since the mid-1990s among involved professional organizations, health policy-making institutions, and national AD societies that genetic testing should not be routinely performed in connection with late-onset Alzheimer disease (although it can be used to confirm a clinical diagnosis).[30] However, the Alzheimer's Association, based in the United States, has recently revised its guidelines to indicate that with careful counseling testing for the APOE gene may be appropriate; the Alzheimer's Society in the United Kingdom states that it does not recommend testing for the APOE gene, noting that this is explicitly in contrast to policy in the United States; the Canadian site also recommends against testing, and the national site in France does not even mention genetic testing for APOE as a possibility.[31] Virtually every one of the clinicians whom I interviewed stated that patients should not be informed about their APOE test results, although a few take issue with this position.[32] All such statements against disclosure of test results are justified on the grounds that there is no known prevention or treatment for AD that is more than minimally effective. Even so, many private companies in the United States now offer testing, and this appears not to be limited by the fact that only one of them, Athena Diagnostics, holds the patent for APOE testing. It is difficult to obtain accurate data, but it appears that public uptake of this testing is greatly on the increase, and has been reported to me as being in the tens of thousands.[33] An "Early Alert Alzheimer's Home Screening Test" kit marketed directly to consumers in the United States in their homes is also available.[34]

In 1999 the National Institutes of Health (NIH) approved what would become for more than a decade a series of randomized controlled trials that go under the umbrella name of REVEAL (Risk Evaluation and Education for Alzheimer's Disease).[35] These trials were initially designed to assess "feasibility, safety, psychological impacts, and behavioral outcomes of APOE genetic testing and disclosure."[36] REVEAL is now in its fourth phase. The discussion below is based on findings from the first two phases in which over 440 healthy individuals were enrolled as subjects. These individuals come from families in which one or more close members have been diagnosed with AD after the age of 60. The first trial was sited at Boston University Medical School, Case

Western Reserve Medical School in Cleveland, and Cornell University Medical School in New York. In its second phase the trial was extended to include Howard University in Washington, D.C. Research subjects were recruited through systematic ascertainment from American AD research registries kept at the respective medical schools, through referral by the study personnel, by responding to an advertisement, through word of mouth, or through community presentations.[37] Upon recruitment into phase 1 of the project, research subjects were randomized into intervention and control groups. In the original three-sited sample, the majority identified themselves as "white." Participants are highly educated, with a mean of 17 years of education. Virtually everyone in the Howard sample described themselves as African American; they also have a high mean education level of 15 years. Ages ranged from 30–86, and 70% of the participants were women. The position taken by the research subjects at each of the sites was that they had entered the trial because they were, above all else, eager to assist with medical research about Alzheimer disease. The majority said that wanting to learn about their genotype was a secondary consideration, although 91% were eager to find out their result.

Upon recruitment, participants attended an educational session about Alzheimer disease in the form of a PowerPoint presentation, which gave emphasis to current ideas about the multicausality of AD and to theories about who is at risk for AD, based on gender, family history, and genotype. After this presentation people were asked to return to the research site at a later date for a blood draw. There were few dropouts at this stage of the trial, but after screening for cognitive impairment and clinically significant depression or anxiety, a small number of subjects were informed that they would no longer be part of the trial. People in the intervention arm were informed about their APOE status a few weeks after having their blood test. Those assigned to be controls were not given this information until after the study was completed. Use of a control group was dropped after completion of the first phase of the trial.

Justifications for carrying out this trial were set out as part of the original research proposal. The first stated that testing for susceptibility genes is likely to become increasingly common in the near future, especially in the private sector, and hence knowledge about how people deal with risk information in situations where predictions cannot be made with a high degree of confidence is urgently needed. The second justification stated that to withhold information from people about their bodies is not appropriate. The third was that in many families where someone had died of AD members of the next generation may well believe that they have a virtually 100% chance of contracting the disease. It was argued that if individuals can be encouraged to realize, even if they are homozygous for APOEε4, that their lifetime risk for getting AD is never more than approximately 52% for men and 58% for women (risk estimates given for dementia among African Americans were somewhat higher than this), then their anxiety levels may well be lowered. The fourth and final justification for the research was to create a pool of APOEε4 individuals whose bloods can be used at any time to enrich clinical trials.

In order to carry out the risk disclosure portion of the study, all subjects were shown one of 12 specially designed risk curves relevant to their particular genotype. The population database from which these curves were created was derived from a meta-analysis involving very large samples of Caucasian (white) subjects;[38] modified curves were created for the Howard sample, and these subjects were given risk estimates for dementia and not AD. The curves are based on differences in age, sex, and the six possible combinations of the APOE genotype.[39] During the "risk disclosure," involved subjects were informed about their APOE genotype and a genetic counselor then interpreted the assigned risk curve privately for each individual.[40] Subjects were allowed to take a copy of their risk curve home, but it was up to them to write down their genotype if they so choose. All subjects were then systematically monitored by means of three follow-up structured interviews conducted by genetic counselors over the course of 12 months, the results of which were compared with the reactions of individuals in the control group. In phase 2 of the project, when the four sites were involved, no control group was used, and clinicians and counselors were both involved in disclosing APOE status to research subjects. Quantitative findings from these interviews suggest that the great majority of research subjects do not experience increased anxiety levels that persist much beyond the time of receiving their APOE results.[41]

Conceptualizing REVEAL

The REVEAL study (figure 7.1) was led by Robert Green, a neurologist who, with 10 years of experience behind him, decided to return to university and obtain a master's degree in public health with an emphasis on epidemiology, statistics, and public policy, in order to "strengthen my analytic and experimental design skills." Green, who had already been involved with designing randomized controlled trials, says, "I remember meeting lots of family members who came to the [neurology] clinic with their relatives diagnosed with AD, and I had long conversations with them. Some of them clearly understood that there is a heritable component to this disease and they wanted to know about their own risk for developing it." These conversations with families had taken place in the mid- to late 1990s, at a time when Green was working as part of MIRAGE, a research project funded by the National Institute on Aging at 15 research centers located primarily in the United States with additional sites in Canada, Germany, and Greece.

The MIRAGE study, headed up by Lindsay Farrer, a medical geneticist, epidemiologist, and biostatistician, commenced in May 1991, resulting in a data set composed of over 2,500 AD patients and their first-degree relatives with a focus on genetic and environmental risk factors for AD. These research subjects were located from research registries or specialized memory clinics. As a result of his experience as part of the MIRAGE team, Green came to the conclusion that applying for a grant to carry out an RCT designed to assess

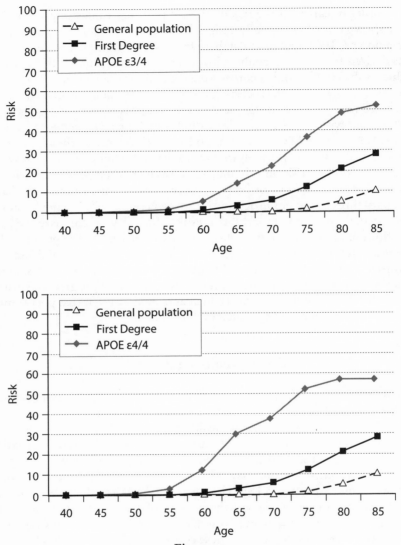

Figure 7.1.

Lifetime risk curves that are part of the education and counseling protocol used in the REVEAL study. Copies of these curves were given to women with APOE genotypes ε3/ε4 and ε4/ε4. Each curve appears in juxtaposition to a curve based on estimates of risk of AD in the general population and to a second curve based on estimates of increased lifetime risk in families where first-degree relatives have been diagnosed with AD. Roberts JS, Cupples LA, Relkin N, et al. Genetic Risk Assessment for Adult Children of People with Alzheimer's Disease: The Risk Evaluation and Education for Alzheimer's Disease (REVEAL) Study. *J Geriatr Psychiatry Neurol* 2005;18:250. © 2005 SAGE publications. Reprinted by permission of SAGE publications.

the responses of individuals to being informed of their APOE status—"an important though not perfect risk marker for AD"—would be timely. In the latter part of the 1990s Green was already thinking and writing about the necessity of detecting AD at a preclinical stage and, because he assumed that treatment for slowing or preventing the disease would be on the market relatively soon, he decided that the time was ripe to assess how information about a susceptibility gene such as APOE could best be imparted to implicated families, and with what effects. During a long conversation with me at his home he said,

> We have models about how to deal with imparting information about deterministic genes. But we have no paradigms for dealing with susceptibility genotyping. We don't have the vocabulary to talk with ordinary people about probabilistic information. REVEAL would be a landmark study to explore ways of talking and communicating susceptibility information.

Green had to submit the grant twice before it was funded, by which time six consensus statements had been published in medical journals and by the Alzheimer's Association, all advising against performing and disclosing genetic testing for APOE to unaffected individuals. He decided to approach several of the key people who had been signatories to one or more of the consensus statements, and recruited them as co-PIs or advisors to the REVEAL project.

Our conversation turned to the question of how exactly the risk curves given to REVEAL subjects had been constructed:

> For REVEAL we took two sources of information; we took the crude MIRAGE study data which offered an absolute sense of underlying risk of first degree relatives, and we took the relative risk data from an APOE meta-analysis that had been done and we put them together. Now this is quite a unique way of doing things! I mean we took data from two different sources, gathered using different techniques, and to my knowledge, for genetic counseling purposes this had never been done before. We superimposed data about absolute risk and data on relative risk. There is a paper that we struggled with about this, because people are going to say we've never seen curves like this.[42] . . . [W]e're trying to help people with the question we think they want answered: "Doc, what are *my* chances of getting Alzheimer disease?" We argued explicitly in the grant that information about susceptibility genes is more like conventional medical risk information and is not the same as deterministic or Mendelian genetic information. People will create a safety net for themselves in the uncertainty of the information they've been given. It's like smoking. Everyone knows that smoking is associated with lung cancer but they can say: "my uncle

smoked until he was 95, and he's never had lung cancer." We're trying to bring probabilistic information to bear on the individual in a way that is meaningful in their life.

When I talked with Lindsay Farrer, the principal investigator of the MIRAGE project, he commented that MIRAGE, "a family-based study," could more aptly be described as a "pooled data analysis" than a meta-analysis, because originally the data "had been collected under different schemes with different assumptions." Like Robert Green, he stressed that a primary concern throughout the development of the REVEAL disclosure methods was that people should not go away with raised anxiety levels and, as he was put it, "jump off a bridge." Above all, the "information we give out needs to be accurate and interpretable," stated Farrer, but the reality is that "for the immediate future most of these estimates are going to be imprecise."

Norman Relkin, a neurologist who was a co–principal investigator for RE-VEAL, freely admitted that he and Robert Green had some differences about the study. In Relkin's opinion, ultimately genetic testing for APOE will not be a consumer-driven market but will be medically driven because testing will have to be routinized for research purposes: "It really doesn't matter whether people want it or not right now. It's all hypothetical."

I talked with Relkin in 2003, and he predicted that detection of prodromal AD in order to prevent AD would soon become commonplace, involving the detection of biomarkers by means of genetic testing and neuroimaging—technologies with which he was at that time already deeply involved. He summarized the differences among the three co-PIs, whom he described as a "good mix," thus: "Peter [Whitehouse] is the most skeptical of us, and he's right, if we can't settle on a clear purpose then we shouldn't be doing this; Bob is right in that there is 'a need to know' [on the part of the public]. And I'm right, APOE testing is likely to become a no-brainer, it will be just like drawing blood for cholesterol—we need this for research, and if we don't, we'll be in dire straits later on. Maybe in five years time all we will need is a finger print." By 2008 Norman Relkin and Peter Whitehouse had become consultants to REVEAL but were no longer PIs. It is of note that two of the three original PIs told me, when asked, that they had not undergone APOE testing, one stating he had no intention of being tested because there is no evidence of AD in his family. The third declined to respond.

Recalling Genotypes

After the REVEAL trial had been in progress for approximately one year, Robert Green suggested that a qualitative component would be a welcome addition to the study. Following considerable reflection I agreed to do this, on the understanding that I would use my own funding to carry this out, and that conclusions drawn from the qualitative findings might well differ to some extent from those based on the quantitative findings. The qualitative project consisted of open-ended interviews carried out by a team of five anthropologists with a subsample of 79 of the REVEAL subjects, 12 months or more after participants had initially received their genotype results.[43]

Based on these interview findings, it was clear that the majority of the people with whom we had talked had transformed the estimates they had been given into accounts that were fused with their experience of having a close relative, sometimes several relatives, diagnosed with Alzheimer disease. In addition to assessments of their own family history, accumulated knowledge about the disease that they had gathered from a variety of sources, including AD societies, neurologists, their family physicians, and the media, all contributed to interpretations of the information subjects had been given during the trial. In other words, it was clear that risk estimates provided in the REVEAL study rarely displaced "lay knowledge" about who in their family is particularly at risk for AD—knowledge that participants had brought with them to the project. Furthermore, the findings from the REVEAL trial follow-up interviews showed that approximately half of the 400 participants, among those who correctly recalled the risk information they had been given, actually believed their risk was significantly different from that which they had been told. As the REVEAL researchers put it, "[T]est participants adhere to baseline risk perceptions even in the face of disconfirming evidence."[44]

Of the 79 participants in the qualitative study, only 27% were able to recall their results correctly, and 23% remembered either incorrectly or nothing at all. The remaining 50% retained only the "gist" of the information, saying, "I have the bad gene," "I have a lower risk than I imagined," "I'm next to worst," and so on. Under the circumstances, it is difficult to imagine how results of genetic testing given out as part of REVEAL might radically transform the way individuals perceive their own risk. Responses such as the one given by Vicki were not uncommon:

> I was just thinking on my way in here today, oh I bet they're going to ask me about which genes I have. And I can't remember. . . . I should have reviewed! (50, APOEε3/3, two affected relatives)

And Jackie said,

> I don't remember much . . . to be truthful, not much. I'm sure I have my risk estimate somewhere, but I don't remember where. (45, control subject, two affected relatives)

Paul also emphasizes the difficulty he has recalling his results:

> Even though she [the genetic counselor] has explained things to me several times, I still couldn't tell you which one of the markers, of the

four, they were watching. . . . [W]e had gotten all this information at the first meeting and we all dutifully took home our notes of this, and then we come back three months later or whatever, and they'd ask us these things again and I said, "Oh, cripe." And I still don't know whether I have a 10 percent or a 20 percent or a 50 percent chance. (53, control subject, one affected relative)

Tessa was one of the subjects who retained what we judged was the gist of the information she had been given:

I keep forgetting. I have problems with it, I know I'm either an ε2 or ε4, but keep forgetting which. The thing that I do remember is, whichever one I am, it's not a factor. (60, APOEε2/2, one affected relative)

This type of response was equally common among those who were given higher risk estimates because they have at least one ε4 allele. Jacqueline said,

You know, I can't even remember. I would come in from one meeting to the next, and I couldn't remember what my risk was. And to this day, I'm not 100 percent sure, but I know that it's elevated. (54, APOEε4/4, two affected relatives)

And Helen's comment reveals considerable confusion:

In fact, when I first came back to have the follow-up study after we found out the results they asked me that percent and whether it was 3/4, 2/2 or whatever. I don't even remember. The number didn't stick . . . to me it was simply like a 50/50 probability . . . okay, it's 3/4— so I put that down. It's more like a parrot thing than a "yes, I know what this means." (52, APOEε3/4, five affected relatives)

Elizabeth has seven relatives with AD and, not surprisingly, despite learning from the REVEAL education session that ε4 does not cause AD, she finds it difficult to come to terms with this information, particularly when she was informed that she is heterozygous for ε4. But Elizabeth made it clear that she already knew for a "fact" that she will get AD, before being genotyped, and claimed that testing in effect changed nothing.

Fifty-year-old Angela, who has five relatives with AD, was surprised not by her APOEε3/4 test results, but rather with the way in which the genetic counselor cushioned the import of the information by underlining the limitations of the test. To Angela, there is nothing ambiguous about her vulnerability: "Call a spade a spade. In my family, we're all children of Alzheimer's victims. Let's face it, it's in the family."

In contrast, because only one relative, her mother, has had Alzheimer disease in 55-year-old Georgina's family, she found it difficult to accept the increased risk she was given based on her APOEε3/4 genotype: "Is it genetic? I guess. I'm not that convinced . . . my mother's siblings died quite a long time ago and we did not hear rumors about senility. We don't have any basis to think that we are particularly vulnerable." The discrepancy between her own idea of risk and the one presented to her based on her genotype and mother's condition forced her to question the predictive powers of susceptibility testing.

Rebecca also questioned the results she was given, but for the opposite reason than did Georgina. Rebecca, was told that her risk is somewhat elevated due to her family history, and she said, somewhat angrily,

> According to that test, I don't have a whole lot of risk, okay? So, technically I should feel better. But I don't believe it. If I had all the confidence in the world in that test, I would say, "Oh maybe it's not going to happen." But I don't believe it. (48, APOEε3/3, four affected relatives)

Rebecca, with four affected relatives, is right to be concerned, but possibly she had forgotten what she was taught in the education session, that approximately 50% of people diagnosed with AD do not have an ε4 allele.

In contrast, some subjects clearly thought they had learned a valuable piece of information, even though they did not necessarily understand exactly what was explained to them:

> I guess I thought before testing I might have a 90 percent chance of having AD. . . . Now I know its fifty–fifty, just like flipping a coin. (55, APOEε4/4, one affected relative)

Given that there is little that can be done to either prevent or treat AD, responses of the following kind were not uncommon:

> Well, I know where I stand, and my children know where they stand—maybe get it, maybe not. (45, APOEε3/3, one affected relative)

For some subjects family concerns clearly took priority over individual testing. Carolyn, a psychiatric nurse aged 52, stated that she was less interested in finding out about her own genotype than that of her sister, who also took part in REVEAL. Carolyn is married but has no children, whereas her sister Linda has two, and Carolyn perceives a considerable difference in the significance of testing for the two of them:

> If Alzheimer's happens to me, it happens to me. But I would be much more concerned if I had children. . . . I would want to know every

> single thing out there. She has two kids, you know. . . . So when my sister learned that the testing was in Boston, I really came along for her, not so much for myself. . . . I mean, it's good knowledge to have for myself, but I wanted to be there for her. . . . To do it together as sisters.

Carolyn learned that she has a 3/3 genotype, whereas her sister is 3/4, and Carolyn's experience of having been the primary caregiver for their father contributed to her response to her sister's results, suggesting possibly that her apparent concern about her sister's children and their possible future was of secondary importance to her:

> In all honesty, I try not to think about these results, because when I think of my sister's risk factors and what I went through with my dad—I really don't want to think about going through it with her, you know.

When asked specifically about her reaction to her own results, Carolyn responded,

> I didn't think one way or the other when I found out my risk factor. . . . I guess I don't recall an awful lot.

And yet she went on to justify her participation in REVEAL as having a desire to know about her own genotype:

> Knowledge is power. I really believe that. I mean, I don't think you can necessarily change your destiny, but certainly to go through life with your eyes only half open doesn't help you at all.

When asked what kinds of actions such power might motivate, Carolyn remained unsure:

> I think [REVEAL] provides useful information. . . . Just don't ask me how I would use it. . . . I honestly don't know.

Laura was informed that she is homozygous for the APOEε4 allele. A 55-year-old from New York City, she had already undergone genetic testing for breast cancer prior to participating in REVEAL—her family has a history of breast cancer and psychiatric disorder. In addition, Laura's mother and one of her mother's cousins had died following diagnoses with AD, and she suspects the disease is also present on her father's side of the family. Not surprisingly, before agreeing to participate in REVEAL, Laura perceived herself to be at risk for Alzheimer disease. She had been in therapy for depression for many years, and her therapist was her major support while her mother was sick with

AD. Laura characterized herself as someone who "wants to know" about her physical condition. During the interview Laura said,

> I have a family where everyone's depressed. I mean, that has been the story of my life. And it's sort of like, you know, if there were 10 depressional genes, I know I would have each one.

Even though Laura appreciated the experience of participating in REVEAL, she remembers few of the specific details:

> You know what, having been to, like, these little workshops, I'm still totally confused. I know I have two of them, whatever these bad things are, or something. And I've got one on my mother's side and on my father's side. So, I do know that by the time I'm, like, 70 I have a 50 percent chance of having it, which doesn't seem so bad except that most people have a 10 percent chance and reach 70. It's not too good.

When asked to explain more about the "bad things" Laura replied,

> I don't know. I don't know what gene it is. . . . It's not the BRCA gene.

She has a heightened concern about becoming demented because she is convinced that she is already suffering from memory loss. More than the information she was given about her APOE status, her perceived poor memory, together with her family history, concern her:

> I can say that I've always felt all my life that I've had some memory issues . . . so, I have this little question, whether it's something that you actually have in some way even when you are very young. . . . Do people wind up getting Alzheimer's who were aware of some memory problem when they were younger, and the connection hasn't been researched yet?

Given that relatively few of the participants recalled with accuracy what they had been taught as part of the REVEAL study, it is perhaps surprising to find that in a follow-up study conducted by REVEAL researchers they found that over 80% of participants reported that they had told someone else about their test results—primarily close family members, followed by friends, and, least of all, health care professionals. The questionnaire simply asked, "Have you told anyone the results of your APOE genetic test?"[45] and then went on to ask specifically to whom the results had been imparted. One is left to wonder what exactly people had told their family and friends. The researchers point out that most REVEAL participants had been very positive about the experience and, on the whole, that they had perceived benefits from participating in the trial. Even so, it is evident that many people were unclear what exactly they would do with the information—a point to be further elaborated.

Imparting Risk Estimates

The REVEAL researchers had struggled greatly at their meetings prior to the start of the trial to decide how best they could impart risk estimates to the research subjects. Above all else they desired a standardized impartation of knowledge across the sites in order to be able to quantify their findings. One of the genetic counselors, Alison Jones (pseudonym), told me that on occasion she worked well into the evenings, meeting with people who had been tested and then had returned later for their follow-up interviews needing, in her opinion, considerably more help in order to understand the risk estimates they had been given. She speculated that if APOE testing was going to be widely used, it simply would not be possible for counselors to take the time to give people the guidance they needed, given that their numbers are not large, and the task would fall to physicians who, quite probably, would take less time imparting information. Furthermore, in Alison's opinion, doctors would be likely to impose their own opinions on patients about what the test results signify.

Alison talked at great length about how, when she saw people six months after receiving their test results, they had forgotten virtually everything they had been taught, and simply wanted to chat to the counselor about their caregiving duties or about family tensions as a result of trying to cope with a parent with AD. She reminded me that normally when genetic counseling is carried out, a letter is written to the patient or client after the disclosure meeting in which the test results—the genotype—is set down on paper, together with associated risk estimates and recommendations for behavioral changes, if appropriate. But she quickly added that in the case of REVEAL there was great concern that these subjects might experience genetic discrimination after testing if their results became known in one or more public settings. No insurance company could legally discriminate against these subjects given that they were participating in a clinical trial, but even so, the matter of genetic discrimination is such a sensitive issue that for this reason people were not given their genotype in writing.

Clearly, as a result of fears about discrimination, the REVEAL protocol assisted with the forgetting of the results, and thereby diminished the likelihood that people would follow up and act on their genotype assessment. Even so, on the basis of spending hours with these research subjects, Alison thought most had initially, at least, understood the import of the graphs they had been shown and had taken home. Several of them were adept with numbers, and had asked Alison questions that she simply could not answer, to her chagrin. And she felt quite uncomfortable because what she thought of as complex data about risk involvement for AD were being reduced on the graphs to simple curves from which precise, rather than approximate, numbers were generated for remaining lifetime risk. Alison admitted that the REVEAL investigators were conflicted among themselves as to how information about risk was being imparted to the trial subjects, and, in her experience, some of the tested individuals simply couldn't get away from the iconic but inaccurate figures they keep reading in the media that "between 85% and 95% of the general public aged 85 and older have Alzheimer disease."

Alison said, "[W]e keep telling them about their remaining risk, because they have lived through a great deal already, but all the time they read how risk just goes up and up with age. It seems ridiculous to most people that age is a risk factor, but the older you get the risk goes down." She also noted that a good number of the subjects whom she had counseled were surprised and somewhat upset that once their follow-up interviews were complete they were no longer going to be followed. And several also expressed unhappiness that those of their parents who were still alive were not going to be genotyped in the hope of improving the quality of the risk estimates.

Alison was pleased to hear that the results of the qualitative investigation had shown that everyone, with perhaps one exception, understood that the APOE gene does not cause AD, but merely contributes to it under certain circumstances. And she agreed that REVEAL had not disposed of the uncertainty in the minds of the research subjects about their future risk for AD, but added that this was just as well, although then "you have to ask, why are we doing this? Maybe it will help future researchers to design better ways of imparting susceptibility data." Given that it appears that undergoing APOE genetic testing had not caused increased anxiety of any duration about a demented future,[46] perhaps Alison's observation about assisting researchers in finding improved ways to impart information about susceptibility genes is appropriate. REVEAL findings are also currently being used as a model for imparting risk information in connection with AD biomarkers (see chapter 8).

The Familiarization of Alzheimer's Risk

It appears that the inherent ambiguities embedded in probabilistic information about susceptibility genes may protect many people from experiencing increased anxiety.[47] But it is apparent that there is an additional reason, because the interviews made clear that very many REVEAL subjects understand their risk primarily in terms of family histories and family resemblances, and genotype results are often eclipsed by statements about the family and its history of AD. For example, Alexis says of her family,

> They'll have a high family risk, but other than that, who knows. . . . I
> don't know what their genetics are, but I do think about what their

chances are—but not having anything to do with the ε4. My sister had a small stroke many years ago and she's at risk. I mean, she has the same family history that I have. (53, APOEε3/4, three affected relatives)

Individuals also emphasize how the information provided by genetic testing is not new to them, but only confirms what they already knew or suspected because of their family history. Edward says of his high-risk estimate,

Looking not just at my father's side, but also at my mother who has lost three siblings in the last two years who had dementia, it's been prevalent on her side of the family too. I think any objective person would look at my family tree and say, if I had to place odds, my sisters and I would all be on the wrong side of those odds for the possibility of getting this disease at some point. (37, APOEε4/4, one affected relative)

Anne reflects on the repercussions of genetic testing for her family:

This information [that I am at increased risk] affects my entire family, my children especially; because I carry a gene it is quite possible that they have this also. But the only case [of AD] I have really known was my Dad . . . it's not as though I have others to compare it to . . . he has a twin sister and I look at this twin sister now and I'm saying, well how did he get Alzheimer's, and she doesn't have Alzheimer's? She's still doing her thing, besides three knee replacements and maybe heart medication, she's still fine! (50, APOEε3/4, one affected relative)

Given her family history, the mismatch between being informed that she is at increased risk and the common assumption that identical twins share the same genes and hence risk leaves Anne particularly skeptical of the REVEAL predictions.

Some individuals resort to "blended inheritance" to account for AD risk. This is a concept created some years ago by Martin Richards, now an emeritus professor at the Cambridge Center for Family Research. Richards documented a frequent understanding held among the British public about a mixing or blending of influences inherited from both parents, rather than one entailing a Mendelian transmission of genes.[48] Such ideas stem from a long tradition of such reasoning evident as early as classical times.[49] Similar to earlier work on single gene disorders, including that of Cox and McKellin cited above, there is a tendency among many REVEAL respondents to resort to "blended inheritance" to identify a family member who in some way resembles the afflicted person as the individual most likely to be at risk for developing the disorder, whether individual genotypes are known or not. Phenotypic resemblances among family members (physical, mental, and/or emotional) signify a shared genotype, including risk for disease. Anita comments,

> I know that Alzheimer's runs in the family. I always assumed my sister will get it, not me. I have more of my father's traits. My sister has many more of my mother's traits. So, I figured if that's the gene [APOE], it goes with her trait. (50, APOEε3/4, five affected relatives)

Christina says of her high-risk result,

> Well, I really wasn't surprised. I mean, I look like my mother. And my uncle and my mother look alike, and they look like their father. So if I just go by history I would not be surprised that I would carry the risk. (52, APOEε3/4, three affected relatives)

And Zoe reminded us,

> I've shown you the picture of me and my dad. We look like clones, practically, physically. And nobody's really said—I don't know whether the information is out there because I haven't read it—whether or not that makes a difference, a person's physical appearance. But I have a suspicion that it does. (56, control subject, three affected relatives)

One participant commented about her brother:

> My brother is very worried. My brother is not very sophisticated scientifically, and he tends to feel that he has inherited a lot of my mother's qualities. He has her hair color and her blue eyes and many of her behavioral traits as well. I don't mean to belittle my brother. (50, ε3/3, three affected relatives)

In addition, sometimes people believe that the genes of one parent are "stronger" than those of the other parent, and hence more likely to transmit a specific disease. Alternatively, because women usually live to be older than men, apparent patterns of AD inheritance observed in the family lead some people to believe that the APOEε4 allele (and hence AD) can be inherited only on the female side of the family.

Conflation of ideas about genetics and heredity is a well-documented feature of "folk" or lay models of heredity and kinship in North America and Europe.[50] This apparent ignorance is sometimes interpreted as due to a deficiency in scientific knowledge, but it can also be understood as a strategy to come to terms with the inherent uncertainties inevitably associated with predictions of risk based on genetics. Duden and Samerki argue that individuals carry out a "superimposition of incompatible spheres of meaning" by "blending abstract and concrete, invisible and visible, statistical and individual."[51] Such superimpositions are relatively easy to come by when genetic information apparently corresponds to previously held beliefs about risk for a condition that "runs in the family."

Beliefs about the Causes of Alzheimer's

Despite the confusion and apparent ambiguity surrounding their own APOE test results, the majority of REVEAL participants agree that genetics plays a role in determining who will develop AD. But, with only one exception, everyone stressed that genetics is just one among several possible causes for late-onset AD (this is what informants have been taught in the education session, and the majority stated that they held such ideas before entering the trial). Contributory causes other than genetics cited by the REVEAL subjects included diet (35%), environment (29%), level of physical activity (19%), contact with aluminum (19%), age (16%), depression (17%), and mental activity (17%). Less frequent but consistently named explanations included stress, head injuries, smoking, and alcohol. In the education session people had been taught that aging is the greatest risk factor for AD, but this idea apparently seemed so evident to the majority that many do not mention it. Aluminum was not named as a risk factor in the education session, but this discarded theory, originally put about long ago by the medical world, persists widely among the public. Given recent highly publicized reports about contact sports and the neurological damage sustained by head injuries, it is likely that today more people would cite head injury as a major risk factor than was the case several years ago.

Muriel juggles with several different ideas about AD causation based on her experience with her mother:

> I mean there's always the diet thing and I do somewhat watch that because I'm certainly aware of the diet connections. There's a lot more that people don't understand about cholesterol as well . . . there is this other thing about keeping your mental activity up as you get older, you know, stimulate the brain; do crossword puzzles, learn new things and keep your brain working. My mother did that though. She watched her diet and she was very careful and doggone it, it didn't keep it from happening. And the genetic side, of course, is also not understood. There's not just one cause. (41, APOEε3/3, one affected relative)

Many interviewees assume that genetic susceptibility for AD can be aggravated, mitigated, or prevented by other factors over which one can exert some degree of control. Christina says,

> [T]here's certain things in terms of diet or exercise, that research has shown, that if you kind of walk tightly around some of these issues,

you may not trigger that genetic potential. You have the genetic potential: no question. Whether it shows up or not has a lot to with what you do and your environment. . . . Before we had blood markers, all we had were family history risks, so to me that's still genetics and what that information tells you is that something is running in your family either inherent to you, or because of something that your family does. I'd like to think that I have something to do with how it manifests. It's sort of like diabetes, which I have a very strong family history for. And knowing that there's certain things in terms of diet or exercise that research has shown may avoid triggering that genetic potential helps. I'd like to think the same way about Alzheimer's. (52, APOEε3/4, three affected relatives)

For Rosie, the role of genetics is ambiguous:

I think [genetics] are a minute aspect of it. People want to say a lot about genetics, and I have to say, we don't know enough. I think that genetics is the big buzz word. . . . Now, with my mom, I think that it could have been a predisposition, but with the stress of having three little ones in her fifties and I guess going through the change of life or whatever, I don't know, all of that could have played a role. Maybe she got depressed and the depression could have led—I don't know. I can't say that I a hundred percent think that it's genetics, even though I did the APOE test. And I forgot what I had! But I refuse to buy into that paradigm. I refuse to believe that there is really an increase [in risk]. I think there's other things that they don't know about . . . and I think that stress and environment helps make a weakness into a disease. I thought it [getting tested] was interesting. But you know, I look at this knowledge not as the answer to anything. And I just take what I want and leave the rest. . . . I don't take anything anybody tells me as God's truth! . . . [I]t's good to know but I'm not going to do anything with it. (51, APOEε3/4, two affected relatives)

And Julia highlights the uncertainties raised by gaps in scientific knowledge:

I think at some point that genes act up. And I don't know what the trigger is, but it's going to send some message that's going to cause something else to happen. I think we can recognize the gene, but I don't know that they know what causes the gene to do the bad stuff . . . so I think there's something larger happening that allows these abhorrent genes or whatever, to run havoc in your body. . . . I don't think they really know how to take a mixture of factors of genetics and gender and whatever, to say, okay, this is what really sparks this clogging of your brain. (57, APOEε3/4, one affected relative)

When imparting information about risk, the REVEAL genetic counselors emphasized the uncertainties associated with the predictions, and it appears that most research subjects listened to these cautionary warnings. The level of uncertainty involved in susceptibility gene testing coupled with the limitations of the predictive power of the APOE test no doubt dissuade many people from giving up complex ideas about AD causation, but, for many, a hope that preventive measures may indeed be protective persists. Furthermore, the probabilistic estimates do not, it seems, have the power to displace preexisting ideas and beliefs about who in the family is at risk. Some people, such as Rosie cited above, are apparently highly skeptical about the value of genotyping for susceptibility genes. On the other hand, Elizabeth is more comfortable with the uncertainties:

> I don't like being kept in the dark—you get a tendency to go off the deep end when you're in the dark, you know? So, the fact that the information doesn't necessarily mean you're going to get it, I mean, that's true about most medical things. Nothing is black and white—everything is like gray. But it's just good to know. I knew about my Dad's AD, although not about his genotype, I knew about my grandfather, I also know that they had a lot of siblings and that it wasn't showing up all that great, you know, but you never really know, okay? (52, APOEε3/4, three affected relatives)

On the whole, people are cautious about the information that circulates in the media and elsewhere about AD prevention, especially when it confounds what they have witnessed in their own family. Rosemary, for example, commented,

> Turmeric was one of the things I remember hearing about, and I started using yellow and red peppers and I season my food with turmeric when I can. I had intended to do further research on the Internet, but you know when you first get the [genetic] result you think you're going to go out and look up all this stuff to find out a way that you can stop this. But then, after a while, you get lackadaisical about it, just like I have. And I just say, "I'm not going to get it. I'm NOT going to get it." Because whatever tests and stuff you may run, I think the ultimate decision rests with God. (62, APOEε4/4, one affected relative)

Jolie is skeptical of the material she has read about mental exercise:

> The one thing that I don't believe will help or hinder Alzheimer's is the fact that they say that you have to keep mentally active and do crossword puzzles or play cards—because my mother was active. She worked at the school lunch program for years, read the newspaper

from cover to cover, so she was mentally stimulated. And you look at people like Reagan you know what I mean? It's not as if his brain wasn't exercising! (56, APOEε3/4, one affected relative)

And many people contextualize their risk for AD by comparing it to other uncertainties in day-to-day life. Rosie says, "There's a risk in walking out the door in the morning, there's a risk in getting out of bed. There's a risk in everything—living is a risk! Existing is a risk!" And Tara, age 57, likens her ε3/3 test result to a weather forecast: "[T]here's always the probability of [AD] happening. It's something like: it's going to snow, but will it snow ten inches or one?"

Even those few people given the highest risk estimates because their genotype proved to be ε4/4, apparently do not experience a radical transformation in their sense of self. Helen was surprised about her own response:

When [the genetic counselor] told me the results, she was saying, "So how do you feel now?" It wasn't anything that surprised me because I felt that there was a strong probability that I would have this genetic component because of the Alzheimer's on my mother's side. And yet, I was hoping that if I got this information I would have this great transformation—I'd lose 50 pounds, I'd live every moment. You know, it didn't happen like that. . . . It was nebulous . . . if it had been real concrete . . . and it spoke to my sense of organization—you know . . . then I'd get things planned . . . but nope. So it's just another part of the way life is, which is quite uncertain. (42, APOEε4/4, three affected relatives)

She added that her recent decision to purchase long-term care health insurance was more to do with her personal experience caring for her mother than with her genetic test results.

Similar reactions were evident during many of the interviews. Interviewees often took over the discussion, talking expansively about the emotional trauma of watching a close relative succumb to Alzheimer's and their struggles with caregiving. Vicki, a 50-year-old African American who was interviewed at Howard University, poured out a heartrending story about her life over the past decades. Her concern about her own risk for AD was eclipsed by her experience of caring for her ailing father while his dementia steadily worsened, with no support from her siblings who claimed they were too busy to get involved. At the same time Vicki struggled with her mother's response to her husband's illness. She was "in denial," said Vicki, because she was resentful of her husband's forgetfulness and erratic behavior. It was Vicki who took her father to doctor's appointments, dealt with his wandering, and found a nursing home when she could no longer care for him. For nearly seven years until his death she went through the ordeal of "caring for this man that was no longer my Dad." Her mother was diagnosed with Lou Gehrig's disease just a year and a half after her father's death, and so she started "retraining again, to learn about Lou Gehrig's." While her parents were alive, Vicki was also caring for an aunt and uncle both diagnosed with AD, struggling to make important financial and legal decisions for them with no assistance from their own children who live far away. Her uncle died three years after Vicki became primarily responsible for his care, but care of her aunt continues, and at the interview Vicki said she was feeling guilty about not spending more time with her.

Vicki, who tested APOEɛ3/3, has diabetes and is concerned about her blood sugar levels and cholesterol. She and her husband have raised three children, and she has worked full-time at a government job virtually all of her adult life with only an occasional leave. She recently took an early-retirement package: "Retirement suits me," she says, and Vicki is now fully occupied with her family, church activities, and a local advocacy group on homelessness. She is excited about the future and says, "If I do [get AD], then I do. . . . I told my husband [about my results] and he's been through as much as I have and he was like, 'well, whatever, we'll deal with it.'"

Based on the follow-up structured interviews carried out by REVEAL researchers exactly 12 months after testing, it was found that people who had been given an ɛ4 genotype reported more often than others in the study that they had made certain dietary and other behavioral changes stimulated by their test results, and some had also made changes in their insurance coverage.[52] However, among a sample of 54 individuals who were part of the qualitative study, interviewed more than 12 months after receiving their test results, 42 reported that they had not made any "use" of their results, including 13 individuals who tested ɛ4 positive. The principal reasons given for inaction were that nothing can be done to prevent or cure AD and, second, that the information they had received has such low predictive power that it was not clear how any action taken would assist with prevention. Others said they had already been taking care with their diet long before entering REVEAL, and a few said that they were simply too busy to make any changes, especially in connection with insurance. Overall, relatively few long-term changes appear to have taken place in the lives of most of the REVEAL subject based on their genotype results.

The above interview excerpts afford only a glimpse into people's lives, a delimited static picture that is frustrating because there is no possibility for future encounters. Whether, as the REVEAL participants find themselves becoming older, knowledge about their genotypes will become more significant is open to question. This will in turn depend upon what other events have happened in their lives, including the outcome of their predictions about who in their families will get the disease based on family history, resemblances, and shared personalities. Before participating in REVEAL, all the research subjects had given permission for their blood that had been drawn to be stored and used in ongoing basic science research in connection with Alzheimer disease. No doubt a good number of these individuals justifiably feel considerable satisfaction having assisted as best they can with scientific research into the condition that places such a burden on their families.

It is clear that many individuals in the REVEAL study did not comfortably accept probability estimates derived from genotyping about neuropathological changes looming on the horizon. Ian Hacking's concept of "interactive kinds" is helpful in considering to what extent the results of biomarker testing such as genotyping and PIB scanning may affect individual subjectivity, and perhaps even ideas about one's identity. Hacking's argument is that once individuals are diagnosed and hence "classified" as having a state or a condition, a "classificatory looping" may well follow in which people undergo a transformation, in effect becoming individuals who are, for example, exemplars of multiple personality disorder, depression, anorexia, being at risk for dementia, and so on. Labeling entails being slotted into a niche where packages of technological practices, regulations, and specific policies are enacted on one. In turn, labeled individuals may transform the niche as a result of responses to such labeling. On the basis of research with focus groups composed of individuals with psychiatric diagnoses, Pickersgill, Cunningham-Burley, and Martin argue that people may best be thought of as "bricoleurs," who "piece together diverse knowledges pertaining to soma, psyche and society; neuroscientific concepts compete with, integrate into, and only occasionally fully supplant, pre-existing notions of subjectivity."[53]

It is generally assumed in the United States, more than elsewhere it seems, that individuals should be informed, even in research settings, about their individual test results, although it is also agreed that people should be clearly informed that biomarkers do not determine individual futures. Findings from the REVEAL study strongly suggest that learning about one's genotype and being labeled at risk for AD do little to change individual subjectivity or sense of self. But, it is quite possible that being given a risk estimate on the basis PIB scanning may have a much more profound effect. If indeed people are beginning to conceptualize their fundamental "essence" as located solely in the brain, as a good number of researchers suggest,[54] rather than in some composite of brain, mind, and social relations, then being informed about amyloid deposits in one's cerebral cortex is likely to be a frightening experience, one that may kick off a biolooping effect involving a profound transformation of self.

Unknown Genotypes

Between 2002 and 2003, 40 interviews were conducted with first-degree relatives (primarily children) of late-onset AD patients in Montréal, Québec. A team of three anthropologists, including myself, contacted participants at clinics and gerontology units in which their relatives were receiving treatment. These participants ranged in age from 29 to 70 years, and 58% were women. In contrast to the REVEAL subjects, the Montréal interviewees had not been exposed to any systematic information about the genetics of AD, and the majority rely on information about AD obtained from other caregivers, the media, family physicians, and advocacy groups. Research has shown that family physicians actively discourage genetic testing for AD,[55] and so too do AD advocacy groups.

Not a single participant spontaneously introduced discussion of APOE into the conversation during the interview, and only half the sample believed that genetics is implicated in AD causation (this finding may no longer hold, nearly a decade later). Carla was among those who stressed the contribution made by genetics:

> I tell myself that the genetics are responsible for the disease. I think it's the family baggage, rather than an aluminum pan or living in a certain area or whatever. . . . It's true that there were hereditary antecedents. My grandmother suffered through Alzheimer's and my aunt also did. We were aware . . . we knew eventually something would happen. (52, three affected relatives)

Interviewees made it clear that physicians downplayed the significance of genetics when discussing Alzheimer disease with them.

Even though half the sample believes that AD has a hereditary or genetic component, just six believe this is the most important or sole factor. Hannah states,

> I think everything is genetically related, but I think that it would be a propensity towards whatever it may be—asthma, heart disease, AD, cancer—and I think that there's other factors that may make it surface or not. (62, one affected relative)

Ideas about AD causation and prevention are primarily informed by family history and personal experience of the disease in the family. Bridget says,

> I've gone to lectures on Alzheimer's and they say, keep your brain working, keep your brain active, try and do things. But my mum worked her whole life, she played bridge, she played Mahjong, you know, she did everything. Her mind was working all the time, so how do you figure that one out? There is no answer. She was very active her whole life. (38, two affected relatives)

And Jane commented,

> I don't know if there's too much prevention that you can do, other than having a healthy lifestyle, exercising, staying active. Which

I do. I definitely feel like I'm a fighter, but I'm sure my Mom was too. . . . So I guess I'm just kind of unclear about it. (28, five affected relatives)

Several of the interviewees questioned the relevance of genetic testing for AD because they were convinced that nothing could be done to prevent the disease. Sophie was clearly disconcerted by the idea. Her real concerns lay with how, if she became demented, it would affect her husband and children:

Dr. C. once said to me, "There's a test," and I said, "do you have a cure?" He said, "no." I said, "Then why should I take the test?" . . . I don't think I want to know, I mean, my greatest fear is that I don't ever want to put my husband or my children through that—looking after me. But I don't see the point of knowing. I also believe that people with AD know there is something wrong, especially at the beginning. I mean, this is no deep dark secret. My mother knew and she said to me early on, "I'm not what I used to be." (58, two affected relatives)

Frank made it clear that he is preoccupied with caregiving and said,

Without wanting to sound trivial, the medical terminology issues are of absolutely no interest to me. . . . [I]f some great new research comes along and looks promising, then I'll be interested in that, but why my mother has it or the genes that she has or how the genes work in the brain. . . . I could really care less. (51, one affected relative)

When asked if she thought her mother could have avoided AD, Mary replied,

My mother is in such a stage of her condition that there is nothing out there that will arrest it, in any way, shape or form OK? . . . [A]nd we have other medical histories to deal with in my family. My father has had colon cancer, my mother had lymphoma, my mother's late brother had lymphoma, so there are clearly some other risk profiles in the family that I'm more concerned about than I am about the AD actually. (53, one affected relative)

Similar to the REVEAL respondents, about half the Montréal interviewees draw on ideas of family likeness to predict who may be marked for AD. When asked whether she worries about Alzheimer's, Hannah maintains that her brother is more likely to develop the condition than is she:

My mother worries more about my brother than me. She thinks his personality is similar to hers. My brother looks like her family and is built like them. We are both intense, but my mother doesn't talk, neither does my brother; I talk about my feelings more and he's a more closed person. (62, one affected relative)

Not everyone rejects the idea of taking a genetic test should it become available, yet most are unclear about how such a test would yield any new information. Even though many believe that the condition "runs in their family," most are preoccupied with caregiving and other family matters and their own risk for AD is not a present concern. These responses are not dissimilar to those of the REVEAL subjects, the majority of whom, even after learning their genotypes, did not report experiencing increased anxiety levels to any degree and were apparently able to put the ambiguous, uncertain information they had been given out their minds and carry on with their busy daily lives.[56]

Sources of Information about Alzheimer Disease

In Montréal, the primary sources of information about AD named by participants were physicians (65%), newspapers (38%), and other caregivers (35%), followed by AD societies and other organizations (33%). For the REVEAL group, the source of information most often cited was AD societies (or other such organizations, at 52%), followed by newspapers (49%), physicians (44%), and the REVEAL study (39%). Other sources of lesser significance for both groups ranged from television and radio to Internet sites and chat rooms, professional medical literature, and work-related experience, such as nursing. It is of note that REVEAL participants ranked their experience as research subjects relatively low as a source of information, suggesting that the experience did not take over their daily lives, nor drive their concerns about the future.

Alzheimer societies are professional advocacy organizations primarily dedicated to alleviating the social effects of Alzheimer disease among patients and families and to increasing public understanding and awareness of the disease. Activities carried out by these societies include the running of peer support groups, public presentations, and educational sessions for patients, families, and health professionals. Since the late 1970s, AD societies have played a significant role in public and political conceptualizations about and perceptions of AD.

This chapter concludes with a brief presentation of research conducted by Miriam Padolsky, a medical sociologist, at the Ottawa (ASO) chapter of the Alzheimer Society of Canada (ASC), founded in 1978.[57] Padolsky's research, carried out in 2005, makes clear that discussion about genetics is highly circumscribed in both ASC and ASO publications and practices, as is the case for other AD societies in the United States, the United Kingdom, France, and elsewhere. In interactions with patients and families ASO staff attempt to contextualize and mitigate fears about inheriting Alzheimer disease. Information about genetics appears in ASC brochures dealing with genetic testing, heredity, and causes of AD; it also appears on websites where "myths" and "common questions" are discussed. The difference between familial (early-onset) and sporadic (late-onset) Alzheimer disease is emphasized, and it is also pointed out that at present, there is no test to determine if a person will or will not develop later onset Alzheimer disease. Moreover, the predictive limitations of the APOE gene are emphasized: "[T]he APOE gene cannot predict Alzheimer disease, but it may be useful in confirming a diagnosis. It should be noted that the absence of the 4 allele reduces the risk of Alzheimer disease."[58] However, materials on genetics represent a very small proportion of the information available on the ASC website. Most staff members report that genetics is rarely raised by their clients as a subject of concern, and in their presentations and individualized educational messages staff members devote only a small amount of time to discussing genetics; they often draw on the concept of "risk factor" as a way of encouraging healthy mental and physical lifestyles, but not to promote genetic testing.

It seems highly unlikely, then, that people are mobilized to be concerned about the genetics of late-onset Alzheimer disease on the basis of what they have read or come across at AD society events.[59] And a systematic review of 1,800 newspaper and magazine articles published between 2001 and 2006 on the topic of Alzheimer disease revealed that little emphasis has been given to genetics,[60] a finding consistent with an earlier review covering the years 1995 to 2001.[61] In only 205 of these articles is mention made of genetics, and an impression that genes determine the disease is avoided almost without exception; in contrast, the complexity of AD causation is highlighted. Only 22 articles mention genetic testing for late-onset AD, and with only one exception, such testing was portrayed as being of no benefit to individuals. It is also evident, as earlier discussion made clear, that family doctors and general practitioners do not encourage patients to search out genetic testing, although no doubt this happens more often in the United States than elsewhere. One Canadian researcher confessed to me that for one of his projects, after obtaining informed consent to take blood for various tests including APOE typing, he told his research subjects that he would inform them of their genotype if they wrote him a personal letter at a later date asking for this information. Thus far no one has sent such a letter of request.

Knowledge is power, Carolyn recites, but knowledge about APOE results in remarkably little action on the part of most people. REVEAL subjects appear on the whole to be pragmatic, primarily because nothing can be done to avert a condition that may well be inscribed in their destiny, and that of very many other human beings. Norman Relkin told me when REVEAL subjects claim that they took part in the trial primarily because they wanted to help research, this was not really the case—this is simply the reply that people think they are expected to select, he claims, when rank ordering a list of reasons from which they must choose one or more. Whether this is so or not, trial subjects were willing to have their blood used for research purposes, knowing that their only tangible reward would be to learn about their APOE genotype. Few if any of their uncertainties were dissipated by going through the trial, and research subjects are left to hope that a cure, perhaps enabled to a small degree by their blood donations, will finally come about.

It is of note that in phase 4 of REVEAL, currently under way, individuals who have been diagnosed with MCI have been enrolled as research subjects and will be informed of their APOE status. These subjects will be followed for six months to assess how they respond to the risk estimates they are given. This project is not designed to assess how many of these subjects eventually "convert" to AD.

Widely cited research in the social sciences brings to our attention social and political activities directly related to emerging genetic knowledge.[62] Such alliances exemplify what has been termed "genetic citizenship," when affected families come together to lobby for financial investment and scientific investigation into the rare single gene disorders that affect them.[63] However, the Colombians who harbor the paisa mutation in their families are perhaps best thought of as practicing a remarkable form of ethnic affiliation when they agree to become research subjects, knowing well that they themselves and their immediate family members may not benefit directly from the research findings. Similarly, individuals who agreed to participate in the REVEAL project profess to having become research subjects primarily to benefit advances in research into Alzheimer disease. These activities—an emerging form of "corporeal citizenship"—do not involve political lobbying but, rather, are expressions of social solidarity and a belief that, as individuals and families, people can contribute to scientific advancement by submitting their bodies to testing and procedures designed to bring about insights that may not benefit them directly.

In the penultimate chapter, following an introduction about recent developments in the AD world, I broaden out discussion to consider the worlds of genomics, epigenetics, and epigenomics, and reflect on what this emerging knowledge means for the ongoing endeavor to better confront the AD conundrum.

Chapter 8
Chance Untamed and the Return of Fate

Life is an adventure in a world where nothing is static; where unpredictable and ill-understood events constitute dangers that must be overcome, often blindly and at great cost; where man himself, like the sorcerer's apprentice, has set in motion forces that are potentially destructive and many someday escape his control . . . complete and lasting freedom from disease is but a dream.

—René Dubos, *Mirage of Health*[1]

A two-day meeting titled "Alzheimer's Disease Research Summit 2012: Path to Treatment and Prevention" was held at the National Institutes of Health in Washington in May of that year. The stated goal of the summit was to find effective prevention and treatment approaches for Alzheimer's disease by 2025, and leading Alzheimer's researchers from around the world were present to give presentations and discuss how this could effectively be carried out.[2] On the first day of the gathering Kathleen Sebelius, the secretary for Health and Human Services, emphasized that 16 million people (in the United States) will be affected with Alzheimer's by 2050. And at the major AD conference of the year held in Vancouver in July, the audience was told that by 2050 there will be 36 million people in the world with AD—such oft-repeated mantras and others like them, ultimately targeted at government, create an urgency in connection with all discussion about Alzheimer's these days, but they are repeated so often that many listeners have become tone-deaf, although it seems that, at last, governments are increasingly paying attention.[3] Sebelius stated that the government had launched a website as a resource to enable people to find out more about the disease and caregiving, but the hope is also to attract people who will be willing to sign up as research subjects. Similarly, the Alzheimer's Association has a website, one stated purpose of which is to attract individuals as research subjects.

Media reporting on the summit ran under the heading of "Obama Administration's War on Alzheimer's,"[4] but it was also remarked that funding for Alzheimer's research in the United States continues to be markedly less than for other major diseases: "Last year, the NIH spent $3 billion on research into AIDS, $4.3 billion on heart disease, and $5.8 billion for cancer," according to the Alzheimer's Association, and they note that these sums are a fraction of what goes to AD. President Obama signed the National Alzheimer's Project Act into law in January 2011. In February 2012, the administration said it would push for a $156 million increase in funding for Alzheimer's research over the next two years. At the time of the summit meeting, it was made clear that the proposed 2013 budget would allow for a further $100 million to support AD research and caregiving, but this funding still falls far short of that allotted to other diseases.

Writing for *Scientific American* in early 2012, Gary Stix comments that "[g]overnment declarations of war on drugs or disease often end in losing battles."[5] Stix cites Samuel Gandy, an Alzheimer researcher at Mount Sinai School of Medicine, who told Reuters, "[I]n my mind . . . [a war] provides the unfortunate sense that we will have 'failed' if we don't have a cure by 2025." However, in July 2012, the *New York Times* reported on a finding published in *Nature* that created a positive stir in the research world, one that suggested that AD research might indeed be on the right track with its dogged persistence in tackling the deposition of amyloid.[6] As part of whole genome sequencing of nearly 1,800 Icelanders, Kari Stefansson's team deCode Genetics, of which he is the executive director, found that about 1 in 100 Icelanders 85 years of age and older carry a coding mutation (A673T) in the APP gene. This mutation apparently protects these individuals against Alzheimer's disease and cognitive decline and, furthermore, it appears to be effective even when people are homozygous for APOEε4. Those who carry this mutation proved to have an approximately 40% reduction in the production in vitro of amyloid producing peptides.

The article in *Nature* claims, "The strong protective effect of the A673T substitution against Alzheimer's disease provides proof of principle for the hypothesis that reducing the β-cleavage of APP may protect against the disease. Furthermore, as the A673T allele also protects against cognitive decline in the elderly without Alzheimer's disease, the two may be mediated through the same or similar mechanisms."[7] The researchers conclude that their finding provides indirect support for the theory that the processes of "normal cognitive aging" may "be shared, at least in part" with that of "the pathogenesis of Alzheimer's disease." And they propose that Alzheimer's disease may represent "the extreme of age-related cognitive decline in cognitive function"[8]— thus strongly supporting an argument that acknowledges an entanglement of aging and dementia.

This same article in *Nature* states that over 30 coding mutations of the APP gene have been found, 25 of them pathogenic, almost all of which are associated with early-onset AD. It has also been shown that the protective mutation appears to be very rare indeed among North Americans, suggesting that it arose only relatively recently in Iceland. Gina Kolata notes in the *Times*, "[T] his find bolsters hopes of drug companies that have zealously developed drugs to reduce amyloid levels with the expectation that they might alter the course of the disease or even prevent it."[9] Samuel Gandy is quoted as claiming that this find is "extraordinarily important"—the most significant in the field since the first mutations associated with dominantly inherited Alzheimer's were discovered 22 years ago. And John Hardy, who formulated the amyloid cascade hypothesis, stated that this research "is obviously right," but added, "as provocative as the discovery of the protective gene mutation is, the strategy of reducing amyloid levels—the ultimate test of the amyloid hypothesis—still must be evaluated in typical [late-onset] Alzheimer's disease. For example, perhaps people need to have lower levels of beta amyloid from birth to really be protected from Alzheimer's disease." This statement suggests that Hardy may be thinking in terms of a developmental approach to AD in which modulating factors present from birth may play a part (a point to which we will return).

At the Vancouver meeting, shortly after the article about the protective mutation had appeared, Hardy stood up at a packed scientific session on genetics to insist that the Iceland finding is not new, and had been made years ago (but presumably was not well publicized). He added that he was a reviewer for the *Nature* article, and not all his critiques of the draft paper had been dealt with satisfactorily. Hardy increasingly voices skepticism about the utility of the amyloid hypothesis, at least in the way it was originally formulated by himself, and virtually everyone would agree today that amyloid functioning continues to be relatively poorly understood, and that AD should be recognized as heterogeneous—insights that have significance for drug development.

In a 2010 article published in *Alzheimer's & Dementia*, Iqbal and Grundke-Iqbal argued categorically that AD is multifactorial, and that at least five subtypes can be recognized. They pointed out that these subtypes may well respond differently to disease-modifying drugs and, for the purposes of clinical trials, should be separated out from one another. Another article that appeared in *Nature* in 2011, insisted, "[R]esearchers don't know enough about the biology of Alzheimer disease to identify the right targets [for treatment]. The disease is the result of a long chain of events, but some of the links in that chain are still a mystery—nobody is certain which link to cut to stop disease progression."[10] Even so, the Icelandic finding of a protective mutation was referred to repeatedly at the Vancouver meeting, as were trial findings about a new drug designed to clear β-amyloid from the brain. This exploratory trial had involved intravenous injections of immunoglobulin in 24 people resulting in stabilization of AD symptom development in the early stages of the disease over the course of three years. In contrast to earlier research involving immunotherapy use, this study did not result in devastating side effects, and hence further research will be carried out with larger samples. This finding, very small though the sample was, created the most media interest among all the data presented at the Vancouver meeting. A cure is what experts and the public want to hear about.

The six-day meeting with over 4,000 participants was a lively affair; at last a breakthrough might be in sight in connection with Alzheimer disease, although most paper presenters were, at best, cautiously optimistic, if that. One plenary session and several other panels focused on ongoing GWAS research. Gerald Schellenberg presented details of the recently formed International Genomics of Alzheimer's Project (IGAP) he now heads up—a consortium designed to replicate earlier GWAS findings. This consortium will also continue the hunt for other susceptibility genes and attempt to document on what pathways these genes are located. IGAP is located in 12 countries, and 20,000 SNPs are being tested. Other GWAS-related papers given at the meeting confirmed that both relatively common and rare variants appear to be implicated in AD risk, none of which have anything like the explanatory value that does APOEε4. In one paper, emphasis was given to the way in which such variants may be protective, and hence influence age of onset of AD, and the Icelandic findings cited above are likely to incite many similar papers. But these data were presented divorced of any contextualizing data that might suggest the social and environmental circumstances under which such genes are protective. If challenged, the involved researchers would no doubt counter that they are not yet at that stage.

A very large number of sessions at the Vancouver meeting were devoted to biomarkers, often described as endophenotypes—a reflection of the move to the molecular prevention of AD. One paper provided evidence that plasma biomarkers could be of use as a standardized diagnostic tool and that would improve on simple clinical assessments of prodromal AD. A whole session was devoted to longitudinal tracking of biomarkers and their age-associated appearance. Claims were made for strong similarities in biomarker changes in dominantly inherited AD and late-onset AD, even though they take place following a different time trajectory. Such findings furnish support for the entrenched belief that early- and late-onset AD are essentially the same condition, even though causation is different, but, as we will see shortly, a few months later second thoughts about this assumption moved to center stage.

A densely packed session was devoted to CAP—the Consortium for Alzheimer's Prevention—composed of three groups of researchers in the United States who are loosely working together to try to make their respective clinical trials sufficiently compatible that generalizations can be made from their findings. One trial is known as "anti-amyloid treatment in asymptomatic Alzheimer's disease," or ε4 for short. This is a secondary prevention trial in which Reisa Sperling, at Harvard University, who has been very active in initiating the move to AD prevention, is involved. The objective of this project is to establish how best to treat older individuals at risk for developing Alzheimer's disease. Risk estimation is based on biomarker evidence of amyloid deposition by means of imaging. The hypothesis being tested is that by decreasing amyloid burden during the preclinical stages of AD, "downstream neurodegeneration" and cognitive decline will be delayed or avoided altogether.

A second trial, is associated with DIAN (Dominantly Inherited Alzheimer Network, discussed briefly in chapter 5) and is described as the largest and most extensive worldwide network investigating mutational Alzheimer's. This network is headed up by John Morris and Randall Bateman at Washington University and is now preparing to launch a prevention trial funded by the Alzheimer's Association and a consortium of 10 pharmaceutical companies (the DIAN Pharma Consortium). The conclusions drawn thus far by this group stated in the Vancouver meeting abstract are as follows:

> Because the clinical and pathological phenotypes of dominantly inherited AD appear similar to those for the far more common late-onset "sporadic" AD, the nature and sequence of brain changes in early-onset AD are also likely relevant for late-onset AD. Clinical studies in AD caused by gene mutations are likely to pioneer the way to prevention trials for all forms of AD. The scientific knowledge gained from secondary prevention trials is likely to inform about the cause of AD, validate biomarkers to accelerate treatment development, and determine the effects of treating AD early.[11]

The third trial discussed in this panel was the one that will make use of subjects enrolled from the Basque families living in Colombia (see chapter 5). The conclusion to the abstract of this paper stated boldly, "We are excited about our progress, plans, current timelines and the chance to work with other researchers, programs, and stakeholders. Together, we have a chance to set the stage of a new era in AD prevention research, develop the resources, biomarker endpoints, and accelerated regulatory approval pathway."[12] In his presentation Eric Reiman concluded by stating that this trial is "giving people at the highest risk an opportunity to fight for themselves and the field." A palpable excitement pervaded the presentations at this panel, but, equally, so did a sense of urgency about the need to prevent AD, and the singular role that will be played by families plagued by this condition who, the audience was told, were finally getting the attention they had longed for. At the conference Reiman reported on a second study in addition to the one in Medellín involving a national registry in the United States of between 20,000 to 50,000 people who have been identified as carrying the APOEε4 gene associated with increased risk for late-onset AD. The aim of this trial, using 400 of these individuals, will be to test yet another new anti-amyloid drug.

One of the plenary sessions at the Vancouver meeting was devoted to amyloid, aging, and neural activity, and emphasis was given to the way in which tracking of cognitively normal people for amyloid deposition could well provide insights into the AD disease mechanism. The abstract for this presentation stressed, "[I]t is unclear why β-amyloid is deposited in some individuals and not others, why some people develop cognitive symptoms while others do not, and how Aβ affects brain structure and function."[13] And a recent article in *Alzheimer's & Dementia* reminds readers that although Aβ neurotoxicity has been clearly demonstrated in vitro, the role of Aβ in neurodegeneration has not been "unequivocally" demonstrated in vivo.[14] In short, stubborn deficits persist in expert understanding of amyloid.

A featured research session had taken place prior to the plenary session, in which Robert Green, the PI of the "evidence-based" REVEAL trial, presented a paper dealing with preliminary findings in connection with disclosure to individuals diagnosed with MCI of their APOE status. In addition, these subjects were given an estimated risk of "conversion" to AD in the following three years.[15] Another paper in this session focused on plans for "ethical" disclosure of their "amyloid status" to clinically normal research subjects. These individuals will be subjects in biomarker studies as part of the ε4 trial noted above. Amyloid-positive individuals will be in both treatment and placebo arms, and amyloid-negative individuals will be placed in the "natural history" arm of the trial. During the presentation it was noted that uncertainties are associated with the "likelihood and timing of clinical progression to AD" even when amyloid can be detected in the brain, and it was stressed that safeguards about disclosure will be embedded in the trial—these safeguards will be based on findings obtained from the REVEAL study. At the same time, the impact on individuals of learning about their amyloid status will be systematically evaluated as part of the trial.[16] Another paper stressed that methods developed in the field of genetic testing "provide a template for how to disclose amyloid imaging results to asymptomatic older adults."[17]

Over 1,400 posters were on display at the meeting. The majority dealt with biomarkers; others dealt with GWAS, PIB scanning, hippocampal structure and volume, and cognitive testing. Several dealt with sleep patterns—a current hot topic; others stressed the importance of physical activity in staving off AD, one emphasized the positive effects of ballroom dancing as protective against AD, and another claimed that the value of meditation had been underestimated. Yet other posters dealt with olfactory changes as diagnostic of AD, and several emphasized the importance of recognizing changes in gait as a signifier of incipient or early stages of AD—changes that can easily be detected in the office of a GP.

I found many skeptics at the meeting about the move to molecular prevention of AD, and a few outraged individuals among the expert audience with whom I mingled. However, judging by the questions at the end of the sessions I attended, many researchers were captivated by the enormous amount of time and energy being invested in order to move the world of Alzheimer's research forward toward its current goal of prevention and/or treatment of AD by 2025—at least in the United States. It is undeniable that achievement of this goal in the manner in which it is at present conceptualized will be dependent upon better elucidation of the activities of amyloid—the capricious actor that thus far has refused to be tamed. One participant at the conference, who has attended this event many times in both Europe and North America, told me that many of the individuals with whom she had talked in Vancouver were commenting that the messages embedded in the talks were much less dogmatic than in previous years, a sure sign that presenters are not assuming that major breakthroughs are imminent, even though an aura of excitement was present in some quarters.

A Worldwide "Call to Arms"

Turn him [the King] to any cause of policy,
The Gordian Knot of it he will unloose,
Familiar as his garter.

—Shakespeare, *Henry V*, Act 1, Scene 1

In the issue of *Alzheimer's & Dementia* that appeared in the same month as the July 2012 Vancouver conference, an exceptionally long article about the development of biomarkers concluded by asserting that what is needed now in connection with AD prevention is a "swift and serious global 'call to arms' AD initiative" that will unite and integrate global interdisciplinary translational research.[18] This article, written by two researchers in the department of Psychiatry at the University of Frankfurt together with Zaven Khachaturian, now heading up the Campaign to Prevent Alzheimer's Disease by 2020, is titled "Development of Biomarkers to Chart All Alzheimer's Disease Stages: The Royal Road to Cutting the Therapeutic Gordian Knot."

This title suggests that these authors continue to think in terms of clearly definable stages in the accumulation of neurological changes associated with AD—an assumption that present research strongly indicates is inappropriate, as do certain of the findings discussed and cited in the authors' own presentation. However, the goal in setting out this call to arms is to circumvent the perceived limitations of research to date—research that has tossed up so many anomalies. Standardization is the order of the day. As readers well know, the legend of the Gordian Knot recounts how Alexander the Great used his sword to cut through a massive knot that no one could remember how to untie. This idiom is usually used today to convey the meaning of cutting through extraneous, obfuscating complexities to get to the point of some argument, debate, or activity. Hampel et al. argue in their article, "[T]o date, systems-based, integrated, comprehensive, validated, and qualified Alzheimer's disease (AD) biomarkers seem to be the much desired golden touch to 'cutting the Gordian therapeutic Alzheimer knot.'"[19] The metaphors are mixed, but one gets the idea—well-honed biomarkers will, sword-like, solve the Alzheimer conundrum.

In their article, after first setting out the obligatory reminder about the "looming medicoeconomic" crisis, noting that certain governments are finally responding to this challenge, and claiming that "a modest delay of 5 years in the onset of disability will reduce the cost and prevalence of the disease by half," Hampel et al. go on to stress the impatience shared by researchers, advocacy groups, and caregivers alike to resolve the AD problem. The challenge now, they insist, is to develop biomarkers as clinical tools that will advance the AD world into the "unknown territory" of reliably identifying asymptomatic people at risk for AD, and they call for a shift to a "systems biology" approach in order to accomplish this. As far as these authors are concerned, to make prevention a realistic goal, the time to develop appropriate therapies must be shortened to three to five years or thereabouts—a goal that will entail the validation of biomarker changes on a massive scale among "large, diverse populations." This cannot be accomplished without the building of a massive infrastructure worldwide that will create a comprehensive longitudinal database including cohorts of "physically and/or mentally normal and/or 'successful' healthy aging people," as well as individuals at elevated risk for AD, and also asymptomatic preclinical/prodromal MCI individuals. The key to this endeavor is the use of high throughput technologies, designed to produce an omnibus database that will enable the global standardization of biomarkers. The limitation of current biomarker research to date, it is noted, is that individuals used as research subjects thus far have been too "heterogeneous," and larger cohorts must be used.[20] Hence, the recruitment for research purposes of vast cohorts of healthy volunteers worldwide is now under way.

The enormity of what is proposed is staggering, and, obviously, the project cannot be global due to the lack of access of millions of people to any form of formal medical assistance in so many parts of the world. Others have to walk for hours to receive attention for any kind of problem, however urgent. And in very many countries, even when some health care is in theory available, the resources and expertise needed to track volunteer research subjects are simply not present. Moreover, the expectation that younger people will volunteer as research subjects in droves for repeated invasive testing, wherever they live, is quite remarkable. Who is going to be willing to submit to a form of corporeal citizenship that demands years of regular spinal taps and neuroimaging, aside, perhaps, from the spouses of affected individuals? The bonds of altruism will be sorely tested. Considerable monetary rewards might change the situation, especially among the poor, malnourished, and destitute, but this raises the specter of exploitative commodification similar to that already seen in connection with organ procurement and with drug trials carried out in economically deprived countries.

In addition, the underlying assumption is that by systematically making use of huge samples, anomalies will no longer be significant, leaving in place "pure" refined biomarkers strung along an orderly temporal scale commencing from age 40 or 50 that leads inexorably forward to Alzheimer's disease. All other implicated variables extraneous to this model are set to one side to expose the newly revised natural history of Alzheimer's, a condition anchored firmly in localization theory that now has a detectable presymptomatic phase involving amyloid deposition—the key to AD prevention. But as John Hardy has noted, it may be that some differences with respect to amyloid among individuals are present from birth—perhaps we do not all have the same levels of amyloid from the outset. Furthermore, it is still not known exactly what role amyloid plays in this conundrum.

A Nail in the Coffin of the Amyloid Hypothesis

At a press conference at the Vancouver meeting William Thies, at that time the Alzheimer's Association chief medical and scientific officer, made a statement designed to up the ante in connection with AD research: "The good news at the Alzheimer's Association International Conference is that we are making advances toward earlier detection of Alzheimer's, greater knowledge of dementia risk factors, and better treatments and prevention." Thies continued, "These advances are critical in order to create a future where Alzheimer's disease is no longer a death sentence but a manageable, treatable, curable, or preventable disease." He then reminded the press, "The soaring global costs of Alzheimer's and dementia care, the escalating number of people living with the disease, and the challenges encountered by affected families all demand a meaningful, aggressive and ambitious effort to solve this problem. . . . The urgency is clear. By midcentury, in the U.S. alone, care for people with Alzheimer's will cost more than $1 trillion. This will be an enormous and unsustainable strain on the health care system, families, and federal and state budgets. The first-ever U.S. National Plan to Address Alzheimer's Disease was unveiled in May, and must be speedily and effectively implemented."[21]

But, on July 23, less than a week later, an article appeared in the *New York Times* under the heading "Alzheimer's Drug Fails Its First Big Clinical Trial." Reporter Andrew Pollack notes that this deals "a blow to the field, to a theory about the cause of the disease, and to the three companies behind the drug." One of the companies is Pfizer, and they reported that the phase 3 trial of their drug, bapineuzumab—produced together with Johnson & Johnson, with further financial input from Elan—improved neither cognition nor daily functioning of the patients to whom it was given, as compared with a placebo. Readers were informed that "most doctors and Wall Street analysts had been expecting the drug not to succeed," because the phase 2 trial had not been statistically significant. At the Vancouver meeting there had been no sign that this failure could be imminent, and the leaders in the field must have been decidedly worried as they gave their upbeat messages. It is significant that in this trial the subjects, 1,100 Americans with mild to moderate AD, were all APOEε4 carriers, but the phase 2 part of the trial had indicated that the drug had a better chance of working with individuals without that genotype. In early August, following further disappointing results, in which noncarriers of the APOEε4 gene were the trial subjects, Pfizer and Janssen announced that they were completely halting development of bapineuzumab for mild to moderate Alzheimer's disease. It is still possible that this drug will be made use of in trials designed to assess its worth in the prevention of AD. In the proposed trial in which Colombian families will participate, it is planned to use a closely related drug—a monoclonal antibody, solanezumab, designed to bind β-amyloid.

Not surprisingly, involved scientists have responded quickly to these findings by arguing that no doubt these drugs are being given too late to individuals already badly affected by AD neuropathology, even though clinical symptoms are mild. Benefit will quite probably result if these drugs are used to prevent the development of unwanted plaque before it commences: "All these symptomatic trials are 25 years too late," Samuel Gandy of Mount Sinai School of Medicine is reported to have said.[22] This failure, and the findings of other similar trials if they too prove to be unsuccessful, or produce only equivocal results, will be made use of to boost the shift to the molecular prevention of AD. Moreover, Eric Reiman impressed on his audience in Vancouver several times that trials using families with dominantly inherited AD can be done much faster than those that make use of older late-onset subjects in whom the disease progresses much more slowly—thus saving not only time but also money.

In August, following the Vancouver meeting, it was reported that solanezumab had failed to slow memory decline in two late-stage studies of about 1,000 patients in each study. But officials at Eli Lilly were "encouraged" even so, because when they combined the results of both studies using a sample composed of patients who were at only a mild stage of AD, then the results could be interpreted as statistically significant. This very tentative finding encouraged those who believe that if one administers the drug very early, long before cognitive symptoms are apparent, then it may well work to deter disease progression.[23] By October, the drugs that would be used in the DIAN study had been announced: Gantenerumab and solanezumab together with another possible drug. This trial involving 160 people is designed to try "to prevent Alzheimer's symptoms from ever occurring," said John Morris. "[T]his will be a new strategy."[24]

Even though the move to prevention is boldly described as a paradigm shift involving a new strategy in the hope, no doubt, of sustaining the attention of both researchers and key funders of the AD enterprise, it is clearly something less that that. For the time being the amyloid hypothesis hangs on, battered and bruised though it is, and remains the key postulate even as prevention moves to the fore. But increasingly researchers are being forced to confront the idea that this hypothesis may well be past its due date, and is in rather urgent need of modification or retirement.

Two articles and associated commentary that appeared in *The Lancet Neurology* in December 2012, based on preliminary findings with the Colombian families, make clear the extent to which some serious reflection has to take place in the AD world. Before the publication of these articles, researchers had been convinced, on the basis of brain-imaging and CSF analysis, that amyloid plaque deposition can commence many years before any signs of clinical symptoms become manifest. This latest research confirmed this finding and, among the Colombian families who carry the early-onset mutation, such changes were detected up to 20 years before any signs of clinical symptoms, and in individuals as young as 18 to 26. These results give support to the idea that early detection of presymptomatic biomarkers is essential if a successful drug regimen is to be developed.[25]

Also of great significance was the finding written up in these articles that among early-onset mutation carriers a buildup of amyloid plaque probably takes place due to excessive production of the Aβ peptide, whereas in contrast, in late-onset AD, amyloid deposition, including cases where the APOEε4 is involved, takes place due to an inability to clear Aβ peptide from the system. This finding, admittedly with a very small cross-sectional sample, suggests that significant differences exist between dominantly inherited and late-onset AD, even in the final molecular pathways, and not simply in terms of causation—thus throwing into jeopardy the assumption that early and late-onset AD are for all intents and purposes the same condition. Equally of interest was the remarkable finding that neurodegeneration involving reduced gray matter and altered synaptic functioning, among yet other changes, may take place in advance of amyloid plaque deposition. Not surprisingly, these findings are regarded as exploratory by the researchers, and the research has limitations. However, if the findings hold up, and particularly if such changes are also well demonstrated in late-onset AD cases, then 100 years of virtually unchallenged neuropathologic criteria for an Alzheimer diagnosis will need modification. Amyloid plaques may not be, after all, the first sign of the AD phenomenon.

These preliminary findings will be followed shortly by many more findings derived from the consortia working on dominantly inherited AD. It is likely that heated debate about the amyloid cascade hypothesis will erupt, pushing the AD world deeper into a serious shake-up of normal science. In the meantime we are informed that people who suffer from disrupted sleep might be on the path to Alzheimer's disease—"increased daytime sleepiness is the biggest predictor [of AD]." Work on mouse models suggests that the beginnings of plaque formation may cause sleep disturbances, but it is acknowledged that the situation may be more complex in humans.[26] And we know that a very large number of people with plaques in their brain never become demented—no mention of this in the article.

In the remainder of this chapter, following a very brief discussion about the emergence of molecular genomics, I turn to a new approach in AD research oriented by postgenomic and epigenetic knowledge. This type of research receives considerably less coverage in professional settings and in the media than does the search for drugs, but if the future impact on the global economy of aging societies is to be confronted, it should be given greater salience than it has received thus far.

Beyond the Dogma of Genetic Determinism

Over the course of the past decade many remarkable changes have taken place in the world of molecular genetics. One outcome has been that genes have been demoted in the minds of many, perhaps the majority of experts in the world of genomics, from "real," substantial entities, to the status of a concept. Not surprisingly, despite their changed status, genes continue to be heuristically very powerful, even though research has made it clear that scientists cannot determine exactly where genes begin or end,[27] nor are genes stable, and they do not, on their own, determine either individual phenotypes or the biological makeup of future generations. Quite simply, genes are not "us," and the gene can no longer pass as the fundamental animating force of human life; it has been dethroned, Evelyn Fox Keller informs us, from its place as "part physicist's atom and part Plato's soul."[28]

It is paradoxical that this definitional disarray of the gene was brought to a head as a result of the Human Genome Project. As is now well known, when mapping the human genome, involved scientists labeled 98% of the DNA they had isolated as "junk" because it did not conform with their idea of how the blueprint for life was assumed to work. In recent years things have changed dramatically, and junk DNA, thrust summarily to one side in order to focus on the task of mapping only those genes that code directly for proteins, can no longer be ignored. This junk, although most of it appears to be nonfunctional and does not code for DNA, is nevertheless clearly implicated in gene expression and regulation at times, and hence is being sifted through systematically.[29] Furthermore, the activities of noncoding RNA are believed to compose the most comprehensive regulatory system in complex organisms.[30] Noncoding RNA has been shown to profoundly affect the timing of processes that take place during development, including stem cell maintenance, cell proliferation, apoptosis (programmed cell death), the onset of cancer, and other complex ailments.[31] Consequently, the research interests of many molecular biologists are no longer confined largely to mapping structure, but have expanded to the elucidation of the mechanisms of cell and organ function throughout the life span of individuals, and also through evolutionary time. Central to this endeavor is to understand gene regulation—above all, how, and under what circumstances, genes are switched "on" and "off"—in other words, what brings about their expression.

Using this approach, the effects of evolutionary, historical, environmental, and cultural variables on developmental processes, health, and disease are brought to the fore. Determinist arguments are in theory no longer appropriate, and both micro- and macro-environmental interactions in connection with cell functioning are key to this type of research. This emerging epigenetic knowledge (as it has come to be known), in theory if not in practice, is a serious challenge to the central dogma on which molecular genetics was founded.

Metaphors associated with the mapping of the human genome—the book of life, the code of codes, the holy grail, and so on—are now outmoded. With the cell at center stage, genetic pleiotropy,[32] gene/gene, gene/protein, and gene/environment interactions cannot conveniently be set to one side—and GWAS researchers have quickly discovered just how challenging this complexity makes their research endeavor. A space has opened up between genotype and phenotype, a zone of endophenotypes—unstable, interim states—partially recognized 100 years ago, but then set to one side until relatively recently.[33] Today, many endophenotypes are recognized as key biomarkers, researched in order to better understand and perhaps prevent complex disease, as was evident at the Vancouver meeting.

The molecularized universe has turned out to be so very much more complicated, and exciting, than most people had imagined. This is a universe entirely in tune with postmodernity, a landscape littered with a pastiche of shape-shifters—smart genes, transcription factors, jumping genes, and so on—an environment of the unexpected, in which boundaries formerly thought of as stable are dissolved. It is evident that some genes code for more than one protein, while many others do not code for proteins at all. Increasingly, it has become clear, with the partial exception only of single gene disorders, that multiple factors, including events both internal and external to the body, serve to enhance or inhibit gene expression. This means that efforts to divine individual futures by means of genetic testing for anything but the rare single gene disorders are precarious indeed.

In her book *The Century of the Gene*, the historian and philosopher of science Evelyn Fox Keller summed up where she believes we stood at the beginning of the 21st century:

> Genes have had a glorious run in the twentieth century, and they have inspired incomparable and astonishing advances in our understanding of living systems. Indeed, they have carried us to the edge of a new era in biology. . . . But these very advances will necessitate the introduction of other concepts, other terms, and other ways of thinking about biological organization, thereby loosening the grip that genes have had on the imagination of the life sciences these many decades.[34]

Fox Keller, although she agrees that the concept of the gene is "good enough" for many experimental purposes, concludes that only the adoption of new concepts will bring about timely insights into the workings of living systems. But the challenge is enormous. At the Vancouver conference Gerard Schellenberg was asked an apparently naïve question by someone in the audience following his presentation on GWAS findings in connection with AD: "Isn't it well known that gene/environment interactions are at work at all times and that genes cannot simply be researched in isolation?" Schellenberg took a deep breath and smiled wryly. In response he said, "That's true, but we have to do the easy stuff first, and then we can move on to the environment." He gave no hint that such a reductionistic approach produces, in effect, decontextualized findings that may well, ultimately, lead investigators astray.

Epigenetics: An Expansion of Horizons

In the beginning was a cell. Well, not in the *very* beginning—only after a mere 1 or 2 billion years of life's story, during which time most our basic principles were already well established.

—Kenneth M. Weiss and Anne V. Buchanan, *The Mermaid's Tale*[35]

The philosopher of biology Paul Griffiths noted a decade ago, "[I]t is a truism that all traits are produced by the interaction of genetic and environmental factors [but] the almost universal acceptance of this view has done little to reduce the prevalence of genetic determinism—the tendency to ignore contextual effects on gene expression and the role of non-genetic factors in development."[36] Space does not permit a detailed summary of current epigenetic theories; suffice it to say that the very word "epigenetics" has more than one meaning. Many argue that the discipline is not new, and existed by the 1940s or even earlier, although others disagree with this position.[37] Most current research into epigenetics focuses on the expression and regulation of genes. For example, questions posed about the phenotype ask why monozygotic twins do not always manifest the same diseases and, when they do, why the age of onset can differ by up to two decades, as is the case for dominantly inherited Alzheimer disease. This narrowly conceptualized epigenetic approach immediately makes the limitations of genetic determinism patently evident.

A broader approach to epigenetics, known by its adherents as "developmental systems theory" (DST), is now well established. The starting point of the DST approach is an ontological reversal of genetic determinism, and it gives priority to dynamic interactions among very many variables, allowing for numerous possible outcomes. Barnes and Dupré, sociologist and philosopher of science respectively, argue that "instead of being spoken about as independent atoms of hereditary material, genes, conceptualized as DNA, are now referred to as parts of the chemical/molecular systems within the cell."[38] In their book *Nature After the Genome*, Parry and Dupré argue that DNA is not simply involved with heredity; we have to ask what DNA *does* throughout the life cycle. It is not appropriate to "conceptualize nature as passive, something upon which humans act, usually with the assistance of technology, but rather as active and lively, as responding to human actors, and as able to resist them."[39]

In a 2010 article, Dupré insists that the dogged idea of individual genetic homogeneity over the life cycle, despite ever increasing evidence to the contrary, is highly misleading. He argues that an assumption of homogeneity persists because too much weight has been given to comparisons of genomic sequences, but "to know what influence a genome will actually have in a particular cellular context one requires a much more detailed and nuanced description of the genome than can be given merely by a sequence."[40] Such epigenetic changes to genomes are brought about by chemical interactions among the molecules within the cell membrane that surround the genome, usually composed of RNA and proteins. Similarly, the molecular biologist Strohman has argued that "there are regulatory networks of proteins that sense or measure changes in the cellular environment and interpret those signals so that the cell makes an appropriate response."[41] This regulatory system, a "dynamic-epigenetic network," has a life of its own that is not specified by DNA. Understanding about the significance of DNA has been radically altered using this new approach, and contingency now takes center stage. For commentators concerned with ontology, the idea, implicit in so many arguments, of DNA as having "agency" is thoroughly anthropomorphic and inappropriate.[42]

Increasingly complex research at the molecular level proceeds apace, resulting in many remarkable findings. For example, a recent study is grounded in the idea of competing and integrated biological pathways within cells that control the "biogenesis, folding, trafficking and turnover of proteins present within and outside the cell."[43] Without doubt this type of research may have relevance for an increasingly fine-grained understanding of Alzheimer-related neuropathology, and possibly be of significance for drug development; but at this level the question of AD causation is severely truncated and limited to late-stage molecular changes.

Buzzing Confusion

The molecular biologist Richard Lewontin penned a radical critique several years ago about current approaches to biology in which he questions the common assumption that laws of nature determine biology, as is true for Newtonian physics. Writing about levels of organization with respect to organisms as a whole, he states,

> Unlike planets, which are extremely large, or electrons, which are extremely small and internally homogenous, living organisms are intermediate in size and internally heterogeneous. They are composed of a number of parts with different properties that are in dynamic interaction with one another and the parts are, in turn, composed of yet smaller parts with their own interactions and properties. Moreover, they change their shapes and properties during their lifetimes, developing from a fertilized egg to a mature adult, ending finally sans

teeth, sans hair, sans everything. In short: organisms are a changing nexus of a large number of weakly determining interacting forces.[44]

Lewontin wonders if biology is inevitably a story of "different strokes for different folks," a collection of exquisitely detailed descriptions of diverse forms and functions down to the molecular level—or, from "this booming, buzzing confusion," can a biologist perhaps derive some general claims that are freed from the "dirty particulars" of each case? "Not laws," he quickly adds, but at least some widely shared characteristics? He agrees with Fox Keller that both history and epistemology seem to speak against this, and as far as making sense of life—of biology—is concerned, all our models, metaphors, and machines, while they have contributed much to our understanding, provide neither unity nor completeness. On the contrary—facing up to complexity is the order of the day.

Lewontin is well aware, of course, that not all biologists are comfortable with the emphasis he gives to complexity and the "confusion" he associates with the functioning of living organisms. But it is becoming increasingly hard to ignore the reality that organisms of all kinds can and do adapt to new environments, toxic insults, and manipulations of various kinds with surprising rapidity. Whether it be the beak size of the finches studied by Charles Darwin on the Galapagos Islands,[45] the dramatic change in the reproductive life of cod in response to contemporary overfishing,[46] the development of resistance on the part of micro-organisms in response to antibiotics, and so on, it is no longer appropriate to think of biological change as a very slow unfolding of events, and the same is now beginning to be recognized in connection with human biology.

The biological anthropologists Kenneth Weiss and Anne V. Buchanan, in their article titled "Is Life Law-Like?" spell out some of the specific implications associated with an assumption that nature is "governed by universal, unexceptional laws." They argue, "A law of nature is a process or mechanism of cause and effect," and "we pursue the laws of nature through what has become known as 'the scientific method.'" This involves, of course, systematic, controlled observation—an empirical approach—in which reductionism, replication, prediction, and the ability to deduce new facts are central. And notably in the 20th century, the idea of predictability has been increasingly extended to include "the law-like distributions of probabilistic processes." The effects of this approach have had the result of conceptualizing "nature as, in effect, deterministic." Weiss and Buchanan agree that a very small fraction of DNA related causality is highly predictable, but they insist that recent technological developments are "enabling us to see that, in important ways, life might not be law-like in the Enlightenment sense, or even that we may not know when we have found such laws."[47] In Weiss and Buchanan's opinion, such findings do not challenge empiricism, but they do question current empirical approaches to causation and inference: "Much of life seems to be characterized by *ad hoc*, ephemeral, contextual probabilism without proper underlying distributions."[48] In other words, John Stuart Mill's idea of "Western fatalism"—a belief that we can know how things will turn out because the scientific order follows regular patterns," should, it seems, be rescinded, and fate, rather than fatalism, can reclaim some recognition.[49]

The micro-mechanisms of basic biological processes are not in contention when biologists make such arguments—rather it is assumptions about the regularity of causal processes that are at issue, with enormous implications in connection with creating generalizations based on population-based data to be applied to individuals wherever they reside.

Intimations of the Future?

It is perhaps the case that certain members of the younger generation are already disaffected with what they think of as scientific reductionism. In 2007 informal, exploratory conversations were held with 30 university-educated individuals aged between 25 and 39 who reside in Montréal in connection with their exposure to, understanding of, and interest in genetics.[50] Only one or two of these people had experience of serious disease in their families, and none reported single gene disorders.

All 30 of these individuals acknowledged the significance of genes in disease causation, yet virtually every one of them also cited social, environmental, and behavioral variables, including upbringing, education, economic status, environmental pollutants, diet, and personality, as contributors. As Keith, age 31, put it, "[G]enes don't give us the whole picture." He went on,

> A condition might come from genetics, but it's also your mind, and your education, and family, and all these things . . . that is what makes it hard to understand how genes affect each and every individual. It's so complex, how your body works, how your mind works and how other factors affect you.

In many cases these interviewees appeared to think that external factors can in effect assail genes. Candice, age 28, argued,

> If your genes aren't "strong enough" to fight a lot of the chemicals and external things . . . then you could get cancer because your body wasn't able to withstand the chemical intrusion.

And Joyce, 31, noted,

> I could be at high risk of some crazy, rare disease. Maybe I'll just never know because by coincidence I have avoided anything that would turn

that gene on or off. It just sits there silently and nothing ever happens to it.

To mitigate external threats, these individuals engage in everyday prevention strategies, such as eating organic food, taking vitamins, exercising, and controlling their weight. They believe such activities are essential in maintaining healthy gene/environment bodily interactions, and many produced vivid accounts of mediating "toggles" and "triggers" that influence gene expression by turning "switches" "on" or "off":

> —I imagine in a disease process it's just like tripping the switch. So if you smoke, you're just flipping a lot of switches which otherwise would never have been switched. . . . [D]oing things like eating lots of vegetables and consuming antioxidants and things keep the switches on or off, whatever they need to be, that's protective. (Joyce, 31)

> Something in the environment, whether it's too much food, or working too much, could cause a short-circuit, an overload in the capacity. Something will set the gene off, whether cigarette smoke, or pollutants, a toxin in food, whether it's food coloring, whether it's a certain chemical. Something will set it off. (Larry, 34)

Some individuals suggested that because they have "greater access to resources and knowledge" than their parents, they are less at risk than the previous generation. Larry comments,

> [Y]ou can have a predisposition to having diabetes, but if the off-spring's diet, and activity level are high, and they're conscious of what they're eating, of their habits, and their stress level, this is something that possibly can be avoided.

Joyce reflects on the history of breast cancer in her family:

> My great-grandmother, grandmother, and mother had it—but also let's look at their lifestyle. None of them breast-fed. My grandmother was a smoker. They all have really high fat diets, low exercise, a lot of alcohol. Maybe I inherited the predisposition, but my lifestyle's so different.

Individuals such as Joyce actively try to distance themselves from family medical histories by emphasizing that it is possible to have a degree of control over genes. Another person surmised that he might avoid serious illness, unlike his grandfather, who was raised on a farm, performed heavy physical labor, and "lived a life that's so different from me and in a different environment from me." The majority shun the idea of a determined genetic history, focusing instead on individual responsibility and decision making as the key to positive health outcomes. Some insist that although they share genetic substance with their parents, they themselves have "transformed" their genes due to generational differences in lifestyle.

These conversations suggest that these young people think of genes as, in effect, unstable entities, subject to modification by environment and human behavior, and the impression is given that a good number believe that through their own exertions they may well be able to contain or subvert "risky" aspects of their genetic heritage that may have been passed along to them. It seems that neither fatalism nor fate applies here, but that dominant values of middle-class youth of today—self-control and discipline—are at work. Some of these individuals are destined to have their beliefs shattered, of course, and one wonders how they will manage the chronic uncertainty associated with increased genetic and biomarker testing, if and when that takes place. But, in the case of Alzheimer's, dietary care and exercise have been shown repeatedly to stand individuals in good stead in order to sustain the cardiovascular system and thus protect from stroke and mixed dementias. Without doubt it is all for the best to cultivate such everyday habits, whatever the driving values behind such behaviors. What must be kept in mind is that by far the majority of people living in the world today are in no position to cultivate their health and well-being, a situation becoming rapidly worse with unrelenting efforts at exploitative economic growth and development, the gains of which are only very rarely indeed redistributed fairly among local populations.

Epigenomics and the Life Experiences of Individuals

> People sometimes say that the human brain is the most complex item in the universe. But the whole person of whom the brain is part is necessarily a much more complex item than the brain alone. And whole people can't be understood without knowing a good deal both about their inner lives and about the other people around them. Indeed they can't be understood without a fair grasp of the whole society that they belong to, which is presumably more complex still.
>
> —Mary Midgley, *Science and Poetry*[51]

Systematic research into epigenetics and the larger field of epigenomics (the global analyses of epigenetic changes across the entire genome) has recently exploded. Many researchers recognize what might be termed a "distributed molecular agency" with the cell at center stage in which the focus of the majority of basic scientists, with some notable exceptions, remains concentrated on microenvironments within the body. It is now well established that, at the molecular level, environmental variables can bring about effects on cellular processes at specific DNA sites, mediated by several processes, the best known of which is methylation. It has also been shown that, under certain circumstances, methylation and demethylation and related processes can be reversed.[52] And some researchers claim that these epigenetic changes can be inherited independently of DNA.[53] This "epigenomic" research has opened the door to what has been described by some as neo-Lamarckianism, although these findings are only associational and are by no means fully accepted as yet.[54]

Research into the epigenetics of Alzheimer disease is under way, although to date it is relatively sparse. One study showed links between memory formation and epigenetics.[55] In another project an analysis of DNA methylation across 12 potential Alzheimer's susceptibility loci was carried out. It was found that "age-specific epigenetic drift" from a previously established norm was apparent in brain tissue taken from individuals who had been diagnosed with AD, as compared to normal controls. The authors of this study argue, "The epigenome is particularly susceptible to deregulation during early embryonal and neonatal development, puberty, and especially old age."[56] They also found that certain genes that contribute to β-amyloid processing showed "significant interindividual epigenetic variability," a finding that they argued may be associated with susceptibility for AD.[57] Another study found that the brain tissue obtained from identical twins had markedly different levels of DNA methylation. One twin was diagnosed with AD at aged 60 and died 16 years after the diagnosis, and the other died aged 79, with no signs of dementia. The twin who was demented had significantly lower DNA methylation in his brain tissue than that of his twin. It was established that the twin who had become demented had been exposed extensively to high pesticide levels as a result of his earlier employment, but it is not yet known if such exposure can be considered as a contributor to the onset of AD.[58] It is argued that these differences can be explained by "epigenetic drift" caused by environmental exposure, lifestyle, diet, drug abuse, or merely stochastic fluctuations. Wang and coauthors point out that

> epigenetic modifications may exert only subtle effects on the regulation of specific genes. Thus, abnormal DNA methylation may only cause a disease phenotype when several loci are affected at the same time.[59]

They add that people predisposed to AD as a result of other variables, including the APOE genotype, may be pushed over a critical threshold after which the brain starts to malfunction. They conclude that late-onset AD

> may represent merely an extreme form of normal aging, which would imply that every human being has a certain predisposition to developing Alzheimer's. In our model, the epigenetic effects can accumulate

throughout life, especially from the time-point when the epigenetic machinery suffers from old age, but also from early embryonal stages or even trans-generational [effects], influenced by epigenetic events in the parents.

The part of the epigenetic system in which methylation and related activities take place is known as the chromatin-marking system. Chromatin is the "stuff of chromosomes"—the DNA, plus proteins, and other molecules associated with it.[60] The structure of chromatin is crucial in the activation of genes, and enables states of gene activity or inactivity to be perpetuated in cell lineages. Alternative hereditable differences in chromatin have come to be known as "chromatin marks," among which DNA methylation is the best recognized—a feature found in all vertebrates, plants, and also many invertebrates, fungi, and even some bacteria. It has been known for decades that methylation processes are crucial to normal development, and several publications have appeared over the years demonstrating methylation effects on neurodevelopment in experimental animals. The findings of Wang and colleagues are insightful but not unprecedented, and earlier findings similarly suggest that links exist between methylation patterns early in life and AD incidence in later life.[61] A recent article titled "The Aging Epigenome" provides an excellent technical summary of this topic.[62] However, based on these findings, the AD story will continue to be one of embodied molecular interactions, with the addition only of a time dimension provided by molecularized developmental processes. What remains poorly investigated in a systematic manner are those variables external to the body that no doubt influence the aging process both directly and indirectly. Thus far association studies alone exist, and the time is ripe to design research projects in which macro and micro variables are examined in tandem, challenging though this may be.

Meanwhile well-publicized findings that appear at regular intervals about newly discovered genetic mutations tend to drive a reductionistic approach, one strongly fueled by drug company interests, that without fail captures the imagination of the public. One such finding first appeared online in the *New England Journal of Medicine* in November 2012 detailing research that had shown that heterozygous rare variants of the gene TREM2 are associated with a significant increase in risk for Alzheimer's disease.[63] Two groups of researchers, one part of an international consortium known as the Alzheimer Genetic Analysis Group, and a second associated with deCode Genetics, independently published the same finding in the same issue of the journal. Both studies concluded that individuals with this rare mutation have a threefold to fivefold lifetime increase in risk of developing AD as compared to the general population—a risk that compares with that of APOEε4. This mutation affects the action of white blood cells in the brain, thus interfering with the immune system in such a way that toxic amyloid cannot be effectively eliminated. However, the TREM2 mutations occur in less than 2% of Alzheimer patients, and hence screening patients is not realistic. Even so, a senior researcher at Washington University involved in the research, Alison Goate, noted for the *New York Times*, "The field is in desperate need of new therapeutic agents . . . this will give us an alternative approach," that is, presumably, an approach directed at maintaining the functioning of the immune system.[64]

It is highly likely that as GWA studies proceed apace more rare variants will be found at regular intervals—Rudolph Tanzi's predictions about the value of GWAS appear to be proving correct. Should more rare variants be found, it will provide further stunning evidence of the enormous complexity at work in the Alzheimer syndrome.

In the final chapter that follows I revisit the tensions set out in Orientations and reconsider them in light of the so-called paradigm shift to Alzheimer prevention that has permeated much of the discussion in previous chapters.

Chapter 9
Transcending Entrenched Tensions

The world is indeed complex; so too should be our representations and analyses of it. Yet science has traditionally sought to reduce the "booming, buzzing, confusion" to simple, universal, and timeless underlying laws.

—Sandra D. Mitchell, *Unsimple Truths*[1]

Competing Ideas about Causality

The preceding chapters have made clear that the interrelated tensions highlighted in the opening Orientations revolve to a greater or lesser degree around matters relating to ideas about AD causation. Is Alzheimer's best modeled as though this condition is purely a "material" matter? In other words, does AD take hold and evolve entirely in the brain as a fully localized phenomenon? If yes, then efforts to treat this condition should be focused on the presumed molecular pathologies associated with the final common pathway(s) that result from the amyloid cascade hypothesis, or some modified version of it. This localized approach is by definition reductionistic, a product of a consciously parsimonious approach, and is oriented primarily to the development of effective drugs. The shift in emphasis toward prevention based on detection of AD biomarkers does not dislodge this reductionism; it means only that if effective drugs are discovered, then they will be administered before the behavioral changes associated with AD can be discerned by means of cognitive testing.

In practice, reductionism is being further embedded it seems because, as a result of the shift to prevention, Alzheimer-associated neuropathology has for the first time been formally bifurcated from clinical symptomatology in the new diagnostic guidelines—thus sundering mind and body. An intermediate stage between normality and dementia is explicitly recognized, in which biomarker changes can be detected but individuals show no signs of behavioral changes (see chapter 4). A good number of people with amyloid biomarkers will develop AD-like behavioral symptoms several years after the biomarkers are first observed, but others will not. Those individuals who test biomarker positive, but who remain cognitively normal until death, are presumed by most researchers to be repositories of the disease that would have eventually become manifest had death not intervened.

Versions of this dominant model of AD grounded in localization theory have held sway for many years. Even so, adherents do not deny that many factors external to the body may well predispose individuals to, and increase susceptibility for, dementia; although the majority make no attempt to research variables other than those internal to the brain. The result has been that serious ruptures have long pervaded the AD world because everything extraneous to the body has been pushed to one side by the majority of experts and given short shrift. It is probably safe to assume that dementia is universally distributed though time and space,[2] but contemporary research, including the emerging work on epigenetics, has made it clear that it is not equally distributed within or among families, communities, or populations, strongly supporting the position that environments—sociopolitical and ecological—influence the development and functioning of the brain. But findings such as these do not captivate molecular researchers working to cure neuropathologies, except those few who have moved in the direction of epigenetic investigations.

A small number of experts over the past 100 years have never been enamored by arguments about plaques and tangles. These individuals have often stressed the significance of the social and environmental milieus in which aging takes place, and range from the work of the psychiatrist David Rothschild writing in the 1940s,[3] to the contemporary writings of the recently deceased psychologist Tom Kitwood and his followers in Great Britain whose motto is "the person comes first."[4] Research in cross-cultural epidemiology has also highlighted the significance of environmental milieus,[5] as does the work of Carol Brayne and colleagues based in the United Kingdom, and now evolving research on epigenetics raises numerous questions for further investigation into developmental and life course experiences and susceptibility to dementia. A concerted effort to reduce obesity and diabetes worldwide would without doubt lower the incidence of mixed dementias. But, in effect, this will only amount to an extended application of localization theory to other bodily systems and their effects on brain functioning, unless the social and political dimensions that account for the prevalence of these diseases—above all, poverty, inequalities, and the promotion and marketing of fast food—are systematically tackled— something that is beginning to take place in certain towns and cities, such as the restrictions in size of soda bottles suggested by Michael Bloomberg, the current mayor of New York City.

Contextualizing the Brain

The move to biomarker-based diagnostic criteria for prodromal dementia will cement a bifurcation of mind and body. But the very fact that biomarker signifiers are based on probabilities involving estimates of risk and susceptibility leaves the door open for a critical discourse, one that asks why some people get sick and others do not,[6] thus forcing a move upstream to consider possible contributions made by mind, individuals, families, and environments as either protective, neutralizing, or causative agents.

Certain readers may have wondered as they perused the preceding chapters why "mind" simply does not appear in discussions. After all, is not the mind of an individual a mediator between stimuli external and internal to the human body? But this position is not accepted by many vocal philosophers of the day. Patricia Churchland, for example, believes that it is "the brain rather than some non-physical stuff, that feels, thinks, decides."[7] And by far the majority of neurologists and cognitive psychologists assume that when they test patients using the Mini-Mental State Examination, MoCA, and so on, they are measuring mental activities that take place fully inside the brain, and are, in effect, a direct reflection of brain structure and function. However, in his book *Out of Our Heads*, the philosopher Alva Noë argues persuasively, "[Y]ou are not your brain." Noë concedes that prominent philosophers of neuroscience counter such a position, but in turn he responds that at present, we simply have no evidence as to how a vast assembly of nerve cells and their associated molecules might give rise to consciousness.

For Noë and philosophers who share his view, to have a mind is to be conscious, whereas to have a brain is to have a certain kind of bodily organ or part. The brain is only one key element in the achievement of consciousness—consciousness is a phenomenon that requires the "joint operations of brain, body, and world. Indeed consciousness is an achievement of the whole animal in its environmental context. I deny, in short, that you are your brain. But I don't deny that you have a brain. And I certainly don't deny that you have a mind."[8] Explicitly following the philosopher Susan Hurley, Noë argues that the bony skull is not a "magical membrane" that delimits trafficking, and mental processes do not have to be explained entirely in terms of internal processes. Hence, he argues, we would do well to understand the phenomena of mind, and therefore of consciousness and experience, as one of "boundary crossing." Noë reminds readers about the plasticity of the human brain, evident by its long-term evolutionary adaptations, as well as environmental adaptations over shorter durations, and also over the life course of individuals—a never-ending relay of responses to stimuli, external and internal to the body, that takes place over a vast time dimension.

Noë comments that establishment neuroscience is committed to the Cartesian doctrine: "[A] thing within us that thinks and feels. Where the neuroscientific establishment breaks with Descartes is in supposing that the thinking thing is the brain." Neuroscience and certain philosophers have, in effect, obliterated the ghost in the machine, and Descartes's dualism is collapsed into a monism, but this move has left us with a piece of meat—the brain—that simply cannot account for consciousness and subjective human experience in all its variety. Noë conceptualizes mind, as did Hurley, as horizontally modular architecture not bounded by the confines of the human body. He concludes that neuroscience must come to terms with the way in which we are "dynamically coupled with the world."[9] And his position is not unlike that of the biologist Scott Gilbert, who argues that new insights about individual biological development suggest that "self" is permeable and that "[w]e are each a complex community, indeed, a collection of ecosystems."[10]

Noë has nothing to say about dementia, but his writing clearly suggests that he would reject localization theory as, at best, a partial account of dementia-like phenomenon and other neuropathological conditions (see the book *Dementia: Mind, Meaning, and the Person* for a full explication of this position).[11] Of course, mind, in its popular understanding as confined fully in the brain and operating as the seat of brain-bounded self and subjectivity, cannot cause the terrible neurological devastation of AD—there is nothing psychosomatic about AD, although psychiatric disorder may be a contributor at times. But mind as consciousness and as part of everyday lived experience that involves not only shared social life, education, access to adequate nutrition, and so on, but also, at times, enormous trauma, can undoubtedly make an indirect contribution to neuropathology. By definition, this boundary-traversing mind is a concept that mediates between external and internal environments and stimuli. And this mind permits us to ask why some people become demented and others remain cognitively "normal" even when "neuropathology" is detected. The chain of AD causality is partially shifted from proximate to more distal causes, to use Ernst Mayr's terminology.[12]

Deeply held assumptions in the neurological world that account for the tenacity of localization theory ensured that AD became understood as a condition confined to the brain, in terms of both cause and effect. This facilitated the recognition of AD as a bona fide disease to which moral recriminations should not be attached. But, by reducing causal explanations for AD to molecular changes alone, medicalization sidelined socioeconomic, political, and public health arguments in connection with risk for AD. One matter that would be brought to the fore by shifting an explanatory model for AD causation to spheres more expansive than neuropathology is the question of allocation of responsibility for Alzheimer's. I will return to this point in the closing paragraphs.

Is Alzheimer Disease on a Continuum with Normal Aging?

Turning to the second tension, that of AD as normal or pathological aging, Wang and colleagues argue that it is appropriate to think of every individual as being "predisposed" to developing Alzheimer disease. And the biologist of aging Tom Kirkwood points out that "aging is a continuum, affecting all of us all the time."[13] His position is that aging is inextricably entangled with conditions that become manifest in later life, including dementia. Using epidemiological methods, Carol Brayne and colleagues concluded that factors other than the simple presence of Alzheimer neuropathology must determine the clinical expression of dementia, notably in those 75 years of age and older. Their argument (see also chapter 1) states, "[C]ognitive dysfunction in later life is a life-span issue and is affected by genetic, developmental, and lifestyle factors, accumulated neural insults, innate and acquired cerebral reserve and compensatory mechanisms, and age-related decline."[14] Moreover, decades of accumulated autopsy evidence have demonstrated that amyloid plaques, tangles, and other assumed signifiers of Alzheimer's can be found to varying degrees in all aging brains.

In a 2007 article Brayne reminds readers that age itself is, by a long way, the greatest risk factor for dementia, and individuals aged 90 and older have more than 25 times the risk of developing dementia than those aged 65 to 69.[15] She is critical of research attempts to isolate and study "pure" Alzheimer's, because in such studies individuals with age-related pathologies such as vascular disease have been so often excluded as research subjects, although in recent years this has changed somewhat. Brayne points out that if, for example, one were to exclude people with hypertension, then about 30% of many older populations could not be subjects for dementia research, thus introducing a major bias, and raising an important epistemological question also raised explicitly by David Bennett and Howard Chertkow: what *is* "normal" in an elderly population? Alzheimer researchers have a perennial problem notes Brayne, "deciding where normal aging stops and dementia starts," and such decisions depend in large part on expectations about aging that are profoundly influenced by culture and politics.[16]

In common with a good number of researchers today, Brayne believes that Alzheimer's is not a single disease but a syndrome, with at least two and possibly several subtypes, each of which is closely associated with the aging process. Such heterogeneity should not be standardized into oblivion, Brayne argues, but on the contrary, acknowledged and researched with care. Carol Brayne is the principal investigator in a research program that commenced in 1985 that focuses on longitudinal population-based studies of people aged 65 and older. When I talked with her in 2010, she had this to say:

> The great bulk of people with dementia are 75 and up and most of them are labeled as AD . . . and I actually think that for these older groups going right back and thinking about senile dementia, but obviously trying to find a new, more acceptable label for this overarching condition, would be better for them; and more honest, in a way. . . . I think we need to look at a person and see where they are on a cognitive spectrum and how they manage in daily life—this makes Alzheimer's a spectrum condition—it's like saying it's a disability, and you have to assess what is a person's level of capacity for day to day functions— this approach is much more patient oriented than most, and pays

attention to comorbidities, including heart conditions and diabetes and so on, not just to Alzheimer's. This is where domicile visits and old age psychiatrists are crucial. The phrase I use these days is "capacity to benefit."

Brayne's position is strongly influenced by the previous findings of her group noted in earlier chapters in which, among a sample of over 450 individuals, a very high proportion of older people exhibit so-called neuropathology in their brains at death but, even so, do not show signs of dementia while alive. Brayne agrees that early-onset, dominantly inherited Alzheimer disease conforms closely to gold standard plaque and tangle pathology, and she concedes that this may also be the case for late-onset cases of AD diagnosed in individuals in their 60s and perhaps early 70s. But for the majority of people clinically diagnosed with AD, those who are 75 and older, she argues that something different is at work—that the plaques and tangles in the brains of such individuals are almost without exception present together with a heterogeneous mix of other signs of neuropathology. She has two difficulties with this situation: first that the majority of the research carried out in connection with AD has been done with people in their 60s and 70s, individuals whose symptoms resemble something closer to so-called pure Alzheimer disease. Brayne argues that the findings from such research may well not apply to the majority of diagnosed cases who are among older age groups. Second, it is necessary to account for why so many older people whose brains are "littered" with "pathology" do not exhibit dementia in daily life. With this in mind, Brayne and her team undertook a remarkable study in which they examined closely the oft-repeated claim that education has a potentially protective role against dementia.

Large population-based cohort studies, involving over 870 individuals in all, were tracked in this project, from either 1985 or 1991.[17] The number of years of education completed by these people was recorded at the outset, and they were interviewed repeatedly over the years until death, at which time dementia-related pathologies were systematically assessed at autopsy. At death, 56% of these individuals had been clinically diagnosed with dementia. The findings showed that longer years of education were associated with lower rates of dementia, and with greater brain weight, but no relationship was found between dementia and either neurodegenerative pathologies or vascular pathologies. Furthermore, the "dose" of education was significant, in that more education reduced dementia risk independently of the severity of pathology.

These researchers emphasize that further research is called for, and they agree that the relationship among education, neuropathology, and dementia remains unclear, but suggest that their sample provides "sufficient power" to begin to tackle three questions effectively: Does education protect against the accumulation of pathologies in the brain? Does education appear to compensate in some way for the usual cognitive impairment associated with pathology? And, third, does such compensation vary with pathological severity?

Their conclusion is that "the number of years of education has no protective effect against the accumulation of neurodegenerative or vascular pathologies in the brain at death." However, education militates against "an association of pathological burden and cognitive decline" even when, for the majority, there had been intervals of more than 70 years between years in school and death. Furthermore, it appears that those individuals who had experienced more education early in life were better protected.

This research supports a "brain reserve" or "cognitive reserve" hypothesis; however, the researchers are cautious about the significance of the findings, and note the shortcomings of previous research in connection with this concept, most often due to sampling bias. They are alert to the strong possibility that education may well be a proxy for socioeconomic status, exposure to toxins, and/or poor diet. Their findings make clear that greater brain weight is associated with more education—but acknowledge that this finding could be either cause or effect. They also recognize that education, and not simply number of years in school, no doubt has an effect over the entire life span. Brayne and colleagues conclude, "Those with more education do appear to have heavier brains and maintain cognition in the face of a burden of neuropathology compared to those with less education. Education attenuates dementia risk but does not mitigate it altogether," as the notable case of the philosopher and novelist, Iris Murdoch, made evident. The group recognizes that considerable individual variation exists, despite population-based findings, demanding further investigation.

Anthropologists are likely to suggest that some fine-grained research is in order among populations where little or no formal education is the norm and literacy rates are low. Research, fairly sparse though it is, suggests that such populations do not appear to have high rates of dementia, although it is by no means absent. These findings raise interesting questions about the care of demented persons and the possibility that the actions of families and communities may slow down or ameliorate the cognitive changes associated with dementia.[18]

The epistemological question of the relationship of aging and dementia remains unanswered, of course, and is unlikely to be settled any time soon. M. Marsel Mesulam, a Chicago-based neurologist, trained also in psychiatry, argued more than two decades ago against a unifying theory of Alzheimer causation. He pointed out that the idea of "neuroplasticity," a concept that refers, as he put it, "to processes of vital importance for the structural upkeep of the brain and for the functional adaptation of the organism to the environment," decreases with age, and this explains why age is the single most important and universal risk factor for AD. According to this formulation, Mesulam argued, Alzheimer's associated with old age may not be a disease at all, but simply the inevitable manifestations of a failure to keep up with the "increasingly burdensome work of plasticity."[19] His position is that an aging brain is normal, but certain variables—genetics, environment, and lifestyles (figure 9.1)—may result in perturbations that accelerate the aging process (or, alternatively, as the work of Brayne and colleagues and other researchers suggests, protect against the ravages of aging). Mesulam concluded,

> [I]mportant insights related to the pathophysiology and prevention of AD may come from the fields of developmental biology. One of the most important goals will be to understand the processes that influence plasticity in the adult human brain and to determine whether their vulnerability to aging and to the other AD-causing factors can be modified.[20]

Mesulam's argument is not one of a radical discontinuity between aging and dementia, and he implies that dementia may well be best understood as a "normal" endpoint in very elderly people. The position taken by Martin Roth (see chapter 1), that of a marked discontinuity between aging and dementia once a crucial threshold has been passed, is not entirely at odds with that of Mesulam. Roth's argument may be appropriate for Alzheimer's diagnosed among younger people, and particularly in cases of dominantly inherited AD. When Roth stated "AD cannot be accounted for in terms of a continuous and predictable extension of normal mental aging," he based his opinion on research findings obtained primarily from samples of people aged largely 70 and younger, and at a time when PIB scanning was not available. The research of Brayne's group and others, also in line with that of Mesulam, strongly suggests that such a statement does not apply to the burgeoning numbers of older people.[21]

Roth acknowledged that genetics and environment no doubt play a role, and account for why some individuals are apparently more vulnerable to AD than are others. However, his assumption, and that of most other researchers, including those searching for effective drugs, has been, and continues to be, that plaques are irrefutable evidence of Alzheimer's in the making, and, with very few exceptions indeed, tangles are the clinching pathological signs of the disease, at whatever age it strikes individuals. Such a diagnosis is decontextualized from everyday life and rests on the belief that localization theory will provide the final answers. But the reality is that new findings are regularly made about both amyloid and tau. The picture becomes ever more complex as technologies enable deeper probing, causing a growing number of researchers to question whether amyloid is indeed driving the neural dysfunction associated with AD. Furthermore, emerging biomarker research is showing that the progression of prodromal dementia does not necessarily take place in the orderly manner set out by the researchers who proposed the move to prevention of AD (see chapter 4); brain structure changes have been shown to take place as early as or even earlier than the commencement of amyloid deposition in some cases.[22]

A great deal remains to be explained, and if drug failures continue to plague the field, then the amyloid hypothesis must be severely modified or dropped entirely, as even those researchers who are the most committed to a molecular approach now concede.[23] A move under way to pay closer attention to cell loss, the significance of which has always been thought of as secondary due to the attention given to plaques and tangles, may rather rapidly bring about some modification of the cascade hypothesis. Paradoxically, biomarker findings has been a major reason for deeper investigation into the onset and significance of cell loss, including the finding that some individuals can apparently adapt to such a loss better than others.

Not only is there still a very long way to go in pinning down the molecularized universe of AD-like dementias, but, furthermore, half the story continues to be barely recognized and researched. The question of what it is that protects so many individuals from dementia even at a great age, including many of those who are assumed to be at risk due to their genetic heritage, has rarely been examined systematically, although work such as that of Brayne and colleagues is an exception. This issue, noted in passing by so many researchers when talking with me, foregrounds the manner in which the biology of aging itself is embedded in social, political, and economic matters. The politics of AD research has been, above all else, to emphasize the horror of dementia, and its unwanted repercussions on the economy, in order first and foremost to raise money for research to cure the disease. Although healthy lifestyles are promoted by AD societies and other organizations dealing with the elderly,

Figure 9.1.
The Tunbridge Wells Rambler's Association, Kent, UK. Annual Bluebell Walk led by the author's father at age 90. Photograph in possession of Margaret Lock; photographer unknown.

the vital question about what it is that raises the odds of having a dementia-free old age is nevertheless marginalized.

In order to assess why and how so many individuals remain healthy and active even in old age, a move must be made away from AD causation conceptualized as in the brain alone, to the boundary-traversing mind, to persons, and to social and political milieus. Of course, a localized approach to the AD conundrum and medicalization of AD will nevertheless continue.

Embedded Bodies

The epigenetic findings by Wang and his colleagues referred to in the previous chapter appear to support and advance the ideas put forward much earlier by Mesulam. And comments by Evelyn Fox Keller, the philosopher of science, cited earlier are also relevant, as we turn to a consideration of the third tension in the AD world. For Fox Keller separation of the terms nature and nurture—the insertion of an "and" in the phrase—is to make these concepts into a false dichotomy. She argues, "From its very beginning, development depends on the complex orchestration of multiple courses of action that involve interactions among many different kinds of elements." This "entanglement between genes and environment" results from "an immensely complex web of interactions between environmental stimuli (both internal and external to the cell) and the structure, conformation, and nucleotide sequence of the DNA molecule."[24] Such a position grounds the science of epigenetics and is applicable throughout individual life spans, Fox Keller insists, and not merely during early development.

Fox Keller limits her argument to what she describes as environmental stimuli, but both epidemiological and emerging epigenetic findings in connection with Alzheimer's suggest that researchers would do well to go further and examine social and political factors. This would include variables mediated by the boundary-traversing mind, such as education, traumatic social and familial events, social isolation, deprivation, and so on, and, in addition, variables that affect neurodevelopment directly, including exposure to toxins, prions, other disease-causing entities, inadequate diet, and others. Clearly, this is a very tall order that would almost certainly entail sophisticated longitudinal studies demanding cooperation across disciplinary boundaries. It is apparent that this will not be easy, particularly because it would require specialists in various disciplines who normally compete for a finite amount of research money to cooperate with one another. But perhaps the disruption in normal science that the AD world is undergoing will result in the kind of "fruitful tension" noted by Fleck, out of which new congeries and novel approaches may arise.

At a conference on epigenetics held in 2011 at McGill University, one of the presenters titled his PowerPoint presentation "DNA Methylation: A

Molecular Link between Nature and Nurture." In his talk, Moshe Szyf, a leading figure in environmental epigenetics in North America, presented a discussion of research findings made by his team involving 25 subjects who had been abused as children, and later committed suicide. The families had agreed to donate the brains of their relatives for research. Upon examination, these brains showed a significantly different pattern of DNA methylation than did those of individuals who had committed suicide where no known abuse had taken place, and also with a control group.[25] Szyf cautions, "Although epidemiological data provides evidence that there is an interaction between nature (genetics) and social and physical environments (nurture) in human development; the main open question remains the mechanism."[26] But he is also the first to admit that obtaining large enough samples to carry out this type of research to great effect is extraordinarily difficult. And a recent review article of epigenetics makes it very clear how provisional are many epigenetic findings to date.[27]

The anthropologist Jörg Niewöhner cautions us about an inappropriate direction in which epigenetics may be moving. He argues that aspects of the social world are liable to be reduced in this type of research to a "quasi-natural experimental system" in which "the molecularization of biography and milieu" may result in standardized models of social change that travel between labs and from thence out into the wider public discourse.[28] Unless such findings can be contextualized in the larger arenas of culture, political economy, and family life, including family intergenerational histories, highly truncated accounts about suicide causation are likely to result, in which parental behavior thought to be wanting may well be interpreted as causal of defective methylation processes that then contribute to suicidal behavior.

Niewöhner has made a preliminary attempt to set out how variables external to the body may be systematically incorporated into lab-created epigenetic findings. He focused on the heterogeneous subfield known as "environmental epigenetics" in which macro-environments are beginning to be given due attention by molecular biologists.[29] Niewöhner argues that environmental epigenetic interactions result in "embedded bodies." Such a body, he writes, is "heavily impregnated by its own past and by the social and material environment within which it dwells. It is a body that is imprinted by evolutionary and trans-generational time, by 'early-life,' and a body that is highly susceptible to changes in its social and material environment."[30] Niewöhner calls for a "molecularization of biography" in which events of significance in people's lives are researched and documented, to a limited extent standardized, and then systematically examined for associations with bodily epigenetic changes.[31] He emphasizes that environmental epigenetics is not, as yet, a well-established discipline, and that there is a long way to go before, for example, the public health implications of its findings could be put to work, but, Niewöhner insists, a serious effort should nevertheless be made to include social and political variables that appear to be implicated in epigenetic effects. The concept of "local biologies" put forward by Lock in the early 1990s represents an earlier effort to come to grips with the crucial importance of extrasomatic variables as key contributors to embedded bodies.[32]

The discussions in previous chapters about Alzheimer genetics, particularly the variation among and across each of the APOE polymorphisms, and about the distribution of plaque in human brains and its less than reliable association with the behavioral changes of dementia, are graphic examples that substantiate a concept of embedded bodies. Such evidence suggests that if improved insights into AD causality with the objective of bringing about its prevention are desired, to simply focus on prodromal AD established by the presence of standardized biomarkers will serve to push the story back to an earlier stage of proximate causation, but no more. A move such as this will not engage with "ultimate" causation—that is, with the numerous extrasomatic factors that are clearly implicated in aging and dementia. Nor will it succeed in creating risk estimates that can predict with accuracy who among us are going to become demented—increasing recognition that the world of biology is, at best, only "law-like," supports this contention. As clinicians repeated to me time and again, family history continues to predict risk for AD far more reliably than does any other variable, and it seems that biomarker usage may not change this situation. The number of times the word "uncertainty" has been used in this book by so many researchers whom I interviewed, is remarkable.

Readers may recall what was cited earlier by one of the leading researchers in the AD enterprise: "The array of challenges for the mission to prevent AD within a decade are no less daunting than those faced by similar national endeavors such as the Apollo space program, the Manhattan Project, or the Human Genome Project."[33] If the AD "epidemic" is to be confronted, a public health approach orchestrated at global and local levels is called for, one that targets variables than can readily be manipulated without resort to expensive technologies, high-powered facilities, and highly trained personnel.[34] This, surely, is the most humanitarian and effective way to approach the problem, one that does not, of course, rule out support of a well-funded research program designed to tackle the molecular puzzles that are so challenging.

Ultimately, however, in addition to further researching molecular changes in the brain and rigorously detecting statistical differences in AD incidence among populations and groups, another level of inquiry is called for. Appreciation of the inextricable entanglement of individual biology with evolutionary, historical, environmental, social, cultural, and political variables, a never-ending process that can usefully be thought of as "biosocial differentiation," demands attention.[35] Only thus can the extent of the variation of AD incidence, and the reasons that contribute to its complexity, begin to be understood. Without the painstaking collection of such knowledge a shift to prevention is liable to run aground. One aspect of this research involves investigation into neuroprotection, but to do this effectively, the brain must first be decentered. Fascinating findings about centenarians and their genetics,[36] and about environments and social arrangements that appear to support long, healthy lives,[37] are steps in the right direction.

Although those who work to increase funding for AD wish to sustain a frightening image of the global future ravaged by an unstoppable AD epidemic, the reality is that there will be no silver bullet for Alzheimer disease—the heterogeneity of the condition that so many researchers surely recognize, ensures that this will be so. The challenge presented by an aging brain exceeds that posed by all other organs of body. Even so, perhaps it will be expedient to retain an overarching AD category for political purposes—the AD label would be used as a heuristic device, a good-enough category, designed to keep societies and governments focused on the enormity of the problem. If and when AD is systematically broken down into subtypes, knowledge about this fine-tuning will filter out into the public domain without necessarily reducing the symbolic power of AD as a tragedy that must be countered by every possible means, including concerted research efforts, outstanding clinical care, and improved family support.

The hype associated with the current "shake-up" in the AD world may lead people in wealthy countries to believe that at last we are on track for a cure. But every indication is that we should not hold our breath in light of the persistent failure of drug trials. The work of Thomas McKeown discussed briefly in the introductory pages of this book showed that during the long interim in which scientists were struggling to develop antibiotics in the early part of the 20th century, improved nutrition and hygiene brought about a major increase in longevity due to a substantial drop in infectious diseases, long before drugs were available. We would do well to keep this in mind—breakthroughs in big science take time, and with a "pandemic" of aging fast approaching, as we are led to believe, a great deal can be done relatively simply to alleviate the situation through improved care of and social support for the elderly, particularly when individuals become dependent and frail.

The positive contribution that so many older people make to society goes consistently unnoticed and demands attention. A recent study in the United Kingdom, for example, showed that far from being a burden on the economy, older people are net contributors. Including tax payments, spending power, caring responsibilities, and the volunteering effort of people aged 65 and older. In the study, it was calculated that this cohort contributes almost £40 billion (approximately $65 billon) more to the U.K. economy than it receives in state pensions, welfare, and health services. This research also suggests that this benefit to the economy will increase in coming years as the "baby boomer" cohort enters retirement. By 2030, it is projected that the net contribution of older people will be worth some £75 billion.[38]

The current move to prevention in the AD world by means of surveillance of biomarkers is presently for research purposes alone, in the hope of eventually bringing drugs to market. In the unlikely event that these technologies are approved for clinical use, drugs developed, and their prescription routinized, most elderly people in the world will not have access to them unless enormous social and political changes first take place—a highly unlikely scenario. And facilities will most certainly not be available for the majority of individuals, regardless of where they live, to have AD biomarkers regularly monitored from middle age on. As the memories and minds of these individuals decline with age, a large number may well continue to be thought of merely as mad, senile, or simply old and faded. Their burden on meager health care systems and local economies will not be that great, although families will indeed be burdened, with the result that large numbers of elderly may well be subjected to neglect and even starvation. Such lives may count for little, except possibly to boost global statistics designed to normalize the approaching pandemic of aging, and very occasionally such individuals will be of use as research subjects.[39]

To improve the lot of this vast majority, therefore, a globally orchestrated political shift to the implementation of public health programs is urgently needed, one that is not ultimately driven by a molecularized approach to AD. This approach would be designed to engage with the reality of aging while attempting to reduce dementia prevalence, or at least slow its progression, no doubt largely through the agency of family and communities that receive governmental support. Such a move would also reduce the stigma that continues to be associated with dementia in so many places. But a public health approach such as this, essential though it is, can have only a limited effect. It is clear that, equally important, the contribution to AD incidence of contemporary global politics must be explicitly acknowledged and, if at all possible, acted upon. Alleviation of poverty and chronic inequalities and universal access to good nutrition, sanitation, and education have long been recognized as crucial to human health; the situation today is somewhat improved in some parts of the world, but for millions this is not the case. The forcible removal of hundreds of people from arable land and their relocation, ever-increasing polluted environments, and numerous other harmful stresses exerted on huge swaths of the world's population due to the demands of endless capital accumulation are increasingly toxic. While these conditions persist and increase, Alzheimer incidence will not be reduced. Nor will the incidence of many other illnesses, several of which, including diabetes and obesity, contribute directly to dementia.

In closing, one further reality must be kept in mind: no amount of preventive measures and no drug will defeat aging (even though certain maverick scientists are attempting just this), nor can dementia be "wiped out" as though it is an infectious disease—aging and dementia cannot be disentangled; all we can strive for is to find ways to stave off or halt the progression of AD at whatever age it strikes.

Afterword
Portraits from the Mind

I noted toward the end of the Orientations that opened this book that I would not be able, for reasons of space, to give more than minimal consideration to the actual experience of Alzheimer's as it affects individuals and their families. I am acutely aware that what is missing almost entirely from this treatise is the ultimate, undeniable reason for the existence of the Alzheimer enterprise— the anguish and suffering that this condition brings to millions of individuals and their families. In closing I want to draw readers' attention to a remarkable set of paintings by the artist William Utermohlen that he produced after being diagnosed with Alzheimer's disease. These paintings (A1–A4) draw one into the terrifying, miserable experience of losing one's mind with heartrending clarity. My attention was first drawn to the work of Utermohlen when I came across an article about him in *The Lancet*, published in 2001. And then at the

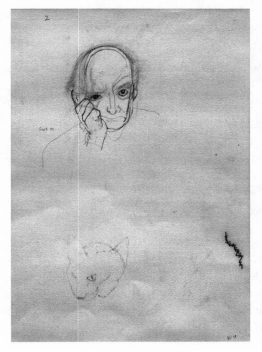

Figure A.1.
Self Portrait with Cat, pencil on paper, 1995, by William Utermohlen.
Reproduced with permission from Galerie Beckel-Odille-Boïcos.

Figure A.2.

Broken Figure, mixed media on paper, 1996, by William Utermohlen. Reproduced with permission from Galerie Beckel-Odille-Boïcos.

Figure A.3.

Mask (Black Marks), watercolor on paper, 1996, by William Utermohlen. Reproduced with permission from Galerie Beckel-Odille-Boïcos.

large international Alzheimer meeting held in Chicago in 2008, I went to an exhibition of Utermohlen's work sponsored by Myriad Genetics, shown at the Chicago Cultural Center. This exhibition was a retrospective in four parts of the artist's work before and after his diagnosis with Alzheimer's disease and included his early portraits (1955–77); the *Dante* Cycle (1964–66) and the *Mummers Parade* Cycle (1969–70); *Conversation Pieces* (1990–91), and, finally, *The Last Portraits* (1995–2000).

William Utermohlen lived from 1933 to 2007 and was diagnosed with AD in 1995. He was born in South Philadelphia, the only son of a first-generation German immigrant, and attended art school in Pennsylvania. He later moved to

England, where he continued his art training and was particularly influenced by the pop art of R. B. Kitaj. He married the art historian Patricia Redmond in 1965 and settled in London, where he lived out his life. After his diagnosis he was taken care of at home by his wife, friends, and caregivers, but his deterioration made it necessary in 2004 to place him in a nursing home, where he eventually died.

The authors of *The Lancet* article inform us that Utermohlen's family had no history of neurological or psychiatric illness and that his medical history was unremarkable except for a car accident at the age of 55 in which he was left unconscious for about 30 minutes. After assessment when he was first referred to a neurologist, he was given at age 61 a diagnosis of "probable Alzheimer's disease." His wife traced the beginning of her husband's problems back to a point several years earlier in his life. The age of onset of the disease was, then, very early indeed, but, with no family history of neurological disease, it could not have been dominantly inherited Alzheimer's disease. By age 65 Utermohlen was showing a "global deterioration in his cognitive state."[1] At this time he agreed with his doctors that he would attempt to continue painting as long as possible in part to assist in helping others to understand what was happening to him neurologically. Patricia Utermohlen commented, "As each small self-portrait was completed, William showed it to his nurse, Ron Isaacs. Ron visited the studio, photographing every new work. Ron's conviction that William's efforts were helping to increase the understanding of the deeply psychological and traumatic aspects of the disease undoubtedly encouraged William to continue."[2]

All his life Utermohlen had experienced a strong desire to draw people, as is strikingly illustrated by his colorful pictures of the Mummers Parade held every New Year's Day in Philadelphia. He had also painted many portraits throughout his career, including self-portraits, and after his diagnosis almost all of his work was concentrated on self-portraits. As *The Lancet* authors note, his self-portraits subsequent to diagnosis "reveal a change commensurate and consistent with the deterioration in his cognitive stage."[3] They point out that self-portraits have been considered by many observers to openly express "a variety of states of mind including terror, sadness, anger, naked pain and resignation."[4] There can be little doubt that themes of sadness and anxiety are readily apparent in Utermohlen's later self-portraits, and even terror, one would surmise, especially in Head 1. It is noted that such themes can readily be found in several of the works painted before any noticeable decline had taken place in Utermohlen's artistic ability. One such picture, painted at the age of 60, shows the artist seated at and gripping onto a table beneath an open skylight. His wife believes that this picture depicts fear and isolation. When asked if this is a reasonable interpretation of the work, the artist himself indicated to his doctors the open skylight and said, "Yes, and I was getting out."[5]

The pictures make it dramatically evident that Utermohlen's processing of spatial and perceptual information had declined greatly, making the portraits remarkably poignant for viewers, and quite possibly raising anxiety and fear for them, perhaps even for their own futures. *The Lancet* authors note that

Figure A.4.

Self Portrait (With Saw), oil on canvas, 1997, by William Utermohlen. Reproduced with permission from Galerie Beckel-Odille-Boïcos.

because Utermohlen used an abstract style for much of his painting, this may have provided an outlet for his continued artistic expression long after his mind was no longer functioning fully, because no restrictions were imposed by the demands of realism, with its need to reflect as accurately as possible what appears before one. They conclude that to be able to continue as an artist at a stage when Alzheimer's disease has "blunted the craftsman's most precious tools offers a testament to the resilience of human creativity."[6] These paintings also indicate that there may be many avenues, as yet only beginning to be explored, that suggest ways in which humans can perhaps be protected from the ravages of this condition by means of lifelong social and cultural activities. Such practices will not inhibit the eventual development of dementia in people who are heavily predisposed, but they may thwart the condition for years.

Notes

Orientations

1. Robert Butler, *The Longevity Revolution: The Benefits and Challenges of Living a Long Life* (New York: Public Affairs, 2008), 121.

2. In developing countries, such as Ghana, the proportion of people aged 60 and older is rising but for the moment remains at less than 10%; however, because the absolute numbers are large, this transformation is characterized as an "aging epidemic." See http://www.modernghana.com/news2/137880/1/ageing-epidemic-looms-in-ghana.html. In China, with an economy that is at once developed and developing, the aging of society is even more dramatic. The Chinese themselves describe the situation as one of "growing old before getting rich." See Hong Zhang, "Vignette 2: China," in Sharon Kaufman, Julie Livingston, Hong Zhang, and Margaret Lock, "Transforming the Concepts of Aging: Three Case Studies from Anthropology," in *Oxford Textbook of Old Age Psychiatry*, ed. Tom Dening and Alan Thomas (Oxford: Oxford University Press, 2013); see also Richard Jackson, Keisuke Nakashima, and Neil Howe, *China's Long March to Retirement Reform: The Graying of the Middle Kingdom Revisited* (Washington, D.C.: Center for Strategic & International Studies, 2009), http://csis.org/files/media/csis/pubs/090422_gai_chinareport_en.pdf.

3. Paul M. V. Martin and Estelle Martin-Granel, "2,500-Year Evolution of the Term Epidemic," *Emerging Infectious Diseases* 12, no. 6 (2006): 976–80.

4. See http://www.google.ca/#hl=en&source=hp&biw=1471&bih=1009&q=Alzheimer%27s+epidemic&aq=f&aqi=g1g-m1g-v2&aql=&oq=&fp=484a924e15169ff5 (accessed March 10, 2011).

5. Meredith Wadman and Nature Magazine, "U.S. Government Sets Out Alzheimer's Plan," *Scientific American*, May 23, 2012, http://www.scientificamerican.com/article.cfm?id=us-government-sets-alzheimers-plan.

6. The term "epigenetics" was created by C. H. Waddington in 1942. When Waddington coined the term the physical nature of genes and their role in heredity were not known; he used it as a conceptual model of how genes might interact with their surroundings to produce a phenotype. The contemporary meaning in scientific discourse is more narrow, referring to heritable traits (over rounds of cell division and sometimes trans-generationally) that do not involve changes to the underlying DNA sequence. See http://en.wikipedia.org/wiki/Epigenetics.

7. Peter J. Whitehouse and Daniel George, *The Myth of Alzheimer's: What You Aren't Being Told about Today's Most Dreaded Diagnosis* (New York: St. Martin's, 2008).

8. Postgenomic research is research carried out since the mapping of the human genome in which emphasis is given to the function of genes in context.

9. Karen Barad, *Meeting the Universe Halfway: Quantum Physics and the Entanglement of Matter and Meaning* (Durham, N.C.: Duke University Press, 2007), 56.

10. Ibid., 33.

11. In addition to ibid., see Lorraine Daston and Peter Galison, *Objectivity* (New York: Zone Books, 2007), and Margaret M. Lock and Vinh-Kim Nguyen, *An Anthropology of Biomedicine* (Oxford: Wiley-Blackwell, 2010).

12. Ian Hacking, "Making up People," in *Reconstructing Individualism: Autonomy, Individuality, and the Self in Western Thought*, ed. T. C. Heller, M. Sosna, and D. E. Wellbery (Stanford: Stanford University Press, 1986), 222–36, http://www.icesi.edu.co/blogs/antro_conocimiento/files/2012/02/Hacking_making-up-people.pdf; Donna J. Haraway, *Simians, Cyborgs, and*

Women: The Reinvention of Nature (New York: Routledge, 1991); Bruno Latour, *Pandora's Hope: Essays on the Reality of Science Studies* (Cambridge, Mass.: Harvard University Press, 1999); Hans-Jörg Rheinberger, "Beyond Nature and Culture: Modes of Reasoning in the Age of Molecular Biology and Medicine," in *Living and Working with the New Medical Technologies: Intersections of Inquiry*, ed. Margaret Lock, Allan Young, and Alberto Cambrosio (Cambridge: Cambridge University Press, 2000), 19–30.

13. Lock and Nguyen, *Anthropology of Biomedicine*, 94.

14. Ludwik Fleck, *Genesis and Development of a Scientific Fact* (Chicago: University of Chicago Press, 1979). See also Charles E. Rosenberg, "What Is Disease? In Memory of Owsei Temkin," *Bulletin of the History of Medicine* 77, no. 3 (2003): 491–505.

15. Warwick Anderson, *The Collectors of Lost Souls: Turning Kuru Scientists into White Men* (Baltimore: Johns Hopkins University Press, 2008); Lock and Nguyen, *Anthropology of Biomedicine*, 103–7; Jacques Pepin, *The Origin of AIDS* (Cambridge: Cambridge University Press, 2011).

16. Bruno Latour, *The Pasteurization of France* (Cambridge, Mass.: Harvard University Press, 1988).

17. Alan Cassels, "Drug Bust," *Common Ground*, November 2010, http://www.commonground.ca/iss/232/cg232_cassels.shtml.

18. See, for example, Michael Lambek, *Knowledge and Practice in Mayotte: Local Discourses of Islam, Sorcery and Spirit Possession* (Toronto: Toronto University Press, 1993); Susan Reynolds Whyte, *Questioning Misfortune: The Pragmatics of Uncertainty in Eastern Uganda* (Cambridge: Cambridge University Press, 1997).

19. For an elaboration of this approach, see, for example, Felicia A. Huppert, Carol Brayne, and Daniel W. O'Conner, eds., *Dementia and Normal Aging* (Cambridge: Cambridge University Press, 1994); and Marcus Richards and Carol Brayne, "What Do We Mean by Alzheimer Disease?," *British Medical Journal* 341 (2010): 865–67.

20. Ian Hacking, *The Taming of Chance* (Cambridge: Cambridge University Press, 1990), 2.

21. Lorraine Daston, *Classical Probability in the Enlightenment* (Princeton: Princeton University Press, 1988).

22. Hacking, *Taming of Chance*, 2.

23. Ulrich Beck, *World at Risk* (Cambridge: Polity Press, 2007), 5.

24. Ibid., 5.

25. Ibid., 216.

26. Ibid., 8.

27. Ibid., 9.

28. Alzheimer's Disease International, *World Alzheimer Report 2009: Executive Summary* (2009), 1, http://www.alz.co.uk/research/files/WorldAlzheimerReport-ExecutiveSummary.pdf.

29. Ibid., 11.

30. "Leading Edge: How Much Is Dementia Care Worth?," *The Lancet Neurology* 9, no. 11 (2010): 1037.

31. "Over the Mountain to Alzheimer's," *Maclean's*, January 4, 2010, http://www2.macleans.ca/2010/01/04/over-the-mountain-to-alzheimers/.

32. "2008 Alzheimer's Disease Facts and Figures," *Alzheimer's & Dementia* 4, no. 2 (2008): 110–33.

33. Carolyn Abraham, "Dementia Researchers Feel Blocked by Ottawa, Big Pharma, Medical Dogma," *Globe and Mail*, September 17, 2010, http://www.theglobeandmail.com/life/health-and-fitness/dementia-researchers-feel-blocked-by-ottawa-big-pharma-medical-dogma/article4389415/.

34. "Northern Ireland Dementia Total More Than Estimated," *BBC News*, February 3, 2010, http://news.bbc.co.uk/2/hi/uk_news/northern_ireland/8494975.stm.

35. Although anthropological research has shown repeatedly that even though many people everywhere accept biomedical accounts of proximate disease causation—that is, accounts about what has gone wrong inside the body—nevertheless a large number of individuals continue to

pose the question of "why me" and seek out explanations for what it is in their past behavior and everyday life that accounts for the problem having arisen in the first place. See, for example, Mark Nichter, *Global Health: Why Cultural Perceptions, Social Representations, and Biopolitics Matter* (Tucson: University of Arizona Press, 2008).

36. Lawrence Cohen, *No Aging in India: Alzheimer's, the Bad Family, and Other Modern Things* (Berkeley: University of California Press, 1998), 184.

37. Margaret Lock, "Centering the Household: The Remaking of Female Maturity in Japan," in *Re-Imagining Japanese Women*, ed. Anne Imamura (Berkeley: University of California Press, 1996), 73–103; Milton Ezrati, "Japan's Aging Economics," *Foreign Affairs*, May/June 1997, http://www.foreignaffairs.com/articles/53050/milton-ezrati/japans-aging-economics.

38. Jennifer Robertson, "Robo Sapiens Japanicus: Humanoid Robots and the Posthuman Family," *Critical Asian Studies* 39, no. 3 (2007): 369–98.

39. United Nations Department of Economic and Social Affairs, Population Division, *World Population Ageing: 1950–2050* (2002), http://www.un.org/esa/population/publications /worldageing19502050/.

40. Yogesh Shah, "Gray Tsunami: Challenges and Solutions of Global Aging," *DMU Magazine*, Summer 2011, http://www.dmu.edu/magazine/summer-2011/my-turn-summer-2011 /gray-tsunami-challenges-and-solutions-of-global-aging/.

41. Richard Jackson and Neil Howe, *The Graying of the Middle Kingdom: The Demographics and Economics of Retirement Policy in China* (Washington, D.C.: Center for Strategic & International Studies, 2004), http://csis.org/files/media/csis/pubs/grayingkingdom.pdf.

42. David Barboza, "China, in a Shift, Takes on Its Alzheimer's Problem," *New York Times*, January 12, 2011.

43. Hong Zhang, "Who Will Care for Our Parents? Changing Boundaries of Family and Public Roles in Providing Care for the Aged in China," *Journal of Long-Term Home Health Care* 25 (2007): 39–46; see also Kaufman et al., "Transforming the Concepts of Aging."

44. See the section by Hong Zhang in Kaufman et al., "Transforming the Concepts of Aging."

45. David L. Kertzer and Peter Laslett, *Aging in the Past: Demography, Society and Old Age* (Berkeley: University of California Press, 1995).

46. Ed Yong, "Life Begins at 100: Secrets of the Centenarians," *Mind Power News*, 2009, http://www.mindpowernews.com/LifeBeginsAt100.htm.

47. Butler, *Longevity Revolution*, 13. See also Thomas McKeown, *The Modern Rise of Population* (New York: Academic Press, 1976).

48. United Nations Department of Economic and Social Affairs, Population Division, *World Population Ageing*.

49. This discussion about global networking among researchers working on Alzheimer disease has much in common with the concept of "biomedical platforms" set out by Peter Keating and Alberto Cambrosio, *Biomedical Platforms: Realigning the Normal and the Pathological in Late-Twentieth-Century Medicine* (Cambridge, Mass.: MIT Press, 2003). Keating and Cambrosio argue that this concept "sheds new light on the articulation and the regulation of the practices focused on the normal and the pathological that characterize contemporary biomedicine" (22). Their purpose is to demonstrate that biomedical platforms are "neither science nor technology but a way of articulating the two" (326). Above all, this hybrid space that has emerged since the end of World War II today defines the standards according to which biomedical actions are evaluated.

50. For development of this theme, see Alexander Peine, "Challenging Incommensurability: What We Can Learn from Ludwik Fleck and the Analysis of Configurational Innovation," *Minerva* 49 (2011): 489–508, 493.

51. See also Susan Leigh Star, "Cooperation without Consensus in Scientific Problem Solving: Dynamics of Closure in Open Systems," in *CSCW: Cooperation or Conflict?*, ed. Steve Easterbrook (London: Springer, 1993), 93–106; Peter Galison and David Stump, eds., *The Disunity of Science—Boundaries, Contexts, and Power* (Stanford: Stanford University Press, 1996).

52. See, for example, Cohen, *No Aging in India*; E. Herskovits, "Struggling over Subjectivity: Debates about the 'Self' and Alzheimer Disease," *Medical Anthropology Quarterly* 9 (1996): 146–64; Charlotte Ikels, "The Experience of Dementia in China," *Culture, Medicine, and Psychiatry* 22 (1998): 257–83; Annette Leibing and Lawrence Cohen, eds., *Thinking about Dementia: Culture, Loss, and the Anthropology of Senility* (New Brunswick, N.J.: Rutgers University Press, 2006); Daniel R. George, "Overcoming the Social Death of Dementia through Language," *Lancet* 376 (2010): 586–87; Judith Levine, "Managing Dad," in *Treating Dementia: Do We Have a Pill for It?*, ed. Jesse F. Ballenger, Peter Whitehouse, Constantine G. Lykestos, Peter V. Rabins, and Jason H. T. Karlawish (Baltimore: Johns Hopkins University Press, 2009), 116–24; John W. Traphagan, *Taming Oblivion: Aging Bodies and the Fear of Senility in Japan* (Albany: State University of New York Press, 2000).

Chapter 1
Making and Remaking Alzheimer Disease

1. Jonathan Franzen, "My Father's Brain," *The New Yorker*, September 10, 2001, 85.

2. Richard M. Torack, *The Pathologic Physiology of Dementia, with Indications for Diagnosis and Treatment* (Berlin: Springer-Verlag, 1978).

3. *King Lear* of course is not a comedy. He is often described by literary critics as demented, but Lear's extreme behavioral changes, including powerful hallucinations and complete derangement followed by an apparent return to normal, should be understood as literary device on the part of Shakespeare and not interpreted via our medicalized minds in a search for truth. The statement by Jacques, on the other hand, does represent the dominant idea of the day that persisted well into the 19th century about the ages of man or the wheel of life. See M. Lock, *Encounters with Aging: Mythologies of Menopause in Japan and North America* (Berkeley: University of California Press, 1993), 305–6, http://www.ucpress.edu/book.php?isbn=9780520201620.

4. Torack, *Pathologic Physiology*, 1.

5. Eric Engstrom, "Researching Dementia in Imperial Germany: Alois Alzheimer and the Economies of Psychiatric Practice," *Culture, Medicine, and Psychiatry* 31, no. 3 (2007): 405–13.

6. H. Braak and E. Braak, "Neuropathological Stageing of Alzheimer-Related Changes," *Journal of Neuropathology & Experimental Neurology* 59 (2000): 733–48.

7. Konrad Maurer and Ulrike Maurer, *Alzheimer: The Life of a Physician and the Career of a Disease* (New York: Columbia University Press, 1986), 41.

8. Ibid., ix.

9. Theodore M. Brown, "Mental Diseases," in *Companion Encyclopedia of the History of Medicine*, vol. 1, ed. W. F. Bynum and Roy Porter (London: Routledge, 1993), 442.

10. Ibid., 444.

11. Michel Foucault, *The Birth of the Clinic: An Archaeology of Medical Perception*, trans. A. M. Sheridan Smith (New York: Vintage Books, 1973), 131.

12. Charles Hughes, "Insanity Defined on the Basis of Disease," *The Alienist and Neurologist* 20 (1887): 170–74, 173.

13. G. E. Berrios, "Alzheimer Disease: A Conceptual History," *International Journal of Geriatric Psychiatry* 5 (1990): 355–65, 356.

14. Cohen, *No Aging in India*, 63.

15. Maurer and Maurer, *Alzheimer*, 2, 4.

16. Ibid., 151–52.

17. Berrios, "Alzheimer Disease."

18. Atwood Gaines and Peter Whitehouse, "Building a Mystery: Alzheimer Disease, Mild Cognitive Impairment, and Beyond," *Philosophy, Psychiatry, & Psychology* 13 (2006): 61–74, 63.

19. Braak and Braak, "Neuropathological Stageing."

20. Cathy Gere, "'Nature's Experiment': Epilepsy, Localization of Brain Function and the Emergence of the Cerebral Subject," in *Neurocultures: Glimpses into an Expanding Universe*, ed. Francisco Ortega and Fernando Vidal (Frankfurt: Peter Lang, 2011), 235–47. Gere's essay tracks

the use of epileptics as research subjects in neurology in the early 20th century, culminating in the experiments carried out by Wilder Penfield at the Montréal Neurological Institute in the 1950s with over 1,100 individuals with epilepsy whose brains were experimented upon using local anesthesia.

21. Berrios, "Alzheimer Disease," 358.

22. Maurer and Maurer, *Alzheimer*, 163.

23. Ibid., 142.

24. Hans-Jurgen Möller and Manuel B. Graeber, "Johann F.: The Historical Relevance of the Case for the Concept of Alzheimer Disease," in *Concepts of Alzheimer Disease: Biological, Clinical, and Cultural Perspectives*, ed. Peter J. Whitehouse, Konrad Maurer, and Jesse F. Ballenger (Baltimore: Johns Hopkins University Press, 2000), 30–46.

25. Maurer and Maurer, *Alzheimer*, 217.

26. Berrios, "Alzheimer Disease," 360.

27. I am indebted to Jesse Ballenger for this thought.

28. Berrios, "Alzheimer Disease," 362.

29. Wolfgang Jilek, "Emil Kraepelin and Comparative Sociocultural Psychiatry," *European Archives of Psychiatry and Clinical Neuroscience* 245 (1995): 231–38.

30. David J. Libon, Catherine C. Price, Kenneth M. Heilman, and Murray Grossman, "Alzheimer's 'Other Dementia,'" *Cognitive and Behavioral Neurology* 19, no. 2 (2006): 112–16; Torack, *Pathologic Physiology*.

31. Konrad Maurer, Stephan Volk, and Hector Gerbaldo, "Auguste D: The History of Alois Alzheimer's First Case," in Whitehouse, Maurer, and Ballenger, *Concepts of Alzheimer Disease*, 20–29, 20.

32. Berrios, "Alzheimer Disease."

33. T. Simchowicz, "Histologische Studien über die senile Demenz," in *Histologische und histopathologische Arbeiten*, vol. 4, ed. F. Nissl and A. Alzheimer (Jena: Fischer, 1913), 267–444.

34. N. Gellerstedt reported this finding in 1933; see W. A. Lishman, "The History of Research into Dementia and Its Relationship to Current Concepts," in Huppert, Brayne, and O'Conner, *Dementia and Normal Aging*, 41–56.

35. Ibid., 44.

36. Thomas G. Beach, "The History of Alzheimer Disease: Three Debates," *Journal of the History of Medicine* 42 (1987): 327–42.

37. Ibid.

38. M. Leale, "The Senile Degenerations, Their Symptom-Complex and Treatment," *International Clinics* 4 (1911): 37–47.

39. C. Mercier, *Sanity and Insanity* (London: Walter Scott, 1895).

40. Martha Holstein, "Alzheimer Disease and Senile Dementia, 1885–1920: An Interpretive History of Disease Negotiation," *Journal of Aging Studies* 11 (1997): 1–13, 7; see also T. Cole, *The Journey of Life: A Cultural History of Aging in America*. Cambridge: Cambridge University Press, 1991.

41. Ignatz Nascher, "Senile Mentality," *International Clinics* 4 (1911): 48–59.

42. Jesse F. Ballenger, *Self, Senility, and Alzheimer's Disease in Modern America: A History* (Baltimore: Johns Hopkins University Press, 2006), 30.

43. Ibid., 119.

44. Cited in ibid., 48.

45. See ibid., 56–80.

46. Martin Roth, Bernard E. Thomlinson, and Gary Blessed, "Correlation between Scores for Dementia and Counts of Senile Plaques in Cerebral Grey Matter of Elderly Subjects," *Nature* 209 (1966): 109–10, 109.

47. Ballenger, *Self, Senility, and Alzheimer's Disease*, 91.

48. Alex Comfort, *A Good Age* (New York: Simon & Schuster, 1976), 47.

49. Robert Katzman, "The Prevalence and Malignancy of Alzheimer Disease: A Major Killer," *Archives of Neurology* 33 (1976): 217–18.

50. Konrad Dillman, "Epistemological Lessons from History," in Whitehouse, Maurer, and Ballenger, *Concepts of Alzheimer Disease*, 129–57, 141.

51. Robert Katzman and Katherine L. Bick, "The Rediscovery of Alzheimer Disease in the 1960s and 1970s," in Whitehouse, Maurer, and Ballenger, *Concepts of Alzheimer Disease*, 104–14.

52. Ballenger, *Self, Senility, and Alzheimer's Disease*, 98.

53. Roger H. Segelken, "Alzheimer Activist," *New York Times*, September 24, 2008.

54. Patrick Fox, "From Senility to Alzheimer Disease: The Rise of the Alzheimer Disease Movement," *Milbank Quarterly* 67 (1989): 58–102.

55. R. Binstock and S. G. Post, eds., *Too Old for Health Care? Controversies in Medicine, Law, Economics, and Ethics* (Baltimore: Johns Hopkins University Press, 1991).

56. Whitehouse, Maurer, and Ballenger, *Concepts of Alzheimer Disease*, 206–7.

57. Medical Research Council, *Senile and Presenile Dementias: A Report of the MRC Subcommittee, Compiled by W. A. Lishman* (London: Medical Research Council, 1977).

58. Patrick Fox, "The Role of the Concept of Alzheimer Disease," in Whitehouse, Maurer, and Ballenger, *Concepts of Alzheimer Disease*, 209–33.

59. M. J. Delvecchio Good, B. J. Good, C. Schaffer, and S. E. Lind, "American Oncology and the Discourse on Hope," *Culture, Medicine, and Psychiatry* 14, no. 1 (1990): 59–79.

60. Fox, "From Senility to Alzheimer Disease"; Claudia Chaufan, Brooke Hollister, Jennifer Nazareno, and Patrick Fox, "Medical Ideology as a Double-Edged Sword: The Politics of Cure and Care in the Making of Alzheimer's Disease," *Social Science & Medicine* 74 (2012): 788–95.

61. Chaufan et al., "Medical Ideology."

62. Lishman, "History of Research," 50.

63. Ibid., 50–51.

64. As my colleague Allan Young puts it.

65. Lishman, "History of Research," 52.

66. Martin Roth, "Dementia and Normal Aging of the Brain," in Huppert, Brayne, and O'Conner, *Dementia and Normal Aging*, 57–76, 65.

67. Foucault, *Birth of the Clinic*.

68. See ibid., chap. 8.

69. Hacking, *Taming of Chance*. See also Georges Canguilhem, *The Normal and the Pathological* (New York: Zone Books, 1991).

70. Lock, *Encounters with Aging*.

71. Hacking, *Taming of Chance*, 164.

72. Canguilhem, *Normal and the Pathological*, 228.

73. David Snowdon, *Aging with Grace: What the Nun Study Teaches Us about Leading Longer, Healthier and More Meaningful Lives* (New York: Bantam Books, 2001).

74. R. H. Swartz, S. E. Black, and P. St George-Hyslop, "Apolipoprotein E and Alzheimer's Disease: A Genetic, Molecular and Neuroimaging Review," *Canadian Journal of Neurological Sciences* 26, no. 2 (1999): 77–88.

75. David Snowdon, "Aging and Alzheimer Disease: Lessons from the Nun Study," *The Gerontologist* 35 (1997): 150–56, 150.

76. George M. Savva, Stephen B. Wharton, Paul G. Ince, Gillian Forster, Fiona E. Matthews, and Carol Brayne, "Age, Neuropathology, and Dementia," *New England Journal of Medicine* 360, no. 22 (2009): 2302–9, 2306.

77. Ibid., 2308.

78. Carol Brayne and Paul Calloway, "Normal Ageing, Impaired Cognitive Function, and Senile Dementia of the Alzheimer's Type: A Continuum?," *The Lancet* 331 (1988): 1265–67.

79. Julie A. Schneider, Zoe Arvanitakis, Sue E. Leurgans, and David A. Bennett, "The Neuropathology of Probable Alzheimer Disease and Mild Cognitive Impairment," *Annals of Neurology* 66, no. 2 (2009): 200–208; Eric E. Smith and Steven M. Greenberg, "Beta-Amyloid, Blood Vessels and Brain Function," *Stroke* 40, no. 7 (2009): 2601–6; Neuropathology Group of the Medical Research Council Cognitive Function and Ageing Study, "Pathological Correlates of

Late-Onset Dementia in a Multicentre, Community-Based Population in England and Wales," *The Lancet* 357 (2001): 169–75.

80. Howard A. Crystal, Dennis Dickson, Peter Davies, David Masur, Ellen Grober, and Richard B. Lipton, "The Relative Frequency of 'Dementia of Unknown Etiology' Increases with Age and Is Nearly 50% in Nonagenarians," *Archives of Neurology* 57, no. 5 (2000): 713–19.

81. Carol F. Lippa and John C. Morris, "Alzheimer Neuropathology in Nondemented Aging Keeping Mind over Matter," *Neurology* 66, no. 12 (2006): 1801–2.

82. Richards and Brayne, "What Do We Mean?"

83. Ibid., 865.

84. Ibid., 866.

85. Robert N. Butler, Richard A. Miller, Daniel Perry, Bruce A. Carnes, T. Franklin Williams, Christine Cassel, Jacob Brody, et al., "New Model of Health Promotion and Disease Prevention for the 21st Century," *British Medical Journal* 337 (2008): a399.

86. Tiago Moreira and Paolo Palladino, "Ageing between Gerontology and Biomedicine," *BioSocieties* 4, no. 4 (2009): 349–65, 351–52.

87. House of Lords, *Science and Technology Committee: 6th Report* (London: Parliament, 2006), http://www.publications.parliament.uk/pa/ld200506/ldselect/ldsctech/146/14603.htm.

88. Moreira and Palladino, "Ageing between Gerontology and Biomedicine," 351–52.

89. Tom Kirkwood, *Time of Our Lives: The Science of Human Aging* (Oxford: Oxford University Press, 1999), cited in House of Lords, *Science and Technology Committee*; see also Moreira and Palladino, "Ageing between Gerontology and Biomedicine," 363.

Chapter 2
Striving to Standardize Alzheimer Disease

1. Zaven Khachaturian, "Plundered Memories," *The Sciences* 37, no. 4 (1997): 20–23.

2. Charles E. Rosenberg, "The Tyranny of Diagnosis: Specific Entities and Individual Experience," *Milbank Quarterly* 80, no. 2 (2002): 237–60, 237.

3. Ibid., 240.

4. See also Marc Berg, *Rationalizing Medical Work: Decision-Support Techniques and Medical Practices* (Cambridge, Mass.: MIT Press, 1997); N. A. Christakis, "The Ellipsis of Prognosis in Modern Medical Thought," *Social Science & Medicine* 44, no. 3 (1997): 301–15; Lock and Nguyen, *Anthropology of Biomedicine*; Harry M. Marks, *The Progress of Experiment: Science and Therapeutic Reform in the United States, 1900–1990* (Cambridge: Cambridge University Press, 1997).

5. Aimee Pasqua Borazanci, Meghan K. Harris, Robert N. Schwendimann, Eduardo Gonzalez-Toledo, Amir H. Maghzi, Masoud Etemadifar, Nadejda Alekseeva, James Pinkston, Roger E. Kelley, and Alireza Minagar, "Multiple Sclerosis: Clinical Features, Pathophysiology, Neuroimaging and Future Therapies," *Future Neurology* 4, no. 2 (2009): 229–46.

6. In medicine the concept of sensitivity measures the percentage of sick people correctly identified as having the condition in question, whereas specificity measures the proportion of negatives correctly identified, that is, the percentage of healthy people who are correctly identified as not having the condition.

7. The neuropathologist in charge of the brain bank located at one of the teaching hospitals of McGill University informed me that this is the case in Montréal.

8. See also Siddhartha Mukherjee, *The Emperor of All Maladies: A Biography of Cancer* (New York: Scribner, 2010).

9. Canadian Study of Health and Aging Working Group, "The Incidence of Dementia in Canada," *Neurology* 55, no. 1 (2000): 66–73.

10. Kathleen M. Hayden, Peter P. Zandi, Constantine G. Lyketsos, Ara S. Khachaturian, Lori A. Bastian, Gene Charoonruk, JoAnn T. Tschanz, et al., "Vascular Risk Factors for Incident Alzheimer Disease and Vascular Dementia: The Cache County Study," *Alzheimer Disease &*

Associated Disorders 20, no. 2 (2006): 93–100; David A. Bennett, Philip L. De Jager, Sue E. Leurgans, and Julie A. Schneider, "Neuropathologic Intermediate Phenotypes Enhance Association to Alzheimer Susceptibility Alleles," *Neurology* 72, no. 17 (2009): 1495–1503; Neuropathology Group of the Medical Research Council Cognitive Function and Ageing Study, "Pathological Correlates."

11. See Carol Brayne, "Research and Alzheimer's Disease: An Epidemiological Perspective," *Psychological Medicine* 23, no. 2 (1993): 287–96 and Carol Brayne and Daniel Davis, "Making Alzheimer's and Dementia Research Fit for Populations," *The Lancet* 380 (2012): 1441–43.

12. Zaven S. Khachaturian, "Diagnosis of Alzheimer's Disease," *Archives of Neurology* 42, no. 11 (1985): 1097–1105.

13. The Mini-Mental State Examination is the most widely used standardized instrument in the world designed to screen for what is known as "cognitive impairment." It was created in 1975 and has been translated into ten languages. It is described as "mini" because it was reduced in size from earlier tests and is not designed to assess mood or psychological states. The test, which is rarely administered in its full form and usually takes about ten minutes or so to complete, has been criticized by some as too long and others as too brief. An updated and revised version was made available in 2010. Psychological Assessment Resources has a copyright on the test, which can be bought today as an Android phone application. It is recommended not only for use by general practitioners, psychiatrists, and neurologists, but also for personal and home use. See, M. F. Folstein, S. E. Folstein, and P. R. McHugh, "'Mini-mental State': A Practical Method for Grading the Cognitive State of Patients for the Clinician," *Journal of Psychiatric Research* 12, no. 3 (1975): 189–98; E. L. Teng and H. C. Chui, "The Modified Mini-Mental State (3MS) Examination," *Journal of Clinical Psychiatry* 48, no. 8 (1987): 314–18.

14. T. Erkinjuntti, T. Ostbye, R. Steenhuis, and V. Hachinski, "The Effect of Different Diagnostic Criteria on the Prevalence of Dementia," *New England Journal of Medicine* 337, no. 23 (1997): 1667–74.

15. Janice E. Graham, Kenneth Rockwood, B. Lynn Beattie, Ian McDowell, Robin Eastwood, and Serge Gauthier, "Standardization of the Diagnosis of Dementia in the Canadian Study of Health and Aging," *Neuroepidemiology* 15, no. 5 (1996): 246–56; Z. Nagy, M. M. Esiri, K. A. Jobst, et al., "The Effects of Additional Pathology on the Cognitive Deficit in Alzheimer Disease," *Journal of Neuropathology & Experimental Neurology* 56 (1997): 163–70; K. Ritchie, D. Leibovici, B. Lessert, and J. Touchon, "A Typology of Sub-clinical Senescent Cognitive Disorder." *British Journal of Psychiatry* 168 (1966): 470–76; Gustavo C. Román and Donald R. Royall, "Executive Control Function: A Rational Basis for the Diagnosis of Vascular Dementia," *Alzheimer Disease & Associated Disorders* 13 (1999): S4–S8.

16. Graham et al., "Standardization of the Diagnosis."

17. See Danny George and Peter Whitehouse, "The Classification of Alzheimer Disease and Mild Cognitive Impairment," in Ballenger et al., *Treating Dementia*, 5–24.

18. Fadi Massoud, Gayatri Devi, Yaakov Stern, Arlene Lawton, James E. Goldman, Yan Liu, Steven S. Chin, and Richard Mayeux, "A Clinicopathological Comparison of Community-Based and Clinic-Based Cohorts of Patients with Dementia," *Archives of Neurology* 56, no. 11 (1999): 1368–73.

19. C. Duyckaerts, P. Delaère, J.-J. Hauw, A. L. Abbamondi-Pinto, S. Sorbi, I. Allen, J. P. Brion, et al., "Rating of the Lesions in Senile Dementia of the Alzheimer Type: Concordance between Laboratories. A European Multicenter Study under the Auspices of EURAGE," *Journal of the Neurological Sciences* 97, nos. 2–3 (1990): 295–323.

20. John V. Bowler, David G. Munoz, Harold Merskey, and Vladimir Hachinski, "Fallacies in the Pathological Confirmation of the Diagnosis of Alzheimer's Disease," *Journal of Neurology, Neurosurgery & Psychiatry* 64, no. 1 (1998): 18–24.

21. C. J. Gilleard, "Is Alzheimer's Disease Preventable? A Review of Two Decades of Epidemiological Research," *Aging and Mental Health* 4 (2000): 101–18, 102.

22. Ibid., 102; see also S. S. Erlich and R. L. Davis, "Alzheimer's Disease in the Very Aged," *Journal of Neuropathology & Experimental Neurology* 39, no. 3 (1980): 352–54; H. M. Wisniewski,

A. Rabe, W. Silverman, and W. Zigman, "Neuropathological Diagnosis of Alzheimer Disease: A Survey of Current Practices," *Alzheimer Disease & Associated Disorders* 2, no. 4 (1988): 396–414.

23. John C. Morris, "The Relationship of Plaques and Tangles to Alzheimer Disease Phenotype," in *Pathobiology of Alzheimer's Disease*, ed. Alison M. Goate and Frank Ashall (London: Academic Press, 1995), 193–218.

24. John C. S. Breitner, "Dementia—Epidemiological Considerations, Nomenclature, and a Tacit Consensus Definition," *Journal of Geriatric Psychiatry and Neurology* 19, no. 3 (2006): 129–36, 136.

25. Jesse F. Ballenger, "DSM-V: Continuing the Confusion about Aging, Alzheimer's and Dementia," *H-madness: History of Psychiatry*, March 19, 2010, http://historypsychiatry .com/2010/03/19/dsm-v-continuing-the-confusion-about-aging-alzheimer%e2%80%99s -and-dementia/.

26. Carol Brayne, "The Elephant in the Room—Healthy Brains in Later Life, Epidemiology and Public Health," *Nature Reviews Neuroscience* 8, no. 3 (2007): 233–39.

27. Catalina Amador-Ortiz, Wen-Lang Lin, Zeshan Ahmed, David Personett, Peter Davies, Ranjan Duara, Neill R. Graff-Radford, Michael L. Hutton, and Dennis W. Dickson, "TDP-43 Immunoreactivity in Hippocampal Sclerosis and Alzheimer's Disease," *Annals of Neurology* 61, no. 5 (2007): 435–45.

28. Khachaturian, "Plundered Memories," 21.

29. Brayne and Davis, "Making Alzheimer's and Dementia Research Fit," 1441.

30. Whitehouse and George, *Myth of Alzheimer's*.

31. John A. Hardy and Gerald A. Higgins, "Alzheimer's Disease: The Amyloid Cascade Hypothesis," *Science* 256 (1992): 184–85.

32. Ibid., 184.

33. Ibid., 184, emphasis added.

34. Ballenger et al., *Treating Dementia*; see also Whitehouse and George, *Myth of Alzheimer's*, esp. chap. 5, "Waiting for Godot."

35. George Perry, Akihiko Nunomura, Arun K. Raina, and Mark A. Smith, "Amyloid-β Junkies," *The Lancet* 355 (2000): 757.

36. John A. Hardy and Dennis J. Selkoe, "The Amyloid Hypothesis of Alzheimer's Disease: Progress and Problems on the Road to Therapeutics," *Science* 297 (2002): 353–56.

37. See, for example, Stephanie J. Soscia, James E. Kirby, Kevin J. Washicosky, Stephanie M. Tucker, Martin Ingelsson, Bradley Hyman, Mark A. Burton, et al., "The Alzheimer's Disease-Associated Amyloid β-Protein Is an Antimicrobial Peptide," *PLoS ONE* 5, no. 3 (2010): e9505.

38. Amritpal Mudher and Simon Lovestone, "Alzheimer's Disease—Do Tauists and Baptists Finally Shake Hands?," *Trends in Neurosciences* 25, no. 1 (2002): 22–26.

39. Ibid.

40. Ibid., 25.

41. Virginia M.-Y. Lee, "Tauists and Baptists United—Well Almost!," *Science* 293 (2001): 1446–47.

42. Benjamin Yang, "A United Disease Theory Brings Two Groups of Alzheimer's Disease Researchers Together," *Discovery Medicine*, May 10, 2009, http://www.discoverymedicine.com /Benjamin-Yang/2009/05/10/a-united-disease-theory-brings-two-groups-of-alzheimer -researchers-together/.

43. The Editors, "Alzheimer Disease," *Nature Medicine* 12, no. 7 (2006): 746–84.

44. Apoorva Mandavilli, "The Amyloid Code," *Nature Medicine* 12, no. 7 (2006): 747–51.

45. Ibid., 747.

46. Edith G. McGeer and Pat L. McGeer, "Neuroinflammation: Alzheimer Disease, and Other Aging Disorders," in *Pharmacological Mechanisms in Alzheimer's Therapeutics*, ed. A. C. Cuello (New York: Springer, 2007), 149–66.

47. C.P.J. Maury, "The Emerging Concept of Functional Amyloid," *Journal of Internal Medicine* 265, no. 3 (2009): 329–34.

48. Rudy J. Castellani, Hyoung-gon Lee, Sandra L. Siedlak, Akihiko Nunomura, Takaaki Hayashi, Masao Nakamura, Xiongwei Zhu, George Perry, and Mark A. Smith, "Reexamining Alzheimer's Disease: Evidence for a Protective Role for Amyloid-β Protein Precursor and Amyloid-β," *Journal of Alzheimer's Disease* 18, no. 2 (2009): 447–52.

49. Jesse F. Ballenger, "Necessary Interventions: Antidementia Drugs and Heightened Expectations for Aging in Modern American Cultures and Society," in Ballenger et al., *Treating Dementia*, 189–209, 199.

50. Ibid., 201.

51. Tiago Moreira, "Truth and Hope in Drug Development and Evaluation in Alzheimer Disease," in Ballenger et al., *Treating Dementia*, 210–30.

52. See, for example, Elaine K. Perry, Peter H. Gibson, Garry Blessed, Robert H. Perry, and Bernard E. Tomlinson, "Neurotransmitter Enzyme Abnormalities in Senile Dementia. Choline Acetyltransferase and Glutamic Acid Decarboxylase Activities in Necropsy Brain Tissue," *Journal of the Neurological Sciences* 34, no. 2 (1977): 247–65; Elaine K. Perry, Robert H. Perry, Peter H. Gibson, Garry Blessed, and Bernard E. Tomlinson, "A Cholinergic Connection between Normal Aging and Senile Dementia in the Human Hippocampus," *Neuroscience Letters* 6, no. 1 (1977): 85–89.

53. Peter J. Whitehouse, Donald L. Price, Arthur W. Clark, Joseph T. Coyle, and Mahlon R. DeLong, "Alzheimer Disease: Evidence for Selective Loss of Cholinergic Neurons in the Nucleus Basalis," *Annals of Neurology* 10, no. 2 (1981): 122–26.

54. Ballenger, "Necessary Interventions"; Moreira, "Truth and Hope"; Peter J. Whitehouse, "Can We Fix This with a Pill? Qualities of Life and the Aging Brain," in Ballenger et al., *Treating Dementia*, 168–82.

55. Alison Abbott, "The Plaque Plan," *Nature* 456 (2008): 161–64.

56. Robert Langreth, "Eli Lilly Alzheimer's Disease Failure Bolsters Amyloid Theory Skeptics—Forbes," *Forbes*, August 17, 2010, http://www.forbes.com/sites/robertlangreth/2010/08/17/eli-lilly-alzheimers-failure-bolsters-skeptics-on-amyloid-theory/.

57. Clive Holmes, Delphine Boche, David Wilkinson, Ghasem Yadegarfar, Vivienne Hopkins, Anthony Bayer, Roy W. Jones, et al., "Long-Term Effects of Aβ42 Immunisation in Alzheimer's Disease: Follow-Up of a Randomised, Placebo-Controlled Phase I Trial," *The Lancet* 372 (2008): 216–23, 222.

58. Ballenger, *Self, Senility, and Alzheimer's Disease*, 99.

59. See, for example, I. J. Bennett, E. J. Golob, E. S. Parker, and A. Starr, "Memory Evaluation in Mild Cognitive Impairment Using Recall and Recognition Tests," *Journal of Clinical and Experimental Neuropsychology* 28, no. 8 (2006): 1408–22.

60. Bruno Dubois, Howard H. Feldman, Claudia Jacova, Steven T. DeKosky, Pascale Barberger-Gateau, Jeffrey Cummings, André Delacourte, et al., "Research Criteria for the Diagnosis of Alzheimer's Disease: Revising the NINCDS-ADRDA Criteria," *The Lancet Neurology* 6, no. 8 (2007): 734–46.

61. Ibid., 734.

62. Ibid., 743.

63. Abbott, "The Plaque Plan."

64. Ibid., 163.

65. Ibid., 164.

66. John A. Hardy, "The Amyloid Cascade Hypothesis Has Misled the Pharmaceutical Industry," *Biochemical Society Transactions* 39, no. 4 (2011): 920–23.

Chapter 3
Paths to Alzheimer Prevention

1. John Bayley, *Iris: A Memoir of Iris Murdoch* (London: Abacus, 2002), 216–17.

2. Patricia Jasen, "Breast Cancer and the Language of Risk, 1750–1950," *Social History of Medicine* 15, no. 1 (2002): 17–43.

3. François Ewald, "Insurance and Risk," in *The Foucault Effect: Studies in Governmentality*, edited by Graham Burchell, Colin Gordon, and Peter Miller (Hemel Hempstead: Harvester Wheatsheaf, 1991), 197–210, see 208.

4. Ibid., 208.

5. Mary Douglas, "Risk as a Forensic Resource," *Daedalus* 119, no. 4 (1990): 1–16.

6. Michel Foucault, *The History of Sexuality*, vol. 1 (New York: Vintage Books, 1980).

7. Robert Castel, "From Dangerousness to Risk," in Burchell, Gordon, and Miller, *The Foucault Effect*, 281–98, see 289.

8. It is my understanding that the terms "corporeal" and "embodied" risk were first coined by Anne M. Kavanagh and Dorothy H. Broom, "Embodied Risk: My Body, Myself?," *Social Science & Medicine* 46, no. 3 (1998): 437–44.

9. Mark Schiffman, Nicolas Wentzensen, Sholom Wacholder, Walter Kinney, Julia C. Gage, and Philip E. Castle, "Human Papillomavirus Testing in the Prevention of Cervical Cancer," *Journal of the National Cancer Institute* 103, no. 5 (2011): 368–83; Marc B. Garnick, "The Great Prostate Cancer Debate," *Scientific American* 306, no. 2 (2012): 38–43.

10. See http://medical-dictionary.thefreedictionary.com/biomarker.

11. Of course, in connection with reproduction, insurance companies may seek to assign responsibility to individuals for passing on their so-called "defective" genes, and in the days when eugenics was practiced genes were loaded with moral valence. As we will see, especially in chapter 6, the more that is uncovered about gene structure and function in the postgenomic era, the more it becomes difficult to associate genes and human responsibility.

12. Pseudonyms were given to all interviewed patients and family members.

13. Also known as the Verdun Protestant Hospital, the Protestant Insane Asylum initially was a philanthropic institution set in farmland, which the patients tended. Today known as the Douglas Mental Health University Institute or, more informally, the Douglas, patients assist with hospital administration and care of its beautiful grounds.

14. Jeremia Heinik, "V. A. Kral and the Origins of Benign Senescent Forgetfulness and Mild Cognitive Impairment," *International Psychogeriatrics* 22, no. 3 (2010): 395–402, 398.

15. James Golomb, Alan Kluger, and Steven H. Ferris, "Mild Cognitive Impairment: Historical Development and Summary of Research," *Dialogues in Clinical Neuroscience* 6, no. 4 (2004): 351–67, 352.

16. Glenn E. Smith, Ronald C. Petersen, Joseph E. Parisi, Robert J. Ivnik, Emre Kokmen, Eric G. Tangalos, and Stephen Waring, "Definition, Course, and Outcome of Mild Cognitive Impairment," *Aging, Neuropsychiatry and Cognition* 3 (1996): 141–47.

17. Ronald C. Petersen, Rachelle Doody, Alexander Kurz, Richard C. Mohs, John C. Morris, Peter V. Rabins, Karen Ritchie, Martin Rossor, Leon Thal, and Bengt Winblad, "Current Concepts in Mild Cognitive Impairment," *Archives of Neurology* 58, no. 12 (2001): 1985–92, 1991.

18. Cited in Tiago Moreira, Carl May, and John Bond, "Regulatory Objectivity in Action Mild Cognitive Impairment and the Collective Production of Uncertainty," *Social Studies of Science* 39, no. 5 (2009): 665–90, 205 of the FDA report, www.fda.gov/ohrms/dockets/AC/01/transcripts/3724t1.pdf.

19. Moreira, May, and Bond, "Regulatory Objectivity," 684.

20. Ronald C. Petersen, ed., *Mild Cognitive Impairment: Aging to Alzheimer's Disease* (Oxford: Oxford University Press, 2003), 12.

21. Moreira, May, and Bond, "Regulatory Objectivity," 685.

22. Ibid., 672.

23. B. Winblad, K. Palmer, M. Kivipelto, V. Jelic, L. Fratiglioni, L.-O. Wahlund, A. Nordberg, et al., "Mild Cognitive Impairment—Beyond Controversies, towards a Consensus: Report of the International Working Group on Mild Cognitive Impairment," *Journal of Internal Medicine* 256, no. 3 (2004): 240–46.

24. Moreira, May, and Bond, "Regulatory Objectivity," 666. Here Moreira and colleagues follow the lead of Keating and Cambrosio, *Biomedical Platforms*.

25. Serge Gauthier, Barry Reisberg, Michael Zaudig, Ronald C. Petersen, Karen Ritchie, Karl Broich, Sylvie Belleville, et al., "Mild Cognitive Impairment," *Lancet* 367 (2006): 1262–70.

26. Ibid., 1268.

27. "Amnestic" means a partial or total loss of memory. Amnestic cognitive impairment is today the most commonly diagnosed form of MCI.

28. Ronald C. Petersen, "The Current Status of Mild Cognitive Impairment—What Do We Tell Our Patients?," *Nature Clinical Practice Neurology* 3, no. 2 (2007): 60–61, 60.

29. Ibid., 61.

30. Ronald C. Petersen, "Mild Cognitive Impairment Is Relevant," *Philosophy, Psychiatry, & Psychology* 13, no. 1 (2006): 45–49.

31. Peter J. Whitehouse, "Mild Cognitive Impairment—A Confused Concept?," *Nature Clinical Practice Neurology* 3, no. 2 (2007): 62–63, 63.

32. Ziad S. Nasreddine, Natalie A. Phillips, Valérie Bédirian, Simon Charbonneau, Victor Whitehead, Isabelle Collin, Jeffrey L. Cummings, and Howard Chertkow, "The Montreal Cognitive Assessment, MoCA: A Brief Screening Tool for Mild Cognitive Impairment," *Journal of the American Geriatric Society* 53, no. 4 (2005): 695–99.

33. Memory clinics are springing up in the United Kingdom, even though the specialty of old age psychiatry takes primary responsibility for most cases of early-onset dementia there. In France there are as many as 150 memory clinics staffed by psychiatrists. Approaches to MCI and the early stages of dementia vary considerably among countries depending in large part on which clinical specialty, neurology or psychiatry, takes primary responsibility.

34. This research underwent an ethics clearance at McGill University and was supported by the director of the memory clinic where the interviews were conducted. All interviewed patients signed an informed consent form before the interview commenced and were at liberty to terminate the interview at any time. As noted above, pseudonyms are used throughout when citing interviews.

35. It has been argued that MoCA is superior to the MMSE in testing for the "intermediate category" known as MCI, and that there are no other screening tools that reliably and quickly distinguish those with MCI from "normal control research subjects"; see Nasreddine et al., "Montreal Cognitive Assessment."

36. Erin Andersen, "Isolated by Affliction, Isolated by Language," *Globe and Mail*, December 18, 2010.

37. Gina Kolata, "Finding Suggests New Target for Alzheimer's Drugs," *New York Times*, September 1, 2010.

38. Gina Kolata, "New Scan May Spot Alzheimer's," *New York Times*, July 13, 2010.

39. Gina Kolata, "In Spinal-Fluid Test, an Early Warning on Alzheimer's," *New York Times*, August 9, 2010.

40. Arthur Brisbane, "The Trouble with Absolutes," *New York Times*, July 13, 2010, Public Editor's Journal sec., http://publiceditor.blogs.nytimes.com/2010/08/24/the-trouble-with-absolutes/.

41. Alliance for Human Research Protection, "NY Times Corrects Gina Kolata Re: Alzheimer's," September 16, 2010, http://www.ahrp.org/cms/content/view/726/1/. See also Health News Review, http://www.healthnewsreview.org/2012/06/historian-writes-on-medical-journalism-in-the-war-on-alzheimers, and Science Journalism Tracker, http://ksj.mit.edu/tracker/2011/01/ny-times-strangely-quiet-alzheimers-test.

42. Gina Kolata, "Guidelines Seek Early Detection of Alzheimer's," *New York Times*, July 14, 2010, http://query.nytimes.com/gst/fullpage.html.

43. Ibid.

44. Gina Kolata, "Drug Trials Test Bold Plan to Slow Alzheimer's," *New York Times*, July 16, 2010.

45. Gina Kolata, "Insights Give New Hope for New Attack on Alzheimer's," *New York Times*, December 13, 2010.

46. Quoted in Gina Kolata, "Early Tests for Alzheimer's Pose Diagnosis Dilemma," *New York Times*, December 17, 2010.

47. Gina Kolata, "F.D.A. Sees Promise in Alzheimer's Imaging Drug," *New York Times*, January 20, 2011.

48. Bruno Dubois, Howard H. Feldman, Claudia Jacova, Jeffrey L. Cummings, Steven T. Dekosky, Pascale Barberger-Gateau, André Delacourte, et al., "Revising the Definition of Alzheimer's Disease: A New Lexicon," *The Lancet Neurology* 9, no. 11 (2010): 1118–27, 1118.

49. Ibid., 1118.

50. Ibid., 1124–25.

51. Ibid., 1125.

Chapter 4
Embodied Risk Made Visible

1. John Berger, *Ways of Seeing* (London: Penguin, 1972), 7.

2. Sanjay W. Pimplikar, "Alzheimer's Isn't Up to the Tests," *New York Times*, July 19, 2010, Opinion sec., http://www.nytimes.com/2010/07/20/opinion/20pimplikar.html.

3. Garnick, "Great Prostate Cancer Debate."

4. Whitehouse and George, *Myth of Alzheimer's*, 4.

5. Bradley T. Hyman, Creighton H. Phelps, Thomas G. Beach, Eileen H. Bigio, Nigel J. Cairns, Maria C. Carrillo, Dennis W. Dickson, et al., "National Institute on Aging–Alzheimer's Association Guidelines for the Neuropathologic Assessment of Alzheimer's Disease," *Alzheimer's & Dementia* 8, no. 1 (2012): 1–13.

6. See Clifford R. Jack, Jr., Marilyn S. Albert, David S. Knopman, Guy M. McKhann, Reisa A. Sperling, Maria C. Carrillo, Bill Thies, and Creighton H. Phelps, "Introduction to the Recommendations from the National Institute on Aging-Alzheimer's Association Workgroups on Diagnostic Guidelines for Alzheimer's Disease," *Alzheimer's & Dementia* 7, no. 3 (2011): 257–62, 258.

7. See ibid., 259.

8. For the guidelines in connection with the new neuropathologic assessment of Alzheimer disease, see Hyman et al., "National Institute on Aging."

9. See Jack et al., "Introduction to the Recommendations."

10. See ibid.

11. Harald Hampel, Simone Lista, and Zaven S. Khachaturian, "Development of Biomarkers to Chart All Alzheimer's Disease Stages: The Royal Road to Cutting the Therapeutic Gordian Knot," *Alzheimer's & Dementia* 8, no. 4 (2012): 312–36.

12. Thomas Kuhn, *The Structure of Scientific Revolutions* (Chicago: University of Chicago Press, 1962).

13. Ibid. See 50th Anniversary ed. (Chicago: University of Chicago Press, 2012), 91.

14. Ian Hacking, "Introduction," in Kuhn, *The Structure of Scientific Revolutions*, 50th Anniversary ed., vii–xxxvii.

15. Reisa A. Sperling, Paul S. Aisen, Laurel A. Beckett, David A. Bennett, Suzanne Craft, Anne M. Fagan, Takeshi Iwatsubo, et al., "Toward Defining the Preclinical Stages of Alzheimer's Disease: Recommendations from the National Institute on Aging-Alzheimer's Association Workgroups on Diagnostic Guidelines for Alzheimer's Disease," *Alzheimer's & Dementia* 7, no. 3 (2011): 280–92, 282.

16. Ibid., 282.

17. Ibid., 282.

18. Ibid., 283.

19. Ibid., 284.

20. Ibid., 287.

21. John C. Morris, "Revised Criteria for Mild Cognitive Impairment May Compromise the Diagnosis of Alzheimer Disease Dementia," *Archives of Neurology* 69, no. 6 (2012): 700–708, 700.

22. Ibid., 700.

23. Ibid., 705.

24. Ibid., 706.

25. Leonard F. M. Scinto and Kirk R. Daffner, eds., *The Early Diagnosis of Alzheimer's Disease* (Totowa, N.J.: Humana Press, 2000).

26. A. Zara Herskovits and John H. Growdon, "Sharpen That Needle," *Archives of Neurology* 67, no. 8 (2010): 918–20, 918.

27. Geert De Meyer, Fred Shapiro, Hugo Vanderstichele, Eugeen Vanmechelen, Sebastiaan Engelborghs, Peter Paul De Deyn, Els Coart, et al., "Diagnosis-Independent Alzheimer Disease Biomarker Signature in Cognitively Normal Elderly People," *Archives of Neurology* 67, no. 8 (2010): 949–56, 949.

28. Ibid., 954, emphasis added.

29. Haiqun Lin, Charles E. McCulloch, Bruce W. Turnbull, Elizabeth H. Slate, and Larry C. Clark, "A Latent Class Mixed Model for Analysing Biomarker Trajectories with Irregularly Scheduled Observations," *Statistics in Medicine* 19, no. 10 (2000): 1303–18.

30. William E. Klunk, Henry Engler, Agneta Nordberg, Yanming Wang, Gunnar Blomqvist, Daniel P. Holt, Mats Bergström, et al., "Imaging Brain Amyloid in Alzheimer's Disease with Pittsburgh Compound-B," *Annals of Neurology* 55, no. 3 (2004): 306–19, 317.

31. G. G. Glenner, "Alzheimer's Disease: The Commonest Form of Amyloidosis," *Archives of Pathology & Laboratory Medicine* 107, no. 6 (1983): 281–82.

32. Kerryn E. Pike, Greg Savage, Victor L. Villemagne, Steven Ng, Simon A. Moss, Paul Maruff, Chester A. Mathis, William E. Klunk, Colin L. Masters, and Christopher C. Rowe, "B-amyloid Imaging and Memory in Non-demented Individuals: Evidence for Preclinical Alzheimer's Disease," *Brain* 130, no. 11 (2007): 2837–44.

33. Howard Jay Aizenstein, Robert D. Nebes, Judith A. Saxton, Julie C. Price, Chester A. Mathis, Nicholas D. Tsopelas, Scott K. Ziolko, et al., "Frequent Amyloid Deposition without Significant Cognitive Impairment among the Elderly," *Archives of Neurology* 65, no. 11 (2008): 1509–17.

34. Kerim Munir, Suzanne Coulter, John H. Growdon, Ann MacDonald, Patrick J. Skerrett, and Jane A. Leopold, "How to Solve Three Puzzles," *Newsweek* 151, no. 3 (January 21, 2008): 64–66, 65.

35. M. A. Mintun, G. N. Larossa, Y. I. Sheline, C. S. Dence, S. Y. Lee, R. H. Mach, W. E. Klunk, C. A. Mathis, S. T. DeKosky, and J. C. Morris, "[11C]PIB in a Nondemented Population: Potential Antecedent Marker of Alzheimer Disease," *Neurology* 67, no. 3 (2006): 446–52; Clifford R. Jack, Val J. Lowe, Stephen D. Weigand, Heather J. Wiste, Matthew L. Senjem, David S. Knopman, Maria M. Shiung, et al., "Serial PIB and MRI in Normal, Mild Cognitive Impairment and Alzheimer's Disease: Implications for Sequence of Pathological Events in Alzheimer's Disease," *Brain* 132, no. 5 (2009): 1355–65, 1363.

36. Sperling, "Toward Defining the Preclinical Stages of Alzheimer's Disease," 282.

37. Jack et al., "Serial PIB and MRI," 1363.

38. Ibid., 1363.

39. C. M. Clark and J. A. Schneider, "Use of Florbetapir-PET for Imaging B-amyloid Pathology," *Journal of the American Medical Association* 305, no. 3 (2011): 275–83.

40. Ibid., 280.

41. Monique Breteler, "Mapping Out Biomarkers for Alzheimer Disease," *Journal of the American Medical Association* 305, no. 3 (2011): 304–5, 304.

42. Ibid., 305.

43. Ibid., 305.

44. Michael Carome and Sidney Wolfe, "Florbetapir-PET Imaging and Postmortem B-amyloid Pathology," *Journal of the American Medical Association* 305, no. 18 (2011): 1857–58.

45. Joseph Dumit, "Critically Producing Brain Images of Mind," in *Critical Neuroscience: A Handbook of the Social and Cultural Contexts of Neuroscience*, ed. Suparna Choudhury and Jan Slaby (Chichester, UK: Wiley-Blackwell, 2012), 195–226, 222.

46. W. J. Jagust, S. M. Landau, L. M. Shaw, J. Q. Trojanowski, R. A. Koeppe, E. M. Reiman, N. L. Foster, et al., "Relationships between Biomarkers in Aging and Dementia," *Neurology* 73, no. 15 (2009): 1193–99.

47. Mark Mintun stated, "[M]ultiple papers now exist that show that subjects who appear cognitively normal but are shown to have amyloid in their brain by PET scanning, have statistically decreased cognitive or memory function compared to subjects who do have amyloid." However, he added, these studies are cross-sectional and do not predict the future. Mintun regarded a paper by

Storandt et al., in which it was noted that the PIB-positive group had been going downward on certain longitudinal cognitive measures, as highly relevant. The only caveat, Mintun added, "was that most of the longitudinal data was done prior to the PET PIB amyloid scan (as part of these subject's long-term commitment to the longitudinal research studies). And so it is not a real prospective study." See Martha Storandt, Mark A. Mintun, Denise Head, and John C. Morris, "Cognitive Decline and Brain Volume Loss Are Signatures of Cerebral Aβ Deposition Identified with PIB," *Archives of Neurology* 66, no. 12 (2009): 1476–81. A second source noted by Mintun is by Morris et al., in which it was shown that cognitively normal elderly individuals do "advance" to dementia at a higher rate if they are positive on PET PIB amyloid scans. However, the caveat with this study, Mintun stated, "is that the follow-up was very short and the number of people who actually became cognitively impaired was very small." See John C. Morris, Catherine M. Roe, Elizabeth A. Grant, Denise Head, Martha Storandt, Alison M. Goate, Anne M. Fagan, David M. Holtzman, and Mark A. Mintun, "Pittsburgh Compound B Imaging and Prediction of Progression from Cognitive Normality to Symptomatic Alzheimer Disease," *Archives of Neurology* 66, no. 12 (2009): 1469–75.

48. F. C. Redlich, "The Concept of Health in Psychiatry," in *Explorations in Social Psychiatry*, ed. Alexander H. Leighton, J. N. Clausen, and R. N. Wilson (New York: Basic Books, 1957), 138–64.

49. Zaven Khachaturian, "Perspective on the Alzheimer's Disease Neuroimaging Initiative: Progress Report and Future Plans," *Alzheimer's & Dementia* 6, no. 3 (2010): 199–201, 201.

50. Zaven S. Khachaturian and Ara S. Khachaturian, "Prevent Alzheimer's Disease by 2020: A National Strategic Goal," *Alzheimer's & Dementia* 5, no. 2 (2009): 81–84, 82.

51. Ibid., 84.

52. Michael W. Weiner, Paul S. Aisen, Clifford R. Jack Jr., William J. Jagust, John Q. Trojanowski, Leslie Shaw, Andrew J. Saykin, et al., "The Alzheimer's Disease Neuroimaging Initiative: Progress Report and Future Plans," *Alzheimer's & Dementia* 6, no. 3 (2010): 202–11.

53. Ibid., 204.

54. Ibid., 204.

55. Ibid., 209.

56. Gina Kolata, "Rare Sharing of Data Led to Results on Alzheimer's," *New York Times*, August 12, 2010, Health/Research sec., http://www.nytimes.com/2010/08/13/health/research/13alzheimer.html.

57. Weiner et al., "Alzheimer's Disease Neuroimaging Initiative," 209.

58. Nikolas Rose, *The Politics of Life Itself: Biomedicine, Power, and Subjectivity in the Twenty-First Century* (Princeton: Princeton University Press, 2006), 192.

59. Simon Cohn, "Disrupting Images: Neuroscientific Representations in the Lives of Psychiatric Patients," in Choudhury and Slaby, *Critical Neuroscience*, 179–94.

60. Julie A. Schneider, Zoe Arvanitakis, Woojeong Bang, and David A. Bennett, "Mixed Brain Pathologies Account for Most Dementia Cases in Community-Dwelling Older Persons," *Neurology* 69, no. 24 (2007): 2197–2204; Carol Brayne, Blossom C. M. Stephan, and Fiona E. Matthews, "A European Perspective on Population Studies of Dementia," *Alzheimer's & Dementia* 7, no. 1 (2011): 3–9; Savva et al., "Age, Neuropathology, and Dementia."

61. Khachaturian, "Plundered Memories," 22.

62. Stephen Lunn, "End of Alzheimer's Curse 'a Decade Away,'" *The Australian*, September 19, 2001.

Chapter 5
Alzheimer Genes: Biomarkers of Prediction and Prevention

1. Michael Ignatieff, *Scar Tissue* (London: Chatto & Windus, 1993), 50.

2. Khachaturian, "Plundered Memories," 21.

3. An autosomal dominant mutation inherited from one parent alone is sufficient to cause the condition in question. If one parent carries the mutation, there is, therefore, a 50 percent chance that the offspring will inherit the mutation.

4. A. Goate, M. C. Chartier-Harlin, M. Mullan, J. Brown, F. Crawford, L. Fidani, L. Giuffra, A. Haynes, N. Irving, and L. James, "Segregation of a Missense Mutation in the Amyloid Precursor Protein Gene with Familial Alzheimer's Disease," *Nature* 349 (1991): 704–6.

5. "Alzheimer's Disease: Amyloid Precursor Protein—Good, Bad or Both?," *ScienceDaily*, December 29, 2009, http://www.sciencedaily.com/releases/2009/10/091018171806.htm.

6. Andrea Tedde, Benedetta Nacmias, Monica Ciantelli, Paolo Forleo, Elena Cellini, Silvia Bagnoli, Carolina Piccini, Paolo Caffarra, Enrico Ghidoni, Marco Paganini, Laura Bracco, and Sandro Sorbi, "Identification of New Presenilin Gene Mutations in Early-Onset Familial Alzheimer Disease," *Archives of Neurology* 60, no. 11 (2003): 1541–44.

7. L. Tilley, K. Morgan, and N. Kalsheker, "Genetic Risk Factors in Alzheimer's Disease," *Molecular Pathology* 51, no. 6 (1998): 293–304; I. Raiha, J. Kaprio, M. Koskenvuo, T. Rajala, and L. Sourander, "Alzheimer's Disease in Finnish Twins," *Lancet* 347 (1996): 573–78; L. E. Nee, R. Eldridge, T. Sunderland, C. B. Thomas, D. Katz, K. E. Thompson, H. Weingartner, H. Weiss, C. Julian, and R. Cohen, "Dementia of the Alzheimer Type: Clinical and Family Study of 22 Twin Pairs," *Neurology* 37, no. 3 (1987): 359–63.

8. Carlos Cruchaga, Sumitra Chakraverty, Kevin Mayo, Francesco L. M. Vallania, Robi D. Mitra, Kelley Faber, Jennifer Williamson, Tom Bird, Ramon Diaz-Arrastia, Tatiana M. Foroud, Bradley F. Boeve, Neill R. Graff-Radford, Pamela St. Jean, Michael Lawson, Margaret G. Ehm, Richard Mayeux, and Alison M. Goate, "Rare Variants in APP, PSEN1 and PSEN2 Increase Risk for AD in Late-Onset Alzheimer's Disease Families," *PLoS ONE* 7, no. 2 (2012): e31039.

9. Bradley T. Hyman, "Alzheimer's Disease or Alzheimer's Diseases? Clues from Molecular Epidemiology," *Annals of Neurology* 40, no. 2 (1996): 135–36.

10. Ibid., 136.

11. Gabriel García Márquez, *One Hundred Years of Solitude* (New York: HarperPerennial, 1991), 47.

12. "The Colombian Alzheimer's Family Testing Possible Cures," *BBC News*, May 19, 2011, Latin America & Caribbean sec., http://www.bbc.co.uk/news/world-latin-america-13428265.

13. Pam Belluck, "Alzheimer's Stalks a Columbian Family," *New York Times*, June 1, 2010, Health sec., http://www.nytimes.com/2010/06/02/health/02alzheimers.html.

14. "Colombian Alzheimer's Family Testing."

15. Belluck, "Alzheimer's Stalks"; see also Pam Belluck and Salvador Rodriguez, "Hoping to Crack Alzheimer's, Together as a Family," *New York Times*, October 3, 2011, Health sec., http://www.nytimes.com/2011/10/04/health/04alzheimers.html.

16. Michele G. Sullivan, "Studies Take Aim at Groups at High Risk for Alzheimer's," *WorldCare Clinical*, March 21, 2011.

17. Julie Steenhuysen, "Roche Alzheimer's Drug Picked for Major Test," *Reuters*, May 15, 2012, http://www.reuters.com/article/2012/05/15/us-alzheimers-genentech-idUSBRE84E0UJ20120515.

18. Ibid.

19. Pam Belluck, "New Drug Trial Seeks to Stop Alzheimer's Before It Starts," *New York Times*, May 15, 2012, Health/Research sec., http://www.nytimes.com/2012/05/16/health/research/prevention-is-goal-of-alzheimers-drug-trial.html.

20. Office of Public Affairs, University of California, Santa Barbara, "Clinical Trials for Alzheimer's Disease Preventative Drug to Begin Early 2013," news release, May 21, 2012, http://www.ia.ucsb.edu/pa/display.aspx?pkey=2734.

21. Steenhuysen, "Roche Alzheimer's Drug."

22. Alex John London and Jonathan Kimmelman, "Justice in Translation: From Bench to Bedside in the Developing World," *The Lancet* 372 (2008): 82–85.

23. Margaret Lock, "Interrogating the Human Genome Diversity Project," *Social Science & Medicine* 39, no. 5 (1994): 603–6; Margaret Lock, "The Alienation of Body Tissue and the Biopolitics of Immortalized Cell Lines," *Body & Society* 7, nos. 2–3 (2001): 63–91; Adriana Petryna, "Clinical Trials Offshored: On Private Sector Science and Public Health," *BioSocieties* 2, no. 1 (March 2007): 21–40; Philip Mirowski and Robert Van Horn, "The Contract Research

Organization and the Commercialization of Scientific Research," *Social Studies of Science* 35, no. 4 (2005): 503–48; Prasanna Kumar Patra and Margaret Sleeboom-Faulkner, "Bionetworking: Experimental Stem Cell Therapy and Patient Recruitment in India," *Anthropology & Medicine* 16, no. 2 (2009): 147–63.

24. Natalia Acosta-Baena, Diego Sepulveda-Falla, Carlos Mario Lopera-Gómez, Mario César Jaramillo-Elorza, Sonia Moreno, Daniel Camilo Aguirre-Acevedo, Amanda Saldarriaga, and Francisco Lopera, "Pre-dementia Clinical Stages in Presenilin 1 E280A Familial Early-Onset Alzheimer's Disease: A Retrospective Cohort Study," *The Lancet Neurology* 10, no. 3 (2011): 213–20.

25. John J. Mitchell, Annie Capua, Carol Clow, and Charles R. Scriver, "Twenty-Year Outcome Analysis of Genetic Screening Programs for Tay-Sachs and Beta-Thalassemia Disease Carriers in High Schools," *American Journal of Human Genetics* 59, no. 4 (1996): 793–98; Stefan Beck and Jörg Niewöhner, "Translating Genetic Testing and Screening in Cyprus and Germany: Contingencies, Continuities, Ordering Effects and Bio-Cultural Intimacy," in *The Handbook of Genetics & Society: Mapping the New Genomic Era*, ed. Paul Atkinson, Peter Glasner, and Margaret Lock (New York: Routledge, 2009), 76–93; Barbara Prainsack and Gil Siegal, "The Rise of Genetic Couplehood? A Comparative View of Premarital Genetic Testing," *BioSocieties* 1, no. 1 (2006): 17–36; J. Ekstein and H. Katzenstein, "The Dor Yeshorim Story: Community-Based Carrier Screening for Tay-Sachs Disease," *Advances in Genetics* 44 (2001): 297–310.

26. W. J. Strittmatter, D. Y. Huang, R. Bhasin, A. D. Roses, and D. Goldgaber, "Avid Binding of Beta A Amyloid Peptide to Its Own Precursor," *Experimental Neurology* 122, no. 2 (1993): 327–34; E. H. Corder, A. M. Saunders, W. J. Strittmatter, D. E. Schmechel, P. C. Gaskell, G. W. Small, A. D. Roses, J. L. Haines, and M. A. Pericak-Vance, "Gene Dose of Apolipoprotein E Type 4 Allele and the Risk of Alzheimer's Disease in Late Onset Families," *Science* 261 (1993): 921–23; A. M. Saunders, W. J. Strittmatter, D. Schmechel, P. H. George-Hyslop, M. A. Pericak-Vance, S. H. Joo, B. L. Rosi, J. F. Gusella, D. R. Crapper-MacLachlan, and M. J. Alberts, "Association of Apolipoprotein E Allele Epsilon 4 with Late-Onset Familial and Sporadic Alzheimer's Disease," *Neurology* 43, no. 8 (1993): 1467–72.

27. Allen D. Roses, "Apolipoprotein E and Alzheimer's Disease: A Rapidly Expanding Field with Medical and Epidemiological Consequences," *Annals of the New York Academy of Sciences* 802, no. 1 (1996): 50–57; Ronald C. Petersen, Stephen C. Waring, Glenn E. Smith, Eric G. Tangalos, and Stephen N. Thibodeau, "Predictive Value of APOE Genotyping in Incipient Alzheimer's Disease," *Annals of the New York Academy of Sciences* 802, no. 1 (1996): 58–69.

28. R. M. Corbo and R. Scacchi, "Apolipoprotein E (APOE) Allele Distribution in the World: Is APOEε4 a 'Thrifty' Allele?" *Annals of Human Genetics* 63 (1999): 301–10.

29. Dennis J. Selkoe, "The Pathophysiology of Alzheimer's Disease," in Scinto and Daffner, *The Early Diagnosis of Alzheimer's Disease*, 83–104.

30. Corder et al., "Gene Dose of Apolipoprotein E Type 4."

31. Tilley, Morgan, and Kalsheker, "Genetic Risk Factors."

32. John A. Hardy, "ApoE, Amyloid, and Alzheimer's Disease," *Science* 263 (1994): 454–55."

33. M. I. Kamboh, "Apolipoprotein E Polymorphism and Susceptibility to Alzheimer's Disease," *Human Biology* 67, no. 2 (1995): 195–215.

34. Lars Bertram and Rudolph E. Tanzi, "Alzheimer's Disease: One Disorder, Too Many Genes?," *Human Molecular Genetics* 13, no. 90001 (January 13, 2004): R135–R141, R135.

35. Ibid., R137.

36. Graham et al., "Standardization of the Diagnosis."

37. M. B. Liddell, S. Lovestone, and M. J. Owen, "Genetic Risk of Alzheimer Disease: Advising Relatives," *British Journal of Psychiatry* 178 (2001): 7–11; K. Ritchie and A. M. Dupuy, "The Current Status of APOε4 as a Risk Factor for Alzheimer's Disease: An Epidemiological Perspective," *International Journal of Geriatric Psychiatry* 14, no. 9 (1999): 695–700.

38. The term "heterozygous" refers to the case where a person carries only one APOE ε4 allele (along with either an APOE ε2 or 3). Someone who is homozygous for APOE ε4 has two of these alleles.

39. Martin R. Farlow, "Alzheimer's Disease: Clinical Implications of the Apolipoprotein E Genotype," *Neurology* 48, no. 5, suppl. 6 (1997): S30–S34.

40. Clive Holmes, "The Genetics of Alzheimer's Disease," *Menopause International* 8, no. 1 (2002): 20–23; Farlow, "Alzheimer's Disease."

41. Holmes, "Genetics of Alzheimer's Disease"; Swartz, Black, and St George-Hyslop, "Apolipoprotein E and Alzheimer's Disease."

42. Holmes, "Genetics of Alzheimer's Disease."

43. Deborah Blacker and Rudolph E. Tanzi, "Genetic Testing in the Early Diagnosis of Alzheimer Disease," in Scinto and Daffner, *The Early Diagnosis of Alzheimer's Disease*, 105–26.

44. See John H. Growdon, "Apolipoprotein E and Alzheimer Disease," *Archives of Neurology* 55, no. 8 (1998): 1053–54, for a summary of these early findings.

45. Alan R. Templeton, "The Complexity of the Genotype-Phenotype Relationship and the Limitations of Using Genetic 'Markers' at the Individual Level," *Science in Context* 11, nos. 3–4 (1998): 373–89, 376.

46. G. W. Small, S. Komo, A. La Rue, S. Saxena, M. E. Phelps, J. C. Mazziotta, A. M. Saunders, J. L. Haines, M. A. Pericak-Vance, and A. D. Roses, "Early Detection of Alzheimer's Disease by Combining Apolipoprotein E and Neuroimaging," *Annals of the New York Academy of Sciences* 802 (1996): 70–78, 76.

47. Roses, "Apolipoprotein E and Alzheimer's Disease: A Rapidly Expanding Field."

48. Allen D. Roses, "Apolipoprotein E and Alzheimer's Disease: The Tip of the Susceptibility Iceberg," *Annals of the New York Academy of Sciences* 855, no. 1 (1998): 738–43.

49. The ε2 allele has a cystine at positions 112 and 158 in the receptor-binding region of APOE. The ε3 allele is cystine-112 and argenine-158. The APOEε4 allele has argenine at both positions. See Nader Ghebranious, Lynn Ivacic, Jamie Mallum, and Charles Dokken, "Detection of ApoE E2, E3 and E4 Alleles Using MALDI-TOF Mass Spectrometry and the Homogeneous Mass-Extend Technology," *Nucleic Acids Research* 33, no. 17 (2005): e149.

50. Daniel Glass and Steven E. Arnold, "Some Evolutionary Perspectives on Alzheimer's Disease Pathogenesis and Pathology," *Alzheimer's and Dementia* 8, no. 4 (2012): 343–50.

51. C. E. Finch and R. M. Sapolsky, "The Evolution of Alzheimer Disease, the Reproductive Schedule, and apoE Isoforms," *Neurobiology of Aging* 20, no. 4 (1999): 407–28.

52. Ibid.

53. M. Gearing, G. W. Rebeck, B. T. Hyman, J. Tigges, and S. S. Mirra, "Neuropathology and Apolipoprotein E Profile of Aged Chimpanzees: Implications for Alzheimer Disease," *Neurobiology, Proceedings of the National Academy of Sciences of the United States of America* 91, no. 20 (1994): 9382–86. Dementia in mammals other than humans is exceptionally hard to investigate in the wild, especially given that most animals die before reaching old age.

54. R. W. Mahley, "Apolipoprotein E: Cholesterol Transport Protein with Expanding Role in Cell Biology," *Science* 240 (1988): 622–30.

55. Matthew C. Keller and Geoffrey Miller, "Resolving the Paradox of Common, Harmful, Heritable Mental Disorders: Which Evolutionary Genetic Models Work Best?," *Behavioral Brain Science* 29, no. 4 (2006): 385–452.

56. Yin C. Paradies, Michael J. Montoya, and Stephanie M. Fullerton, "Racialized Genetics and the Study of Complex Diseases: The Thrifty Genotype Revisited," *Perspectives in Biology and Medicine* 50, no. 2 (2007): 203–27.

57. Corbo and Scacchi, "Apolipoprotein E (APOE) Allele Distribution."

58. Stephanie M. Fullerton, Andrew G. Clark, Kenneth M. Weiss, Deborah A. Nickerson, Scott L. Taylor, Jari H. Stengård, Veikko Salomaa, Erkki Vartiainen, Markus Perola, Eric Boerwinkle, and Charles F. Sing, "Apolipoprotein E Variation at the Sequence Haplotype Level: Implications for the Origin and Maintenance of a Major Human Polymorphism," *American Journal of Human Genetics* 67, no. 4 (2000): 881–900.

59. Hugh Hendrie, "Diagnosis of Dementia and Alzheimer's Disease in Indianapolis and Ibadan: Challenges in Cross-Cultural Studies of Aging and Dementia," *Alzheimer's & Dementia*

5, no. 4 (2009): P122; Oye Gureje, Adesola Ogunniyi, and Lola Kola, "The Profile and Impact of Probable Dementia in a Sub-Saharan African Community: Results from the Ibadan Study of Aging," *Journal of Psychosomatic Research* 61, no. 3 (2006): 327–33. See also L. A. Farrer, "Familial Risk for Alzheimer Disease in Ethnic Minorities: Nondiscriminating Genes," *Archives of Neurology* 57, no. 1 (2000): 28–29.

60. Neill R. Graff-Radford, Robert C. Green, Rodney C. P. Go, Michael L. Hutton, Timi Edeki, David Bachman, Jennifer L. Adamson, et al., "Association between Apolipoprotein E Genotype and Alzheimer Disease in African American Subjects," *Archives of Neurology* 59, no. 4 (2002): 594–600

61. L. A. Farrer, "Intercontinental Epidemiology of Alzheimer Disease: A Global Approach to Bad Gene Hunting," *Journal of the American Medical Association* 285, no. 6 (2001): 796–98.

62. The default network is a network of brain regions that are active when the individual is not focused on the outside world and the brain is at wakeful rest. See also Michael D. Greicius, Gaurav Srivastava, Allan L. Reiss, and Vinod Menon, "Default-Mode Network Activity Distinguishes Alzheimer's Disease from Healthy Aging: Evidence from Functional MRI," *Proceedings of the National Academy of Sciences of the United States of America* 101, no. 13 (2004): 4637–42.

63. R. W. Mahley, K. H. Weisgraber, and Y. Huang, "Apolipoprotein E4: A Causative Factor and Therapeutic Target in Neuropathology, Including Alzheimer Disease," *Proceedings of the National Academy of Sciences* 103, no. 15 (2006): 5644–51. The abstract of this article is as follows:

The premise of this review is that apolipoprotein (apo) E4 is much more than a contributing factor to neurodegeneration. ApoE has critical functions in redistributing lipids among CNS cells for normal lipid homeostasis, repairing injured neurons, maintaining synapto-dendritic connections, and scavenging toxins. In multiple pathways affecting neuropathology, including Alzheimer disease, APOE acts directly or in concert with age, head injury, oxidative stress, ischemia, inflammation, and excess amyloid beta peptide production to cause neurological disorders, accelerating progression, altering prognosis, or lowering age of onset. We envision that unique structural features of APOE4 are responsible for APOE4-associated neuropathology. Although the structures of APOE2, APOE3, and APOE4 are in dynamic equilibrium, APOE4, which is detrimental in a variety of neurological disorders, is more likely to assume a pathological conformation. Importantly, APOE4 displays domain interaction (an interaction between the N- and C-terminal domains of the protein that results in a compact structure) and molten globule formation (the formation of stable, reactive intermediates with potentially pathological activities). In response to CNS stress or injury, neurons can synthesize APOE. APOE4 uniquely undergoes neuron-specific proteolysis, resulting in bioactive toxic fragments that enter the cytosol, alter the cytoskeleton, disrupt mitochondrial energy balance, and cause cell death. Our findings suggest potential therapeutic strategies, including the use of "structure correctors" to convert APOE4 to an "APOE3-like" molecule, protease inhibitors to prevent the generation of toxic APOE4 fragments, and "mitochondrial protectors" to prevent cellular energy disruption.

64. John C. Morris, Catherine M. Roe, Chengjie Xiong, Anne M. Fagan, Alison M. Goate, David M. Holtzman, and Mark A. Mintun, "APOE Predicts Aβ but Not Tau Alzheimer's Pathology in Cognitively Normal Aging," *Annals of Neurology* 67, no. 1 (2010): 122–31, 127.

65. Joseph L. Price, Daniel W. McKeel Jr., Virginia D. Buckles, Catherine M. Roe, Chengjie Xiong, Michael Grundman, Lawrence A. Hansen, et al., "Neuropathology of Nondemented Aging: Presumptive Evidence for Preclinical Alzheimer Disease," *Neurobiology of Aging* 30, no. 7 (2009): 1026–36.

66. Morris et al., "APOE Predicts," 127.

67. Ibid., 128.

68. Eric Karran, Marc Mercken, and Bart De Strooper, "The Amyloid Cascade Hypothesis for Alzheimer's Disease: An Appraisal for the Development of Therapeutics," *Nature Reviews Drug Discovery* 10, no. 9 (2011): 698–712, 701, 705.

69. Ibid., 710.

Chapter 6
Genome-Wide Association Studies: Back to the Future

1. Anne V. Buchanan, Samuel Sholtis, Joan Richtsmeier, and Kenneth M. Weiss, "What Are Genes 'for' or Where Are Traits 'from'? What Is the Question?," *BioEssays* 31, no. 2 (2009): 198–208.

2. A polymorphism is a discrete genetic trait that exists in a population in at least two forms. That is, more than one allele exists at a gene locus in any given population. By convention, the frequency of the rarest of the alleles must be no less than 1 percent. The most common type of polymorphism exists as variation at a single base pair. Polymorphisms can also be much larger in size and involve long stretches of DNA. Natural selection, genetic drift, and/or gene flow—that is, human migration—account for allelic frequencies in populations. Mutations are rare and cannot account by themselves directly for changes in allele frequencies.

3. See http://learn.genetics.utah.edu/content/health/pharma/snips/ for further, readily comprehensible details about SNPs. SNPs are the most simple form and most common source of genetic polymorphism in the human genome, approximately 90 percent of all human DNA polymorphisms.

4. There is some argument among experts about the common disease/common variation hypothesis, but the AD geneticists to whom I spoke adhere to it as set out above, and their publications corroborate this position. This hypothesis has been very lucrative in snaring large amounts of money for GWAS research.

5. Lars Bertram and Rudolph E. Tanzi, "Genome-wide Association Studies in Alzheimer's Disease," *Human Molecular Genetics* 18, no. R2 (2009): 270–81.

6. Ibid.

7. Hui Shi, Christopher Medway, James Bullock, Kristelle Brown, Noor Kalsheker, and Kevin Morgan, "Analysis of Genome-Wide Association Study (GWAS) Data Looking for Replicating Signals in Alzheimer's Disease (AD)," *International Journal of Molecular Epidemiology and Genetics* 1, no. 1 (2009): 53–66, 53.

8. Gabrielle Strobel, "Paper Alert: GWAS Hits Clusterin, CR1, PICALM Formally Published," *Alzheimer Research Forum*, September 7, 2009, http://www.alzforum.org/new/detail.asp?id=2233.

9. Julie Williams, 2009, comment in Strobel, "Paper Alert."

10. http://www.news-medical.net/news/20090907/CLU-and-PICALM-genes-associated-with-Alzheimers-disease.aspx.

11. Adam C. Naj, Gyungah Jun, Gary W. Beecham, Li-San Wang, Badri Narayan Vardarajan, Jacqueline Buros, Paul J. Gallins, et al., "Common Variants at MS4A4/MS4A6E, CD2AP, CD33 and EPHA1 Are Associated with Late-Onset Alzheimer's Disease," *Nature Genetics* 43, no. 5 (2011): 436–41, 436.

12. Ibid., 443.

13. Walter Gilbert, "A Vision of the Grail," in *The Code of Codes: Scientific and Social Issues in the Human Genome Project*, ed. Daniel Kevles and Leroy Hood (Cambridge, Mass.: Harvard University Press, 1992), 83–97.

14. Gina Kolata, "Vast Gene Study Yields Insights on Alzheimer's," *New York Times*, April 3, 2011.

15. Unfortunately I did not ask Schellenberg what he thought about the research being planned for Colombian subjects discussed in chapter 5.

16. For an explanation of the statistical concept of *p* value, see http://en.wikipedia.org/wiki/P-value.

17. The odds ratio is a measure of effect size, describing the strength of association or nonindependence between two binary data values. It is used as a descriptive statistic and plays an important role in logistic regression. Unlike other measures of association for paired binary data such as the relative risk, the odds ratio treats the two variables being compared symmetrically and can be estimated using some types of nonrandom samples. See http://en.wikipedia.org/wiki/Odds_ratio.

18. "New Alzheimer's Genes Identified," *CBC News—Health*, April 3, 2011, http://www.cbc.ca/news/health/story/2011/04/01/alzheimer-genes-identified.html, emphasis added.

19. See http://www.youtube.com/watch?v=WKCa5Cv_fDg&feature=relmfu.

20. Meredith Wadman, "Fleshing Out the US Alzheimer's Strategy," *Nature News*, January 19, 2012, http://www.nature.com/news/fleshing-out-the-us-alzheimer-s-strategy-1.9856.

21. The heritability of a population is the proportion of observable differences between individuals that is due to genetic differences. Factors including genetics, environment, and random chance can all contribute to the variation between individuals in their phenotypic characteristics, and heritability analyzes the relative contributions of differences in genetic and nongenetic factors to the total phenotypic variance in a population. The term "missing heritability" is used when researchers are unable to demonstrate the assumed genetic contribution to a condition.

22. Teri A. Manolio, Francis S. Collins, Nancy J. Cox, David B. Goldstein, Lucia A. Hindorff, David J. Hunter, Mark I. McCarthy, et al., "Finding the Missing Heritability of Complex Diseases," *Nature* 461 (2009): 747–53, 747.

23. Ibid., 751; see also Wei Zheng, Jirong Long, Yu-Tang Gao, Chun Li, Ying Zheng, Yong-Bin Xiang, Wanqing Wen, et al., "Genome-wide Association Study Identifies a New Breast Cancer Susceptibility Locus at 6q25.1," *Nature Genetics* 41, no. 3 (2009): 324–28.

24. Samuel P. Dickson, Kai Wang, Ian Krantz, Hakon Hakonarson, and David B. Goldstein, "Rare Variants Create Synthetic Genome-Wide Associations," *PLoS Biology* 8, no. 1 (2010): e1000294.

25. Or Zuk, Eliana Hechter, Shamil R. Sunyaev, and Eric S. Lander, "The Mystery of Missing Heritability: Genetic Interactions Create Phantom Heritability," *Proceedings of the National Academy of Sciences*, December 5, 2011, http://www.pnas.org/content/early/2012/01/04/1119675109.

26. Lars Bertram, Christina M. Lill, and Rudolph E. Tanzi, "The Genetics of Alzheimer Disease: Back to the Future," *Neuron* 68, no. 2 (2010): 270–81.

27. I am indebted to Ken Weiss for this comment.

28. Cruchaga et al., "Rare Variants," emphasis original.

29. Khachaturian, "Plundered Memories," 21.

30. Alzheimer's Association, William Thies, and Laura Bleiler, "2011 Alzheimer's Disease Facts and Figures," *Alzheimer's & Dementia* 7, no. 2 (2011): 208–44, 235.

Chapter 7
Living with Embodied Omens

1. E. E. Evans-Pritchard, *Witchcraft, Oracles and Magic among the Azande* (Oxford: Clarendon, 1937).

2. Nadia C. Seremetakis, *The Last Word: Women, Death, and Divination in Inner Mani* (Chicago: University of Chicago Press, 1991).

3. Epigenomics, meaning literally over and above the genome, will be discussed in the following chapter.

4. Rose, *Politics of Life Itself*.

5. Edward J. Yoxen, "Constructing Genetic Diseases," in *The Problem of Medical Knowledge: Examining the Social Construction of Medicine*, ed. P. Wright and A. Treacher (Edinburgh: University of Edinburgh, 1982), 144–61, see 144.

6. See, for example, Whyte, *Questioning Misfortune*.

7. Yoxen, "Constructing Genetic Diseases."

8. Abby Lippman, "Led (Astray) by Genetic Maps: The Cartography of the Human Genome and Human Care," *Social Science & Medicine* 35, no. 12 (1992): 1469–96, 1470.

9. See, for example, Troy Duster, *Backdoor to Eugenics* (New York: Routledge, 1990); Troy Duster, "Buried Alive: The Concept of Race in Science," in Goodman, Heath, and Lindee, *Genetic Nature/Culture: Anthropology and Science beyond the Two Culture Divide*, ed. Alan H. Goodman, Deborah Heath, and M. Susan Lindee (Berkeley: University of California Press, 2003), 258–77; Keith Wailoo and Stephen Pemberton, *The Troubled Dream of Genetic Medicine: Ethnicity and Innovation in Tay-Sachs, Cystic Fibrosis, and Sickle Cell Disease* (Baltimore: Johns Hopkins University Press, 2006).

10. Rose, *Politics of Life Itself*, 125.

11. Sarah Franklin, "Life," in *The Encyclopedia of Bioethics*, ed. Warren T. Reich (New York: Simon & Schuster, 1995), 456–62.

12. Nina Hallowell, "Doing the Right Thing: Genetic Risk and Responsibility," *Sociology of Health & Illness* 21, no. 5 (1999): 597–621; Anne Kerr, Sarah Cunningham-Burley, and Amanda Amos, "The New Genetics and Health: Mobilizing Lay Expertise," *Public Understanding of Science* 7, no. 1 (1998): 41–60; Susan Michie, Harriet Drake, Theresa Marteau, and Martin Bobrow, "A Comparison of Public and Professionals' Attitudes towards Genetic Developments," *Public Understanding of Science* 4, no. 3 (1995): 243–53; Carlos Novas and Nikolas Rose, "Genetic Risk and the Birth of the Somatic Individual," *Economy and Society* 29, no. 4 (2000): 485–513.

13. Deborah Heath and Karen-Sue Taussig, "Genetic Citizenship," in *A Companion to the Anthropology of Politics*, ed. D. Nguyent and J. Vincent (London: Blackwell, 2004), 152–67; Rayna Rapp, "Cell Life and Death, Child Life and Death: Genomic Horizons, Genetic Diseases, Family Stories," in *Remaking Life and Death: Toward an Anthropology of the Biosciences*, ed. Sarah Franklin and Margaret Lock (Santa Fe, N.Mex.: School of American Research Press, 2004), 23–60.

14. C. M. Condit, "How the Public Understands Genetics: Non-Deterministic and Non-Discriminatory Interpretations of the 'Blueprint' Metaphor," *Public Understanding of Science* 8, no. 3 (1999): 169–80; Margaret Lock, Stephanie Lloyd, and Janalyn Prest, "Genetic Susceptibility and Alzheimer Disease: The Penetrance and Uptake of Genetic Knowledge," in Cohen and Leibing, *Thinking about Dementia*, 123–54.

15. S. Cox and W. McKellin, "'There's This Thing in Our Family': Predictive Testing and the Construction of Risk for Huntington Disease," in *Sociological Perspectives on the New Genetics*, ed. P. Conrad and J. Gabe (London: Blackwell, 1999), 121–48, 140.

16. Kerr, Cunningham-Burley, and Amos, "New Genetics and Health."

17. Diagnosing Huntington Disease, http://www.nhs.uk/Conditions/Huntingtons -disease/Pages/Diagnosis.aspx; Diane Beeson and Theresa Doksum, "Family Values and Resistance to Genetic Testing," in *Bioethics in Social Context*, ed. Barry Hoffmaster (Philadelphia: Temple University Press, 2001), 153–79; Kimberley A. Quaid and Michael Morris, "Reluctance to Undergo Predictive Testing: The Case of Huntington Disease," *American Journal of Medical Genetics* 45, no. 1 (1993): 41–45.

18. Shirley Hill, *Managing Sickle Cell Disease in Low-Income Families* (Philadelphia: Temple University Press, 1994); Rayna Rapp, *Testing Women, Testing the Fetus: The Social Impact of Amniocentesis* (New York: Routledge, 1999).

19. Kira A. Apse, Barbara B. Biesecker, Francis M. Giardiello, Barbara P. Fuller, and Barbara A. Bernhardt, "Perceptions of Genetic Discrimination among At-Risk Relatives of Colorectal Cancer Patients," *Genetics in Medicine* 6, no. 6 (2004): 510–16.

20. Monica Konrad, *Narrating the New Predictive Genetics: Ethics, Ethnography, and Science* (Cambridge: Cambridge University Press, 2005).

21. Alice Wexler, *Mapping Fate: A Memoir of Family Risk and Genetic Research* (Berkeley: University of California Press, 1995), 224.

22. Ibid., 238.

23. Hallowell, "Doing the Right Thing."

24. Ian Hacking, "The Looping Effects of Human Kinds," in *Causal Cognition: A Multidisciplinary Approach*, ed. D. Sperber, D. Premack, and A. J. Premack (Oxford: Oxford Medical Publications, 1995), 351–83.

25. Paul Rabinow, "Artificiality and Enlightenment: From Sociobiology to Biosociality," in *Essays on the Anthropology of Reason* (Princeton: Princeton University Press, 1996), 91–111.

26. Paul Rabinow, "Afterword: Concept Work," in *Biosocialities, Genetics and the Social Sciences: Making Biologies and Identities*, ed. Sarah Gibbon and Carlos Novas (London: Routledge, 2007), 188–92; see also Gibbon and Novas, *Biosocialities, Genetics and the Social Sciences*; Lock and Nguyen, *Anthropology of Biomedicine*.

27. Duana Fullwiley, *The Enculturated Gene: Sickle Cell Health Politics and Biological Difference in West Africa* (Princeton: Princeton University Press, 2012).

28. Jill Waalen and Ernest Beutler, "Genetic Screening for Low-Penetrance Variants in Protein-Coding Genes," *Annual Review of Genomics and Human Genetics* 10 (2009): 431–50.

29. C. R. Scriver and P. J. Waters, "Monogenic Traits Are Not Simple: Lessons from Phenylketonuria," *Trends in Genetics* 15, no. 7 (1999): 267–72.

30. S. G. Post, P. J. Whitehouse, R. H. Binstock, T. D. Bird, S. K. Eckert, L. A. Farrer, L. M. Fleck, et al., "The Clinical Introduction of Genetic Testing for Alzheimer Disease: An Ethical Perspective," *Journal of the American Medical Association* 277, no. 10 (1997): 832–36; Norman R. Relkin, "Apolipoprotein E Genotyping in Alzheimer's Disease," *The Lancet* 347 (1996): 1091–95; Norman R. Relkin, Younga J. Kwon, Julia Tsai, and Samuel Gandy, "The National Institute on Aging/Alzheimer's Association Recommendations on the Application of Apolipoprotein E Genotyping to Alzheimer's Disease," *Annals of the New York Academy of Sciences* 802, no. 1 (1996): 149–76;

31. http://www.alz.org/national/documents/topicsheet_genetictesting.pdf; http://www.alzheimers.org.uk/site/scripts/documents_info.php?documentID=434; http://www.alzheimer.ca/~/media/Files/national/brochures-tough-issues/Tough_Issues_Genetics_2007_e.ashx; http://www.francealzheimer.org.

32. See, for example, Hyman Schipper, who believes that knowledge about their APOE status will encourage individuals to pay more attention to their dietary intake: Hyman M. Schipper, "Presymptomatic Apolipoprotein E Genotyping for Alzheimer's Disease Risk Assessment and Prevention," *Alzheimer's & Dementia* 7, no. 4 (2011): e118–e123.

33. Personal communication with Robert Green, January 2012.

34. Frederick J. Kier and Victor Molinari, "'Do-It-Yourself' Dementia Testing: Issues Regarding an Alzheimer's Home Screening Test," *The Gerontologist* 43, no. 3 (2003): 295–301.

35. The REVEAL Study is funded by the ELSI Branch of the National Human Genome Research Institute (R01 HG/AG02213). Additional support was provided by an NIA Mentoring Award to Dr. Green (K24 AG027841) and by the Boston University General Clinical Research Center (GCRC; M01 RR00533).

36. Ashida Sato, Laura M. Koehly, J. Scott Roberts, Clara A. Chen, Susan Hiraki, and Robert C. Green, "Disclosing the Disclosure: Factors Associated with Communicating the Results of Genetic Susceptibility Testing for Alzheimer's Disease," *Journal of Health Communication* 14, no. 8 (2009): 768–84.

37. L. Adrienne Cupples, Lindsay A. Farrer, A. Dessa Sadovnick, Norman Relkin, Peter Whitehouse, and Robert C. Green, "Estimating Risk Curves for First-degree Relatives of Patients with Alzheimer's Disease: The REVEAL Study," *Genetics in Medicine* 6, no. 4 (2004): 192–96.

38. Robert C. Green, V. C. Clarke, N. J. Thompson, J. L. Woodard, and R. Letz, "Early Detection of Alzheimer Disease: Methods, Markers, and Misgivings," *Alzheimer Disease & Associated Disorders* 11, suppl. 5 (1997): S1–S5, discussion S37–S39.

39. L. A. Farrer, L. A. Cupples, J. L. Haines, B. Hyman, W. A. Kukull, R. Mayeux, R. H. Myers, M. A. Pericak-Vance, N. Risch, and C. M. van Duijn, "Effects of Age, Sex, and Ethnicity

on the Association between Apolipoprotein E Genotype and Alzheimer Disease. A Meta-analysis. APOE and Alzheimer Disease Meta Analysis Consortium," *Journal of the American Medical Association* 278, no. 16 (1997): 1349–56.

40. In REVEAL 2 some individuals were given an abbreviated disclosure session by the involved clinicians, rather than genetic counselors. The purpose was to find out if a briefer form of disclosure would be as satisfactory as the original longer form with the idea that such disclosure sessions would become part of general practice.

41. Susan Larusse, J. Scott Roberts, Theresa M. Marteau, Heather Katsen, Erin L. Linnenbringer, Melissa Barber, Peter Whitehouse, Kimberly Quaid, Tamsen Brown, Robert C. Green, and Norman R. Relkin, "Genetic Susceptibility Testing versus Family History-Based Risk Assessment: Impact on Perceived Risk of Alzheimer's Disease," *Genetic Medicine* 7 (2005): 48–53; J. S. Roberts, K. D. Christensen, and R. C. Green, "Using Alzheimer's Disease as a Model for Genetic Risk Disclosure: Implications for Personal Genomics," *Clinical Genetics* 80, no. 5 (2011): 407–14.

42. Cupples et al., "Estimating Risk Curves"; see also L. A. Farrer, L. A. Cupples, J. L. Haines, B. Hyman, W. A. Kukull, R. Mayeux, R. H. Myers, M. A. Pericak-Vance, N. Risch, and C. M. van Duijn, "Effects of Age, Sex, and Ethnicity on the Association between Apolipoprotein E Genotype and Alzheimer Disease. A Meta-analysis. APOE and Alzheimer Disease Meta Analysis Consortium," *Journal of the American Medical Association* 278, no. 16 (1997): 1349–56.

43. Margaret Lock, Julia Freeman, Gillian Chilibeck, Briony Beveridge, and Miriam Padolsky, "Susceptibility Genes and the Question of Embodied Identity," *Medical Anthropology Quarterly* 21, no. 3 (2007): 256–76.

44. Jill S. Goldman, Susan E. Hahn, Jennifer Williamson Catania, Susan LaRusse-Eckert, Melissa Barber Butson, Malia Rumbaugh, Michelle N. Strecker, et al., "Genetic Counseling and Testing for Alzheimer Disease: Joint Practice Guidelines of the American College of Medical Genetics and the National Society of Genetic Counselors," *Genetics in Medicine* 13, no. 6 (2011): 597–605.

45. Sato et al., "Disclosing the Disclosure."

46. Robert C. Green, J. Scott Roberts, L. Adrienne Cupples, Norman R. Relkin, Peter J. Whitehouse, Tamsen Brown, Susan LaRusse Eckert, et al., "Disclosure of APOE Genotype for Risk of Alzheimer's Disease," *New England Journal of Medicine* 361, no. 3 (2009): 245–54.

47. I am indebted to Gillian Chilibeck for the phrase "familiarization of Alzheimer's risk."

48. Martin Richards, "Lay and Professional Knowledge of Genetics and Inheritance," *Public Understanding of Science* 5, no. 3 (1996): 217–30.

49. J. Turney, "The Public Understanding of Science—Where Next?." *European Journal of Genetics in Society* 1, no. 2 (1995): 5–22, 12.

50. Katie Featherstone, Paul Atkinson, Aditva Bharadwai, and Angus Clarke, *Risky Relations: Family, Kinship and the New Genetics* (Oxford: Berg, 2006); C. Emslie, K. Hunt, and G. Watt, "A Chip Off the Old Block? Lay Understandings of Inheritance amongst Men and Women in Mid-life," *Public Understanding of Science* 12, no. 1 (2003): 47–65; Richards, "Lay and Professional Knowledge."

51. Barbara Duden and Silja Samerki, " 'Pop Genes': An Investigation of 'the Gene' in Popular Parlance," in *Biomedicine as Culture: Instrumental Practices, Technoscientific Knowledge, and New Modes of Life*, ed. R. V. Burri and J. Dumit (New York: Routledge, 2007), 167–90, 167.

52. Goldman et al., "Genetic Counseling and Testing"; Roberts, Christensen, and Green, "Using Alzheimer's Disease."

53. Martyn Pickersgill, Sarah Cunningham-Burley, and Paul Martin, "Constituting Neurologic Subjects: Neuroscience, Subjectivity and the Mundane Significance of the Brain," *Subjectivity* 4 (2011): 346–65, 361.

54. Joseph Dumit, *Picturing Personhood: Brain Scans and Biomedical Identity* (Princeton: Princeton University Press, 2004); Ortega and Vidal, *Neurocultures*; Jan Slaby, "Steps towards a Critical Neuroscience," *Phenomenology and the Cognitive Sciences* 9, no. 3 (2010): 397–416, see "Hermeneutics of Subjectification," 403.

55. Lock et al., "Susceptibility Genes and the Question of Embodied Identity."

56. Roberts, Christensen, and Green, "Using Alzheimer's Disease."

57. Padolsky conducted participant observation research in 2005 at the Alzheimer Society of Ottawa office, at ASO classes and workshops, and at several ASO special events over the course of twelve months. She interviewed nine ASO staff members, including most of the Family Support and Education team, the director of Family Support and Education, the executive director of ASO, and a sample of administrative staff. In addition, ASC and ASO publications were compared to the materials put out by the U.K. and U.S. Alzheimer societies.

58. Alzheimer Society of Canada, Alzheimer Care: Ethical Guidelines; Genetic Testing (2006), http://www.alzheimer.ca/english/care/ethics-genetictest.htm.

59. Lock et al., "Susceptibility Genes and the Question of Embodied Identity."

60. Ten of the newspapers with articles on AD are published in either Canada, the United States, the United Kingdom, or Australia, and the seven magazines are published in either Canada or the United States.

61. Margaret Lock, Julia Freeman, Rosemary Sharples, and Stephanie Lloyd, "When It Runs in the Family: Putting Susceptibility Genes in Perspective," *Public Understanding of Science* 15, no. 3 (2006): 277–300; Lock et al., "Susceptibility Genes and the Question of Embodied Identity"; Briony Beveridge was responsible for the comprehensive analysis of newspaper reporting on AD.

62. Rayna Rapp, Deborah Heath, and Karen-Sue Taussig, "Genealogical Disease: Where Hereditary Abnormality, Biomedical Explanation, and Family Responsibility Meet," in *Relative Matters: New Directions in the Study of Kinship*, ed. Sarah Franklin and Susan MacKinnon (Durham, N.C.: Duke University Press, 2001), 384–412; Rapp, "Cell Life and Death."

63. Michel Callon and Vololona Rabeharisoa, "Gino's Lesson on Humanity: Genetics, Mutual Entanglements and the Sociologist's Role," *Economy and Society* 33, no. 1 (2004): 1–27; Heath and Taussig, "Genetic Citizenship"; Rapp, "Cell Life and Death"; Karen-Sue Taussig, Rayna Rapp, and Deborah Heath, "Flexible Eugenics: Technologies of the Self in the Age of Genetics," in Goodman, Heath, and Lindee, *Genetic Nature/Culture*, 58–76.

Chapter 8
Chance Untamed and the Return of Fate

1. René Dubos, *Mirage of Health* (London: Harper and Row, 1959), 1–2.

2. National Institutes of Health, *Alzheimer Disease Research Summit 2012: Path to Treatment and Prevention* (Bethesda, Md.: National Institutes of Health, 2012), http://www.nia.nih.gov/about/events/2012/alzheimers-disease-research-summit-2012-path-treatment-and-prevention.

3. Even so, certain governments remain strikingly resistant to recognition of the undeniable increase in diagnosed cases of AD, Canada being one intransigent example.

4. See, for example, Howard Gleckman, "The Obama Administration's War on Alzheimer's," *Forbes*, January 11, 2012, http://www.forbes.com/sites/howardgleckman/2012/01/11/the-obama-administrations-war-on-alzheimers/.

5. Gary Stix, "Obama's War on Alzheimer's: Will We Be Able to Treat the Disease by 2025?," *Scientific American*, January 31, 2012, http://blogs.scientificamerican.com/observations/2012/01/31/obamas-war-on-alzheimers-will-we-be-able-to-treat-the-disease-by-2025/.

6. Gina Kolata, "In Preventing Alzheimer's, Mutation May Aid Drug Quest," *New York Times*, July 11, 2012.

7. Thorlakur Jonsson, Jasvinder K. Atwal, Stacy Steinberg, Jon Snaedal, Palmi V. Jonsson, Sigurbjorn Bjornsson, Hreinn Stefansson, et al., "A Mutation in APP Protects Against Alzheimer's Disease and Age-Related Cognitive Decline," *Nature* 488 (2012): 96–99, 96.

8. Ibid, 98.

9. Kolata, "In Preventing Alzheimer's."

10. Lauren Gravitz, "Drugs: A Tangled Web of Targets," *Nature* 475 (2011): S9–S11.

11. Randall Bateman and John Morris, "The Dominantly Inherited Alzheimer's Network Trials: An Opportunity to Prevent Alzheimer's Disease," *Alzheimer's & Dementia* 8, no. 4 (2012): 427.

12. Eric Reiman, Francisco Lopera, Jessica Langbaum, Adam Fleisher, Naparkamon Ayutyanont, Yakeel Quiros, Laura Jakimovitch, Carolyn Langlois, and Pierre Tariot, "The Alzheimer's Prevention Initiative," *Alzheimer's & Dementia* 8, no. 4 (2012): 427.

13. William Jagust, "Aging, Amyloid and Neural Activity," *Alzheimer's & Dementia* 8, no. 4 (2012): 427.

14. Glass and Arnold, "Some Evolutionary Perspectives."

15. Robert Green, Scott Roberts, Jason Karlawish, Thomas Obisesan, L. Adrienne Cupples, Denise Lautenbach, Margaret Bradbury, et al., "Disclosure of APOE Genotype to Persons with Mild Cognitive Impairment (MCI)," *Alzheimer's & Dementia* 8, no. 4 (2012): 423.

16. Reisa A. Sperling and Scott Roberts, "Disclosure of Amyloid Status in Secondary Prevention Trials for Alzheimer's Disease," *Alzheimer's & Dementia* 8, no. 4 (2012): 423.

17. Jason Karlawish, "Disclosing Amyloid Imaging Results," *Alzheimer's & Dementia* 8, no. 4 (2012): 423.

18. Hampel, Lista, and Khachaturian, "Development of Biomarkers," 332.

19. Ibid., 313.

20. Ara S. Khachaturian, Michelle M. Mielke, and Zaven S. Khachaturian, "Biomarker Development: A Population-Level Perspective," *Alzheimer's & Dementia* 8, no. 4 (2012): 247–49.

21. Alzheimer's Association International Conference, "Sleep Duration, Sleep Disorders, and Circadian Patterns Are Risk Factors and Indicators of Cognitive Decline" (press release, 2012), http://www.prnewswire.com/news-releases/sleep-duration-sleep-disorders-and-circadian-patterns-are-risk-factors-and-indicators-of-cognitive-decline-162589416.html.

22. Andrew Pollack, "Alzheimer's Drug Fails Its First Big Clinical Trial," *New York Times*, July 23, 2012.

23. Tom Murphy, "Alzheimer's Drug Fails Study but Flashes Potential," *Associated Press*, August 24, 2012.

24. Michael C. Purdy, "Investigational Drugs Chosen for Major Alzheimer's Prevention Method (Washington University in St. Louis, 2012), http://www.wustel.edu.

25. Eric Reiman, Yakeel T. Quiroz, Adam S. Fleisher, Kewei Chen, Carlos Velez-Pardo, Marlene Jimenez-Del-Rio, Anne M. Fagan, Aarti R. Shah, Sergio Alvarez, Andrés Arbelaez, Margarita Giraldo, Natalia Acosta-Baena, Reisa A. Sperling, Brad Dickerson, Chantal E. Stern, Victoria Tirado, Claudia Munoz, Rebecca A. Reiman, Matthew J. Huentelman, Gene E. Alexander, Jessica B. S. Langbaum, Kenneth S. Kosik, Pierre N. Tariot, and Francisco Lopera, "Brain Imaging and Fluid Biomarker Analysis in Young Adults at Genetic Risk for Autosomal Dominant Alzheimer's Disease in the Presenilin 1 E280A Kindred: A Case-Control Study," *The Lancet Neurology* 11 (2012): 1048–56; Adam S. Fleisher, Kewei Chen, Yakeel T. Quiroz, Laura J. Jakimovich, Madelyn Gutierrez Gomez, Carolyn M. Langois, Jessica B. S. Langbaum, Napatkamon Ayutyanont, Auttawut Roontiva, Pradeep Thiyyagura, Wendy Lee, Hua Mo, Liliana Lopez, Sonia Moreno, Natalia Acosta-Baena, Margarita Giraldo, Gloria Garcia, Rebecca A. Reiman, Matthew J. Huentelman , Kenneth S. Kosik, Pierre N. Tariot, Francisco Lopera, and Eric M. Reiman, "Florbetapir PET Analysis of Amyloid-β Deposition in the Presenilin 1 E280A Autosomal Dominant Alzheimer's Disease Kindred: A Cross-Sectional Study," *The Lancet Neurology* 11 (2012): 1057–65; Nick Fox, "When, Where, and How Does Alzheimer's Disease Start?," *The Lancet Neurology* 11 (2012): 1017–18; William Jagust, "Tracking Brain Amyloid-β in Presymptomatic Alzheimer's Disease," *The Lancet Neurology* 11 (2012): 1018–20.

26. Mo Costandi, "Disrupted Sleep Might Signal Early Stages of Alzheimer's," *Scientific American*, October 18, 2012, http://www.scientificamerican.com/article.cfm?id=disrupted-sleep-might-signal-early-stages-of-alzheimers.

27. Karola Stotz, Adam Bostanci, and Paul Griffiths, "Tracking the Shift to 'Postgenomics,'" *Community Genetics* 9, no. 3 (2006): 190–96.

28. Evelyn Fox Keller, *The Century of the Gene* (Cambridge, Mass.: Harvard University Press, 2000), 277.

29. S. R. Eddy, "Non-coding RNA Genes and the Modern RNA World," *Nature Reviews Genetics* 2, no. 12 (2001): 919–29; John S. Mattick, "Challenging the Dogma: The Hidden Layer of Non-Protein-Coding RNAs in Complex Organisms," *BioEssays* 25, no. 10 (2003): 930–39; John S. Mattick, "The Hidden Genetic Program of Complex Organisms," *Scientific American* 291, no. 4 (2004): 60–67.

30. Mattick, "Challenging the Dogma."

31. A. Petronis, "Human Morbid Genetics Revisited: Relevance of Epigenetics," *Trends in Genetics* 17, no. 3 (2001): 142–46.

32. Pleiotropy means the diverse effects of a single gene or gene pair on several organ systems and function.

33. I. I. Gottesman, "Schizophrenia Epigenesis: Past, Present, and Future," *Acta Psychiatrica Scandinavica. Supplementum* 384 (1994): 26–33.

34. Keller, *Century of the Gene*, 147.

35. Kenneth M. Weiss and Anne V. Buchanan, *The Mermaid's Tale: Four Billion Years of Cooperation in the Making of Living Things* (Cambridge, Mass.: Harvard University Press, 2009), 89.

36. Paul E. Griffiths, "Developmental Systems Theory," in *Nature Encyclopedia of the Life Sciences* (London: John Wiley, 2001), 1; see also Susan Oyama, Paul E. Griffiths, and Russel D. Gray, *Cycles of Contingency: Developmental Systems and Evolution, Life and Mind*. Cambridge, Mass.: MIT Press, 2001.

37. Eva Jablonka and Marion J. Lamb, *Evolution in Four Dimensions: Genetic, Epigenetic, Behavioral, and Symbolic Variation in the History of Life* (Cambridge, Mass.: MIT Press, 2005), 82.

38. Barry Barnes and John Dupré, *Genomes: And What to Make of Them* (Chicago: University of Chicago Press, 2008), 50.

39. Sarah Parry and John Dupré, "Introducing Nature After the Genome," in *Nature After the Genome*, ed. Sarah Parry and John Dupré (Oxford: Blackwell, 2010), 3–16.

40. John Dupré, "The Polygenomic Organism," in *Nature After the Genome*, ed. Sarah Parry and John Dupré (Oxford: Blackwell, 2010), 19–31, 25.

41. R. Strohman, "A New Paradigm for Life: Beyond Genetic Determinism," *California Monthly* 111 (2001): 4–27, 8.

42. Eva M. Neumann-Held and Christopher Rehmann-Sutter, *Genes in Development: Rereading the Molecular Paradigm* (Durham, N.C.: Duke University Press, 2006), 2.

43. Evan T. Powers, Richard I. Morimoto, Andrew Dillin, Jeffery W. Kelly, and William E. Balch, "Biological and Chemical Approaches to Diseases of Proteostasis Deficiency," *Annual Review of Biochemistry* 78 (2009): 959–91.

44. Richard C. Lewontin, "Science and Simplicity," *New York Review of Books* 50 (2003): 39–42, 39.

45. Jonathan Weiner, *The Beak of the Finch* (New York: Vintage Books, 1994).

46. Alec Wilkinson, "The Lobsterman," *The New Yorker*, July 31, 2006, 56–65.

47. Kenneth M. Weiss and Anne V. Buchanan, "Is Life Law-Like?" *Genetics* 188, no. 4 (2011): 761–71, 761; see also P. K. Stanford, *Exceeding Our Grasp: Science, History and the Problem of Unconceived Alternatives* (New York: Oxford University Press, 2006); C. K. Waters, "Causes That Make a Difference," *Journal of Philosophy* 104 (2007): 551–79.

48. Weiss and Buchanan, "Is Life Law-Like?," 761.

49. David Runciman, "Will We Be All Right in the End?," *London Review of Books* 34, no. 1 (2012): 3–5.

50. Gillian Chilibeck, Margaret Lock, and Megha Sehdev, "Postgenomics, Uncertain Futures, and the Familiarization of Susceptibility Genes," *Social Science & Medicine* 72, no. 11 (2011): 1768–75, see 1773–74.

51. Mary Midgley, *Science and Poetry* (London: Routledge, 2001), 120.

52. Nima Mosammaparast and Yang Shi, "Reversal of Histone Methylation: Biochemical and Molecular Mechanisms of Histone Demethylases," *Annual Review of Biochemistry* 79 (2010): 155–79.

53. Patrick O. McGowan, Aya Sasaki, Ana C. D'Alessio, Sergiy Dymov, Benoit Labonté, Moshe Szyf, Gustavo Turecki, and Michael J. Meaney, "Epigenetic Regulation of the

Glucocorticoid Receptor in Human Brain Associates with Childhood Abuse," *Nature Neuroscience* 12, no. 3 (2009): 342–48; Patrick O. McGowan and Moshe Szyf, "The Epigenetics of Social Adversity in Early Life: Implications for Mental Health Outcomes," *Neurobiology of Disease* 39, no. 1 (2010): 66–72; L. H. Lumey, "Decreased Birthweights in Infants after Maternal in Utero Exposure to the Dutch Famine of 1944–1945," *Paediatric and Perinatal Epidemiology* 6, no. 2 (1992): 240–53; J. E. Harding, "The Nutritional Basis of the Fetal Origins of Adult Disease," *International Journal of Epidemiology* 30, no. 1 (2001): 15–23.

54. Moshe Szyf, Patrick McGowan, and Michael J. Meaney, "The Social Environment and the Epigenome," *Environmental and Molecular Mutagenesis* 49, no. 1 (2008): 46–60; Lucia Daxinger and Emma Whitelaw, "Understanding Transgenerational Epigenetic Inheritance via the Gametes in Mammals," *Nature Reviews Genetics* 13, no. 3 (2012): 153–62. See also Margaret Lock, "The Lure of the Epigenome," *The Lancet* 381 (2013) 1986–1897.

55. Ji-Song Guan, Stephen J. Haggarty, Emanuela Giacometti, Jan-Hermen Dannenberg, Nadine Joseph, Jun Gao, Thomas J. F. Nieland, et al., "HDAC2 Negatively Regulates Memory Formation and Synaptic Plasticity," *Nature* 459 (2009): 55–60.

56. Sun-Chong Wang, Beatrice Oelze, and Axel Schumacher, "Age-Specific Epigenetic Drift in Late-Onset Alzheimer's Disease," *PLoS ONE* 3, no. 7 (2008): e2698, citing Dana C. Dolinoy, Radhika Das, Jennifer R. Weidman, and Randy L. Jirtle, "Metastable Epialleles, Imprinting, and the Fetal Origins of Adult Diseases," *Pediatric Research* 61, no. 5, pt. 2 (2007): 30R–37R.

57. Wang, Oelze, and Schumacher, "Age-Specific Epigenetic Drift."

58. Diego Mastroeni, Ann McKee, Andrew Grover, Joseph Rogers, and Paul D. Coleman, "Epigenetic Differences in Cortical Neurons from a Pair of Monozygotic Twins Discordant for Alzheimer's Disease," *PLoS ONE* 4, no. 8 (2009), http://www.ncbi.nlm.nih.gov/pmc/articles/PMC2719870/.

59. Wang, Oelze, and Schumacher, "Age-Specific Epigenetic Drift."

60. Jablonka and Lamb, *Evolution in Four Dimensions*, 128.

61. See, for example, Estelle Sontag, Christa Hladik, Lisa Montgomery, Ampa Luangpirom, Ingrid Mudrak, Egon Ogris, and Charles L. White III, "Downregulation of Protein Phosphatase 2A Carboxyl Methylation and Methyltransferase May Contribute to Alzheimer Disease Pathogenesis," *Journal of Neuropathology & Experimental Neurology* 63, no. 10 (2004): 1080–91; Kimberly D. Siegmund, Caroline M. Connor, Mihaela Campan, Tiffany I. Long, Daniel J. Weisenberger, Detlev Biniszkiewicz, Rudolf Jaenisch, Peter W. Laird, and Schahram Akbarian, "DNA Methylation in the Human Cerebral Cortex Is Dynamically Regulated throughout the Life Span and Involves Differentiated Neurons," *PLoS ONE* 2, no. 9 (2007), http://www.ncbi.nlm.nih.gov/pmc/articles/PMC1964879/; Diego Mastroeni, Andrew Grover, Elaine Delvaux, Charisse Whiteside, Paul D. Coleman, and Joseph Rogers, "Epigenetic Mechanisms in Alzheimer's Disease," *Neurobiology of Aging* 32, no. 7 (2011): 1161–80.

62. Marc Winnefeld and Frank Lyko, "The Aging Epigenome: DNA Methylation from the Cradle to the Grave," *Genome Biology* 13, no. 7 (2012): 165–68.

63. Rita Guerreiro, Aleksandra Wojtas, Jose Bras, Minerva Carrasquillo, Ekaterina Rogaeva, Elisa Majounie, Carlos Cruchaga, Celeste Sassi, John S. K. Kauwe, Steven Younkin, Lilinaz Hazrati, John Collinge, Jennifer Pocock, Tammaryn Lashley, Julie Williams, Jean-Charles Lambert, Philippe Amouyel, Alison Goate, Rosa Rademakers, Kevin Morgan, John Powell, Peter St George-Hyslop, Andrew Singleton, and John Hardy, "*Trem2* Variants in Alzheimer's Disease," *New England Journal of Medicine* 368 (November 14, 2012): 117–27; Thorlakur Jonsson, Hreinn Stefansson, Stacy Steinberg, Ingileif Jonsdottir, Palmi V. Jonsson, Jon Snaedal, Sigurbjorn Bjornsson, Johanna Huttenlocher, Allan I. Levey, James J. Lah, Dan Rujescu, Harald Hampel, Ina Giegling, Ole A. Andreassen, Knut Engedal, Ingun Ulstein, Srdjan Djurovic, Carla Ibrahim-Verbaa, Albert Hofman, M. Arfan Ikram, Cornelia M. van Duijn, Unnur Thorsteinsdottir, Augustine Kong, and Kari Stefansson, "Variant of *TREM2* Associated with the Risk of Alzheimer's Disease," *New England Journal of Medicine* 368 (November 14, 2012): 107–16, doi:10.1056/NEJMoa1211103.

64. Gina Kolata, "Alzheimer's Tied to Mutation Harming Immune Response," *New York Times*, November 14, 2012.

Chapter 9
Transcending Entrenched Tensions

1. Sandra D. Mitchell, *Unsimple Truths: Science, Complexity and Policy* (Chicago: University of Chicago Press, 2009), 11.

2. Martin Prince, Renata Bryce, Emiliano Albanese, Anders Wimo, Wagner Ribeiro, Cleusa P. Ferri, "The Global Prevalence of Dementia: A Systematic Review and Metaanalysis," *Alzheimer's and Dementia* 9 (2013): 63–75.

3. David Rothschild, "The Practical Value of Research in the Psychoses of Later Life," *Diseases of the Nervous System* 8, no. 4 (1947): 123–28.

4. Tom Kitwood, *Dementia Reconsidered: The Person Comes First* (Maidenhead: Open University Press, 1997).

5. Farrer, "Familial Risk for Alzheimer Disease"; Hugh Hendrie, "Diagnosis of Dementia," 122.

6. J. C. Heyman, M. L. Barer Hertzman, and R. G. Evans, eds., *Healthier Societies: From Analysis to Action* (Oxford: Oxford University Press, 2006).

7. Patricia Churchland, *Brain-Wise: Studies in Neurophilosophy* (Cambridge, Mass.: MIT Press, 2002), 1.

8. Alva Noë, *Out of Our Heads: Why You Are Not Your Brain and Other Lessons from the Biology of Consciousness* (New York: Farrar, Straus and Giroux, 2009), 10.

9. Ibid., 181.

10. Scott F. Gilbert, "The Genome in Its Ecological Context: Philosophical Perspectives on Interspecies Epigenesis," *Annals of the New York Academy of Sciences* 981 (2002): 202–18, 213.

11. Julian C. Hughes, Stephen J. Louw, and Steven R. Sabat, *Dementia: Mind, Meaning, and the Person* (Oxford: Oxford University Press, 2006).

12. Ernst Mayr, "Cause and Effect in Biology," *Science* 134 (1961): 1501–6.

13. Kirkwood, *Time of Our Lives*, cited in House of Lords, *Science and Technology Committee*; see also Moreira and Palladino, "Ageing between Gerontology and Biomedicine," 363.

14. Savva et al., "Age, Neuropathology, and Dementia," 2308.

15. Agustin G. Yip, Carol Brayne, and Fiona E. Matthews, "Risk Factors for Incident Dementia in England and Wales: The Medical Research Council Cognitive Function and Ageing Study. A Population-Based Nested Case-Control Study," *Age and Ageing* 35, no. 2 (2006): 154–60.

16. See also Sharon Kaufman, *. . . And a Time to Die: How American Hospitals Shape the End of Life* (New York: Scribner, 2005); D. D. Von Dras and H. T. Blumenthal, "Dementia of the Aged: Disease or Atypical-Accelerated Aging? Biopathological and Psychological Perspectives," *Journal of the American Geriatrics Society* 40, no. 3 (1992): 285–94.

17. Carol Brayne, Paul G. Ince, Hannah A. D. Keage, Ian G. McKeith, Fiona E. Matthews, Tuomo Polvikoski, and Raimo Sulkava, "Education, the Brain and Dementia: Neuroprotection or Compensation? EClipSE Collaborative Members," *Brain* 133, no. 8 (2010): 2210–16.

18. Martin Prince, Daisy Acosta, Helen Chiu, Marcia Scazufca, and Mathew Varghese, "Dementia Diagnosis in Developing Countries: A Cross-Cultural Validation Study," *Lancet* 361 (2003): 909–17.

19. M. Marsel Mesulam, "Neuroplasticity Failure in Alzheimer's Disease: Bridging the Gap between Plaques and Tangles," *Neuron* 24, no. 3 (1999): 521–29, 526.

20. Ibid., 529.

21. Roth, "Dementia and Normal Aging," 65.

22. Hampel, Lista, and Khachaturian, "Development of Biomarkers," 315.

23. Ibid., 315.

24. Evelyn Fox Keller, *The Mirage of a Space between Nature and Nurture* (Durham, N.C.: Duke University Press, 2010), 7.

25. McGowan and Szyf, "Epigenetics of Social Adversity," emphasis added.

26. Moshe Szyf, "The Early Life Social Environment and DNA Methylation: DNA Methylation Mediating the Long-Term Impact of Social Environments Early in Life," *Epigenetics* 8 (2011): 971–78, 971.

27. Daxinger and Whitelaw, "Understanding Transgenerational Epigenetic Inheritance."

28. Jörg Niewöhner, "Epigenetics: Embedded Bodies and the Molecularization of Biography and Milieu," *BioSocieties* 6 (2011): 279–98, 291.

29. Ibid.

30. Jörg Niewöhner, Martin Döring, Michalis Kontopodis, Jeannette Madarász, and Christoph Heintze, "Cardiovascular Disease and Obesity Prevention in Germany: An Investigation into a Heterogeneous Engineering Project," *Science, Technology & Human Values* 36, no. 5 (2011): 723–51.

31. Niewöhner, "Epigenetics," 290–92.

32. Lock, *Encounters with Aging*.

33. Zaven S. Khachaturian and Ara S. Khachaturian, "Prevent Alzheimer's Disease by 2020," 84.

34. See also Danny George, Peter Whitehouse, Simon d'Alton, and Jesse Ballinger, "Through the Amyloid Gateway," *The Lancet* 380 (2012): 1986–87.

35. Lock and Nguyen, *Anthropology of Biomedicine*, 1.

36. Robin McKie, "Discovery of 'Methuselah Gene' Unlocks Secret of Long Life," *The Observer*, February 3, 2002; Lorna M. Lopez, Sarah E. Harris, Michelle Luciano, Dave Liewald, Gail Davies, Alan J. Gow, Albert Tenesa, et al., "Evolutionary Conserved Longevity Genes and Human Cognitive Abilities in Elderly Cohorts," *European Journal of Human Genetics* 20, no. 3 (2012): 341–47.

37. S. Miyagi, N. Iwama, T. Kawabata, and K. Hasegawa, "Longevity and Diet in Okinawa, Japan: The Past, Present and Future," *Asia-Pacific Journal of Public Health* 15, suppl. (2003): S3–S9.

38. David Brindle, "Older People Are an Asset, Not a Drain," *Guardian*, March 2, 2011.

39. See Didier Fassin, "Another Politics of Life Is Possible," *Theory, Culture & Society* 26, no. 5 (2009): 44–60 for a "politics of life" and the differential value placed on lives.

Afterword
Portraits from the Mind

1. Sebastian J. Crutch, Ron Isaacs, and Martin N. Rossor, "Some Workmen Can Blame Their Tools: Artistic Change in an Individual with Alzheimer's Disease," *The Lancet* 357 (2001): 2129–33.

2. Alzheimer's Association, *Portraits From the Mind: The Works of William Utermohlen 1955–2000: A Retrospective of the Artist's Work Before and After His Diagnosis with Alzheimer's Disease* (Salt Lake City, Utah: Myriad Pharmaceuticals, 2008), 22.One of the authors of *The Lancet* article cited above, Ron Isaacs, was William Utermohlen's nurse.

3. Crutch, Isaacs, and Rossor, "Some Workmen Can Blame Their Tools," 2130.

4. Galerie Beckel-Odille-Boïcos, *William Utermohlen: Paintings and Drawings 1955–1997* (Paris, 2000), extract from exhibition notes.

5. Crutch, Isaacs, and Rossor, "Some Workmen Can Blame Their Tools," 2132.

6. Ibid., 2133.

Bibliography

Abbott, Alison. "The Plaque Plan." *Nature* 456 (2008): 161–64.

Abraham, Carolyn. "Dementia Researchers Feel Blocked by Ottawa, Big Pharma, Medical Dogma." *Globe and Mail*, September 17, 2010. http://www.theglobeandmail.com/life /health-and-fitness/dementia-researchers-feel-blocked-by-ottawa-big-pharma-medical -dogma/article4389415/.

Acosta-Baena, Natalia, Diego Sepulveda-Falla, Carlos Mario Lopera-Gómez, Mario César Jaramillo-Elorza, Sonia Moreno, Daniel Camilo Aguirre-Acevedo, Amanda Saldarriaga, and Francisco Lopera. "Pre-dementia Clinical Stages in Presenilin 1 E280A Familial Early-Onset Alzheimer's Disease: A Retrospective Cohort Study." *The Lancet Neurology* 10, no. 3 (2011): 213–20.

"Ageing Epidemic Looms in Ghana." *Modern Ghana*, June 17, 2007. http://www.modernghana .com/news/137880/1/ageing-epidemic-looms-in-ghana.html.

Aizenstein, Howard Jay, Robert D. Nebes, Judith A. Saxton, Julie C. Price, Chester A. Mathis, Nicholas D. Tsopelas, Scott K. Ziolko, et al. "Frequent Amyloid Deposition without Significant Cognitive Impairment among the Elderly." *Archives of Neurology* 65, no. 11 (2008): 1509–17.

Alliance for Human Research Protection. "NY Times Corrects Gina Kolata Re: Alzheimer's." September 16, 2010. http://www.ahrp.org/cms/content/view/726/1/.

Alzheimer's Association, William Thies, and Laura Bleiler. "2011 Alzheimer's Disease Facts and Figures." *Alzheimer's & Dementia* 7, no. 2 (2011): 208–44.

Alzheimer Society of Canada. "Alzheimer Care: Ethical Guidelines; Genetic Testing." 2006. http://www.alzheimer.ca/english/care/ethics-genetictest.htm.

Alzheimer's Association. *Portraits From the Mind: The Works of William Utermohlen 1955–2000: A Retrospective of the Artist's Work Before and After His Diagnosis with Alzheimer's Disease.* Salt Lake City, Utah: Myriad Pharmaceuticals, 2008.

Alzheimer's Association International Conference. "Sleep Duration, Sleep Disorders, and Circadian Patterns Are Risk Factors and Indicators of Cognitive Decline." Press release, 2012. http://www.alz.org/aaic/releases/mon_1030amct_irregular_sleep.asp.

"Alzheimer's Disease: Amyloid Precursor Protein—Good, Bad or Both?" *ScienceDaily*, December 29, 2009. http://www.sciencedaily.com/releases/2009/10/091018171806.htm.

Alzheimer's Disease International. *World Alzheimer Report 2009: Executive Summary.* 2009. http:// www.alz.co.uk/research/files/WorldAlzheimerReport-ExecutiveSummary.pdf.

Amador-Ortiz, Catalina, Wen-Lang Lin, Zeshan Ahmed, David Personett, Peter Davies, Ranjan Duara, Neill R. Graff-Radford, Michael L. Hutton, and Dennis W. Dickson. "TDP-43 Immunoreactivity in Hippocampal Sclerosis and Alzheimer's Disease." *Annals of Neurology* 61, no. 5 (2007): 435–45.

Andersen, Erin. "Isolated by Affliction, Isolated by Language." *Globe and Mail*, December 18, 2010.

Anderson, Warwick. *The Collectors of Lost Souls: Turning Kuru Scientists into White Men.* Baltimore: Johns Hopkins University Press, 2008.

Apse, Kira A., Barbara B. Biesecker, Francis M. Giardiello, Barbara P. Fuller, and Barbara A. Bernhardt. "Perceptions of Genetic Discrimination among At-Risk Relatives of Colorectal Cancer Patients." *Genetics in Medicine* 6, no. 6 (2004): 510–16.

Ballenger, Jesse F. "DSM-V: Continuing the Confusion about Aging, Alzheimer's and Dementia." *H-madness: History of Psychiatry*, March 19, 2010. http://historypsychiatry.com/2010/03/19/dsm-v-continuing-the-confusion-about-aging-alzheimer%e2%80%99s-and-dementia/.

———. "Necessary Interventions: Antidementia Drugs and Heightened Expectations for Aging in Modern American Cultures and Society." In Ballenger et al., *Treating Dementia*, 189–209.

———. *Self, Senility, and Alzheimer's Disease in Modern America: A History*. Baltimore: Johns Hopkins University Press, 2006.

Ballenger, Jesse F., Peter Whitehouse, Constantine G. Lykestsos, Peter V. Rabins, and Jason H. T. Karlawish, eds. *Treating Dementia: Do We Have a Pill for It?* Baltimore: Johns Hopkins University Press, 2009.

Barad, Karen. *Meeting the Universe Halfway: Quantum Physics and the Entanglement of Matter and Meaning*. Durham, N.C.: Duke University Press, 2007.

Barboza, David. "China, in a Shift, Takes on Its Alzheimer's Problem." *New York Times*, January 12, 2011.

Barnes, Barry, and John Dupré. *Genomes: And What to Make of Them*. Chicago: University of Chicago Press, 2008.

Bateman, Randall, and John Morris. "The Dominantly Inherited Alzheimer's Network Trials: An Opportunity to Prevent Alzheimer's Disease." *Alzheimer's & Dementia* 8, no. 4 (2012): 427.

Bayley, John. *Iris: A Memoir of Iris Murdoch*. London: Abacus, 2002.

Beach, Thomas G. "The History of Alzheimer Disease: Three Debates." *Journal of the History of Medicine* 42 (1987): 327–42.

Beck, Stefan, and Jörg Niewöhner. "Translating Genetic Testing and Screening in Cyprus and Germany: Contingencies, Continuities, Ordering Effects and Bio-Cultural Intimacy." In *The Handbook of Genetics & Society: Mapping the New Genomic Era*, edited by Paul Atkinson, Peter Glasner, and Margaret Lock, 76–93. New York: Routledge, 2009.

Beck, Ulrich. *World at Risk*. Cambridge: Polity Press, 2007.

Beeson, Diane, and Theresa Doksum. "Family Values and Resistance to Genetic Testing." In *Bioethics in Social Context*, edited by Barry Hoffmaster, 153–79. Philadelphia: Temple University Press, 2001.

Belluck, Pam. "Alzheimer's Stalks a Columbian Family." *New York Times*, June 1, 2010, Health sec. http://www.nytimes.com/2010/06/02/health/02alzheimers.html.

———. "New Drug Trial Seeks to Stop Alzheimer's Before It Starts." *New York Times*, May 15, 2012, Health/Research sec. http://www.nytimes.com/2012/05/16/health/research/prevention-is-goal-of-alzheimers-drug-trial.html.

Belluck, Pam, and Salvador Rodriguez. "Hoping to Crack Alzheimer's, Together as a Family." *New York Times*, October 3, 2011, Health sec. http://www.nytimes.com/2011/10/04/health/04alzheimers.html.

Bennett, David A., Philip L. De Jager, Sue E. Leurgans, and Julie A. Schneider. "Neuropathologic Intermediate Phenotypes Enhance Association to Alzheimer Susceptibility Alleles." *Neurology* 72, no. 17 (2009): 1495–1503.

Bennett, I. J., E. J. Golob, E. S. Parker, and A. Starr. "Memory Evaluation in Mild Cognitive Impairment Using Recall and Recognition Tests." *Journal of Clinical and Experimental Neuropsychology* 28, no. 8 (2006): 1408–22.

Berg, Marc. *Rationalizing Medical Work: Decision-Support Techniques and Medical Practices*. Cambridge, Mass.: MIT Press, 1997.

Berger, John. *Ways of Seeing*. London: Penguin, 1972.

Berrios, G. E. "Alzheimer Disease: A Conceptual History." *International Journal of Geriatric Psychiatry* 5 (1990): 355–65.

Bertram, Lars, and Rudolph E. Tanzi. "Alzheimer's Disease: One Disorder, Too Many Genes?" *Human Molecular Genetics* 13, no. 90001 (January 13, 2004): R135–R141.

Bertram, Lars, Christina M. Lill, and Rudolph E. Tanzi. "The Genetics of Alzheimer Disease: Back to the Future." *Neuron* 68, no. 2 (2010): 270–81.

Bertram, Lars, and Rudolph E. Tanzi. "Genome-wide Association Studies in Alzheimer's Disease." *Human Molecular Genetics* 18, no. R2 (2009): 270–81.

Binstock, R., and S. G. Post, eds. *Too Old for Health Care? Controversies in Medicine, Law, Economics, and Ethics.* Baltimore: Johns Hopkins University Press, 1991.

Blacker, Deborah, and Rudolph E. Tanzi. "Genetic Testing in the Early Diagnosis of Alzheimer Disease." In Scinto and Daffner, *The Early Diagnosis of Alzheimer's Disease,* 105–26.

Borazanci, Aimee Pasqua, Meghan K. Harris, Robert N. Schwendimann, Eduardo Gonzalez-Toledo, Amir H. Maghzi, Masoud Etemadifar, Nadejda Alekseeva, James Pinkston, Roger E. Kelley, and Alireza Minagar. "Multiple Sclerosis: Clinical Features, Pathophysiology, Neuroimaging and Future Therapies." *Future Neurology* 4, no. 2 (2009): 229–46.

Bowler, John V., David G. Munoz, Harold Merskey, and Vladimir Hachinski. "Fallacies in the Pathological Confirmation of the Diagnosis of Alzheimer's Disease." *Journal of Neurology, Neurosurgery & Psychiatry* 64, no. 1 (1998): 18–24.

Braak, H., and E. Braak. "Neuropathological Stageing of Alzheimer-Related Changes." *Journal of Neuropathology & Experimental Neurology* 59 (2000): 733–48.

Brayne, Carol. "The Elephant in the Room—Healthy Brains in Later Life, Epidemiology and Public Health." *Nature Reviews Neuroscience* 8, no. 3 (2007): 233–39.

———. "Research and Alzheimer's Disease: An Epidemiological Perspective." *Psychological Medicine* 23, no. 2 (1993): 287–96.

Brayne, Carol, and Paul Calloway. "Normal Ageing, Impaired Cognitive Function, and Senile Dementia of the Alzheimer's Type: A Continuum?" *The Lancet* 331 (1988): 1265–67.

Brayne, Carol, and Daniel Davis. "Making Alzheimer's and Dementia Research Fit for Populations." *The Lancet* 380 (2012): 1441–43.

Brayne, Carol, Paul G. Ince, Hannah A. D. Keage, Ian G. McKeith, Fiona E. Matthews, Tuomo Polvikoski, and Raimo Sulkava. "Education, the Brain and Dementia: Neuroprotection or Compensation? EClipSE Collaborative Members." *Brain* 133, no. 8 (2010): 2210–16.

Brayne, Carol, Blossom C. M. Stephan, and Fiona E. Matthews. "A European Perspective on Population Studies of Dementia." *Alzheimer's & Dementia* 7, no. 1 (2011): 3–9.

Breitner, John C. S. "Dementia—Epidemiological Considerations, Nomenclature, and a Tacit Consensus Definition." *Journal of Geriatric Psychiatry and Neurology* 19, no. 3 (2006): 129–36.

Breteler, Monique. "Mapping Out Biomarkers for Alzheimer Disease." *Journal of the American Medical Association* 305, no. 3 (2011): 304–5.

Brindle, David. "Older People Are an Asset, Not a Drain." *Guardian,* March 2, 2011.

Brisbane, Arthur. "The Trouble with Absolutes." *New York Times,* July 13, 2010, Public Editor's Journal sec. http://publiceditor.blogs.nytimes.com/2010/08/24/the-trouble-with-absolutes/.

Brown, Theodore M. "Mental Diseases." In *Companion Encyclopedia of the History of Medicine,* vol. 1, edited by W. F. Bynum and Roy Porter, 438–62. London: Routledge, 1993.

Buchanan, Anne V., Samuel Sholtis, Joan Richtsmeier, and Kenneth M. Weiss. "What Are Genes 'for' or Where Are Traits 'from'? What Is the Question?" *BioEssays* 31, no. 2 (2009): 198–208.

Burchell, Graham, Colin Gordon, and Peter Miller, eds. *The Foucault Effect: Studies in Governmentality.* Hemel Hempstead: Harvester Wheatsheaf, 1991.

Butler, Robert. *The Longevity Revolution: The Benefits and Challenges of Living a Long Life.* New York: Public Affairs, 2008.

Butler, Robert N., Richard A. Miller, Daniel Perry, Bruce A. Carnes, T. Franklin Williams, Christine Cassel, Jacob Brody, et al. "New Model of Health Promotion and Disease Prevention for the 21st Century." *British Medical Journal* 337 (2008): a399.

Callon, Michel, and Vololona Rabeharisoa. "Gino's Lesson on Humanity: Genetics, Mutual Entanglements and the Sociologist's Role." *Economy and Society* 33, no. 1 (2004): 1–27.

Canadian Study of Health and Aging Working Group. "The Incidence of Dementia in Canada." *Neurology* 55, no. 1 (2000): 66–73.

Canguilhem, Georges. *The Normal and the Pathological.* New York: Zone Books, 1991.

Carome, Michael, and Sidney Wolfe. "Florbetapir-pet Imaging and Postmortem B-amyloid Pathology." *Journal of the American Medical Association* 305, no. 18 (2011): 1857–58.

Cassels, Alan. "Drug Bust." *Common Ground*, November 2010. http://www.commonground.ca/iss/232/cg232_cassels.shtml.

Castel, Robert. "From Dangerousness to Risk." In Burchell, Gordon, and Miller, *The Foucault Effect*, 281–98.

Castellani, Rudy J., Hyoung-gon Lee, Sandra L. Siedlak, Akihiko Nunomura, Takaaki Hayashi, Masao Nakamura, Xiongwei Zhu, George Perry, and Mark A. Smith. "Reexamining Alzheimer's Disease: Evidence for a Protective Role for Amyloid-β Protein Precursor and Amyloid-β." *Journal of Alzheimer's Disease* 18, no. 2 (2009): 447–52.

Chaufan, Claudia, Brooke Hollister, Jennifer Nazareno, and Patrick Fox. "Medical Ideology as a Double-Edged Sword: The Politics of Cure and Care in the Making of Alzheimer's Disease." *Social Science & Medicine* 74 (2012): 788–95.

Chilibeck, Gillian, Margaret Lock, and Megha Sehdev. "Postgenomics, Uncertain Futures, and the Familiarization of Susceptibility Genes." *Social Science & Medicine* 72, no. 11 (2011): 1768–75.

Choudhury, Suparna, and Jan Slaby, eds. *Critical Neuroscience: A Handbook of the Social and Cultural Contexts of Neuroscience*. Chichester, UK: Wiley-Blackwell, 2012.

Christakis, N. A. "The Ellipsis of Prognosis in Modern Medical Thought." *Social Science & Medicine* 44, no. 3 (1997): 301–15.

Churchland, Patricia. *Brain-Wise: Studies in Neurophilosophy*. Cambridge, Mass.: MIT Press, 2002.

Clark, C. M., and J. A. Schneider. "Use of Florbetapir-pet for Imaging B-amyloid Pathology." *Journal of the American Medical Association* 305, no. 3 (2011): 275–83.

Cohen, Lawrence. *No Aging in India: Alzheimer's, the Bad Family, and Other Modern Things*. Berkeley: University of California Press, 1998.

Cohn, Simon. "Disrupting Images: Neuroscientific Representations in the Lives of Psychiatric Patients." In Choudhury and Slaby, *Critical Neuroscience*, 179–94.

Cole, T. *The Journey of Life: A Cultural History of Aging in America*. Cambridge: Cambridge University Press, 1991.

"The Colombian Alzheimer's Family Testing Possible Cures." *BBC News*, May 19, 2011, Latin America & Caribbean sec. http://www.bbc.co.uk/news/world-latin-america-13428265.

Comfort, Alex. *A Good Age*. New York: Simon & Schuster, 1976.

Condit, C. M. "How the Public Understands Genetics: Non-Deterministic and Non-Discriminatory Interpretations of the 'Blueprint' Metaphor." *Public Understanding of Science* 8, no. 3 (1999): 169–80.

Corbo, R. M., and R. Scacchi. "Apolipoprotein E (APOE) Allele Distribution in the World: Is APOEε4 a 'Thrifty' Allele?" *Annals of Human Genetics* 63 (1999): 301–10.

Corder, E. H., A. M. Saunders, W. J. Strittmatter, D. E. Schmechel, P. C. Gaskell, G. W. Small, A. D. Roses, J. L. Haines, and M. A. Pericak-Vance. "Gene Dose of Apolipoprotein E Type 4 Allele and the Risk of Alzheimer's Disease in Late Onset Families." *Science* 261 (1993): 921–23.

Costandi, Mo. "Disrupted Sleep Might Signal Early Stages of Alzheimer's." *Scientific American*, October 18, 2012. http://www.scientificamerican.com/article.cfm?id=disrupted-sleep-might-signal-early-stages-of-alzheimers.

Cox, S., and W. McKellin. "'There's This Thing in Our Family': Predictive Testing and the Construction of Risk for Huntington Disease." In *Sociological Perspectives on the New Genetics*, edited by P. Conrad and J. Gabe, 121–48. London: Blackwell, 1999.

Cruchaga, Carlos, Sumitra Chakraverty, Kevin Mayo, Francesco L. M. Vallania, Robi D. Mitra, Kelley Faber, Jennifer Williamson, Tom Bird, Ramon Diaz-Arrastia, Tatiana M. Foroud, Bradley F. Boeve, Neill R. Graff-Radford, Pamela St. Jean, Michael Lawson, Margaret G. Ehm, Richard Mayeux, and Alison M. Goate. "Rare Variants in APP, PSEN1 and PSEN2 Increase Risk for AD in Late-Onset Alzheimer's Disease Families." *PLoS ONE* 7, no. 2 (2012): e31039.

Crutch, Sebastian J., Ron Isaacs, and Martin N. Rossor. "Some Workmen Can Blame Their Tools: Artistic Change in an Individual with Alzheimer's Disease." *The Lancet* 357 (2001): 2129–33.

Crystal, Howard A., Dennis Dickson, Peter Davies, David Masur, Ellen Grober, and Richard B. Lipton. "The Relative Frequency of 'Dementia of Unknown Etiology' Increases with Age and Is Nearly 50% in Nonagenarians." *Archives of Neurology* 57, no. 5 (2000): 713–19.

Cupples, L. Adrienne, Lindsay A. Farrer, A. Dessa Sadovnick, Norman Relkin, Peter Whitehouse, and Robert C. Green. "Estimating Risk Curves for First-degree Relatives of Patients with Alzheimer's Disease: The REVEAL Study." *Genetics in Medicine* 6, no. 4 (2004): 192–96.

Daston, Lorraine. *Classical Probability in the Enlightenment*. Princeton: Princeton University Press, 1988.

Daston, Lorraine, and Peter Galison. *Objectivity*. New York: Zone Books, 2007.

Davis, D. G., F. A. Schmitt, D. R. Wekstein, and W. R. Markesbery. "Alzheimer Neuropathologic Alterations in Aged Cognitively Normal Subjects." *Journal of Neuropathology & Experimental Neurology* 58, no. 4 (1999): 376–88.

Daxinger, Lucia, and Emma Whitelaw. "Understanding Transgenerational Epigenetic Inheritance via the Gametes in Mammals." *Nature Reviews Genetics* 13, no. 3 (2012): 153–62.

Delvecchio Good, M. J., B. J. Good, C. Schaffer, and S. E. Lind. "American Oncology and the Discourse on Hope." *Culture, Medicine, and Psychiatry* 14, no. 1 (1990): 59–79.

De Meyer, Geert, Fred Shapiro, Hugo Vanderstichele, Eugeen Vanmechelen, Sebastiaan Engelborghs, Peter Paul De Deyn, Els Coart, et al. "Diagnosis-Independent Alzheimer Disease Biomarker Signature in Cognitively Normal Elderly People." *Archives of Neurology* 67, no. 8 (2010): 949–56.

Dickson, Samuel P., Kai Wang, Ian Krantz, Hakon Hakonarson, and David B. Goldstein. "Rare Variants Create Synthetic Genome-Wide Associations." *PLoS Biology* 8, no. 1 (2010): e1000294.

Dillman, Konrad. "Epistemological Lessons from History." In Whitehouse, Maurer, and Ballenger, *Concepts of Alzheimer Disease*, 129–57.

Dolinoy, Dana C., Radhika Das, Jennifer R. Weidman, and Randy L. Jirtle. "Metastable Epialleles, Imprinting, and the Fetal Origins of Adult Diseases." *Pediatric Research* 61, no. 5, pt. 2 (2007): 30R–37R.

Douglas, Mary. "Risk as a Forensic Resource." *Daedalus* 119, no. 4 (1990): 1–16.

Dubois, Bruno, Howard H. Feldman, Claudia Jacova, Steven T. DeKosky, Pascale Barberger-Gateau, Jeffrey Cummings, André Delacourte, et al. "Research Criteria for the Diagnosis of Alzheimer's Disease: Revising the NINCDS-ADRDA Criteria." *The Lancet Neurology* 6, no. 8 (2007): 734–46.

Dubois, Bruno, Howard H. Feldman, Claudia Jacova, Jeffrey L. Cummings, Steven T. Dekosky, Pascale Barberger-Gateau, André Delacourte, et al. "Revising the Definition of Alzheimer's Disease: A New Lexicon." *The Lancet Neurology* 9, no. 11 (2010): 1118–27.

Dubos, René. *Mirage of Health*. London: Harper and Row, 1959.

Dumit, Joseph. "Critically Producing Brain Images of Mind." In Choudhury and Slaby, *Critical Neuroscience*, 195–226.

———. *Picturing Personhood: Brain Scans and Biomedical Identity*. Princeton: Princeton University Press, 2004.

Duden, B., and S. Samerki. "'Pop Genes': An Investigation of 'the Gene' in Popular Parlance." In *Biomedicine as Culture: Instrumental Practices, Technoscientific Knowledge, and New Modes of Life*, edited by R. V. Burri and J. Dumit, 167–90. New York: Routledge, 2007.

Dupré, John. "The Polygenomic Organism." In *Nature After the Genome*, edited by Sarah Parry and John Dupré, 19–31. Oxford: Blackwell, 2010.

Duster, Troy. *Backdoor to Eugenics*. New York: Routledge, 1990.

———. "Buried Alive: The Concept of Race in Science." In Goodman, Heath, and Lindee, *Genetic Nature/Culture*, 258–77.

Duyckaerts, C., P. Delaère, J.-J. Hauw, A. L. Abbamondi-Pinto, S. Sorbi, I. Allen, J. P. Brion, et al. "Rating of the Lesions in Senile Dementia of the Alzheimer Type: Concordance between Laboratories. A European Multicenter Study under the Auspices of EURAGE." *Journal of the Neurological Sciences* 97, nos. 2–3 (1990): 295–323.

Eddy, S. R. "Non-coding RNA Genes and the Modern RNA World." *Nature Reviews Genetics* 2, no. 12 (2001): 919–29.

The Editors. "Alzheimer Disease." *Nature Medicine* 12, no. 7 (2006): 746–84.

Ekstein, J., and H. Katzenstein. "The Dor Yeshorim Story: Community-Based Carrier Screening for Tay-Sachs Disease." *Advances in Genetics* 44 (2001): 297–310.

Emslie, C., K. Hunt, and G. Watt. "A Chip Off the Old Block? Lay Understandings of Inheritance amongst Men and Women in Mid-life." *Public Understanding of Science* 12, no. 1 (2003): 47–65.

Engstrom, Eric. "Researching Dementia in Imperial Germany: Alois Alzheimer and the Economies of Psychiatric Practice." *Culture, Medicine, and Psychiatry* 31, no. 3 (2007): 405–13.

Erkinjuntti, T., T. Ostbye, R. Steenhuis, and V. Hachinski. "The Effect of Different Diagnostic Criteria on the Prevalence of Dementia." *New England Journal of Medicine* 337, no. 23 (1997): 1667–74.

Erlich, S. S., and R. L. Davis. "Alzheimer's Disease in the Very Aged." *Journal of Neuropathology & Experimental Neurology* 39, no. 3 (1980): 352–54.

Evans-Pritchard, E. E. *Witchcraft, Oracles and Magic among the Azande*. Oxford: Clarendon, 1937.

Ewald, François. "Insurance and Risk." In Burchell, Gordon, and Miller, *The Foucault Effect*, 197–210.

Ezrati, Milton. "Japan's Aging Economics." *Foreign Affairs*, May/June 1997. http://www.foreignaffairs.com/articles/53050/milton-ezrati/japans-aging-economics.

Farlow, Martin R. "Alzheimer's Disease: Clinical Implications of the Apolipoprotein E Genotype." *Neurology* 48, no. 5, suppl. 6 (1997): S30–S34.

Farrer, L. A. "Familial Risk for Alzheimer Disease in Ethnic Minorities: Nondiscriminating Genes." *Archives of Neurology* 57, no. 1 (2000): 28–29.

———. "Intercontinental Epidemiology of Alzheimer Disease: A Global Approach to Bad Gene Hunting." *Journal of the American Medical Association* 285, no. 6 (2001): 796–98.

Farrer, L. A., L. A. Cupples, J. L. Haines, B. Hyman, W. A. Kukull, R. Mayeux, R. H. Myers, M. A. Pericak-Vance, N. Risch, and C. M. van Duijn. "Effects of Age, Sex, and Ethnicity on the Association between Apolipoprotein E Genotype and Alzheimer Disease. A Meta-analysis. APOE and Alzheimer Disease Meta Analysis Consortium." *Journal of the American Medical Association* 278, no. 16 (1997): 1349–56.

Fassin, Didier. "Another Politics of Life Is Possible." *Theory, Culture & Society* 26, no. 5 (2009): 44–60.

Featherstone, Katie, Paul Atkinson, Aditva Bharadwai, and Angus Clarke. *Risky Relations: Family, Kinship and the New Genetics*. Oxford: Berg, 2006.

Finch, C. E., and R. M. Sapolsky. "The Evolution of Alzheimer Disease, the Reproductive Schedule, and apoE Isoforms." *Neurobiology of Aging* 20, no. 4 (1999): 407–28.

Fleck, Ludwik. *Genesis and Development of a Scientific Fact*. Chicago: University of Chicago Press, 1979.

Fleisher, Adam S., Kewei Chen, Yakeel T. Quiroz, Laura J. Jakimovich, Madelyn Gutierrez Gomez, Carolyn M. Langois, Jessica B. S. Langbaum, Napatkamon Ayutyanont, Auttawut Roontiva, Pradeep Thiyyagura, Wendy Lee, Hua Mo, Liliana Lopez, Sonia Moreno, Natalia Acosta-Baena, Margarita Giraldo, Gloria Garcia, Rebecca A. Reiman, Matthew J. Huentelman, Kenneth S. Kosik, Pierre N. Tariot, Francisco Lopera, and Eric M. Reiman. "Florbetapir PET Analysis of Amyloid-β Deposition in the Presenilin 1 E280A Autosomal Dominant Alzheimer's Disease Kindred: A Cross-Sectional Study." *The Lancet Neurology* 11 (2012): 1057–65.

Folstein, M. F., S. E. Folstein, and P. R. McHugh. "'Mini-mental State': A Practical Method for Grading the Cognitive State of Patients for the Clinician." *Journal of Psychiatric Research* 12, no. 3 (1975): 189–98.

Foucault, Michel. *The Birth of the Clinic: An Archaeology of Medical Perception*. Translated by A. M. Sheridan Smith. New York: Vintage Books, 1973.

——. *The History of Sexuality*. Vol. 1. New York: Vintage Books, 1980.

Fox, Nick. "When, Where, and How Does Alzheimer's Disease Start?" *The Lancet Neurology* 11 (2012): 1017–18.

Fox, Patrick. "From Senility to Alzheimer Disease: The Rise of the Alzheimer Disease Movement." *Milbank Quarterly* 67 (1989): 58–102.

——. "The Role of the Concept of Alzheimer Disease." In Whitehouse, Maurer, and Ballenger, *Concepts of Alzheimer Disease*, 209–33.

Franklin, Sarah. "Life." In *The Encyclopedia of Bioethics*, edited by Warren T. Reich, 456–62. New York: Simon & Schuster, 1995.

Franzen, Jonathan. "My Father's Brain." *The New Yorker*, September 10, 2001.

Fullerton, Stephanie M., Andrew G. Clark, Kenneth M. Weiss, Deborah A. Nickerson, Scott L. Taylor, Jari H. Stengård, Veikko Salomaa, Erkki Vartiainen, Markus Perola, Eric Boerwinkle, and Charles F. Sing. "Apolipoprotein E Variation at the Sequence Haplotype Level: Implications for the Origin and Maintenance of a Major Human Polymorphism." *American Journal of Human Genetics* 67, no. 4 (2000): 881–900.

Fullwiley, Duana. *The Encultured Gene: Sickle Cell Health Politics and Biological Difference in West Africa*. Princeton: Princeton University Press, 2012.

Gaines, Atwood, and Peter Whitehouse. "Building a Mystery: Alzheimer Disease, Mild Cognitive Impairment, and Beyond." *Philosophy, Psychiatry, & Psychology* 13 (2006): 61–74.

Galerie Beckel-Odille-Boïcos. *William Utermohlen: Paintings and Drawings 1955–1997*. Paris, 2000.

——. *William Utermohlen Œuvres: 1955–1997*. Paris, 2002.

Galison, Peter, and David Stump, eds. *The Disunity of Science—Boundaries, Contexts, and Power*. Stanford: Stanford University Press, 1996.

Garnick, Marc B. "The Great Prostate Cancer Debate." *Scientific American* 306, no. 2 (2012): 38–43.

Garrard, Peter, Lisa M. Maloney, John R. Hodges, and Karalyn Patterson. "The Effects of Very Early Alzheimer's Disease on the Characteristics of Writing by a Renowned Author." *Brain* 128, no. 2 (2005): 250–60.

Gauthier, Serge, Barry Reisberg, Michael Zaudig, Ronald C. Petersen, Karen Ritchie, Karl Broich, Sylvie Belleville, et al. "Mild Cognitive Impairment." *Lancet* 367 (2006): 1262–70.

Gearing, M., G. W. Rebeck, B. T. Hyman, J. Tigges, and S. S. Mirra. "Neuropathology and Apolipoprotein E Profile of Aged Chimpanzees: Implications for Alzheimer Disease." *Neurobiology, Proceedings of the National Academy of Sciences of the United States of America* 91, no. 20 (1994): 9382–86.

George, Daniel R. "Overcoming the Social Death of Dementia through Language." *The Lancet* 376 (2010): 586–87.

George, Danny, and Peter Whitehouse. "The Classification of Alzheimer Disease and Mild Cognitive Impairment." In Ballenger et al., *Treating Dementia*, 5–24.

George, Danny, Peter Whitehouse, Simon d'Alton, and Jesse Ballinger. "Through the Amyloid Gateway." *The Lancet* 380 (2012): 1986–87.

Gere, Cathy. "'Nature's Experiment': Epilepsy, Localization of Brain Function and the Emergence of the Cerebral Subject." In Ortega and Vidal, *Neurocultures*, 235–47.

Ghebranious, Nader, Lynn Ivacic, Jamie Mallum, and Charles Dokken. "Detection of ApoE E2, E3 and E4 Alleles Using MALDI-TOF Mass Spectrometry and the Homogeneous Mass-Extend Technology." *Nucleic Acids Research* 33, no. 17 (2005): e149.

Gibbon, Sarah, and Carlos Novas, eds. *Biosocialities, Genetics and the Social Sciences: Making Biologies and Identities*. London: Routledge, 2007.

Gilbert, Scott F. "The Genome in Its Ecological Context: Philosophical Perspectives on Interspecies Epigenesis." *Annals of the New York Academy of Sciences* 981 (2002): 202–18.

Gilbert, Walter. "A Vision of the Grail." In *The Code of Codes: Scientific and Social Issues in the Human Genome Project*, edited by Daniel Kevles and Leroy Hood, 83–97. Cambridge, Mass.: Harvard University Press, 1992.

Gilleard, C. J. "Is Alzheimer's Disease Preventable? A Review of Two Decades of Epidemiological Research." *Aging and Mental Health* 4 (2000): 101–18.

Glass, Daniel J., and Steven E. Arnold. "Some Evolutionary Perspectives on Alzheimer's Disease Pathogenesis and Pathology." *Alzheimer's & Dementia* 8, no. 4 (2012): 343–50.

Gleckman, Howard. "The Obama Administration's War on Alzheimer's." *Forbes*, January 11, 2012. http://www.forbes.com/sites/howardgleckman/2012/01/11/the-obama-administrations-war-on-alzheimers/.

Glenner, G. G. "Alzheimer's Disease: The Commonest Form of Amyloidosis." *Archives of Pathology & Laboratory Medicine* 107, no. 6 (1983): 281–82.

Goate, A., M. C. Chartier-Harlin, M. Mullan, J. Brown, F. Crawford, L. Fidani, L. Giuffra, A. Haynes, N. Irving, and L. James. "Segregation of a Missense Mutation in the Amyloid Precursor Protein Gene with Familial Alzheimer's Disease." *Nature* 349 (1991): 704–6.

Goldman, Jill S., Susan E. Hahn, Jennifer Williamson Catania, Susan LaRusse-Eckert, Melissa Barber Butson, Malia Rumbaugh, Michelle N. Strecker, et al. "Genetic Counseling and Testing for Alzheimer Disease: Joint Practice Guidelines of the American College of Medical Genetics and the National Society of Genetic Counselors." *Genetics in Medicine* 13, no. 6 (2011): 597–605.

Golomb, James, Alan Kluger, and Steven H. Ferris. "Mild Cognitive Impairment: Historical Development and Summary of Research." *Dialogues in Clinical Neuroscience* 6, no. 4 (2004): 351–67.

Goodman, Alan H., Deborah Heath, and M. Susan Lindee, eds. *Genetic Nature/Culture: Anthropology and Science beyond the Two Culture Divide*. Berkeley: University of California Press, 2003.

Gottesman, I. I. "Schizophrenia Epigenesis: Past, Present, and Future." *Acta Psychiatrica Scandinavica. Supplementum* 384 (1994): 26–33.

Graff-Radford, Neill R., Robert C. Green, Rodney C. P. Go, Michael L. Hutton, Timi Edeki, David Bachman, Jennifer L. Adamson, et al. "Association between Apolipoprotein E Genotype and Alzheimer Disease in African American Subjects." *Archives of Neurology* 59, no. 4 (2002): 594–600.

Graham, Janice E., Kenneth Rockwood, B. Lynn Beattie, Ian McDowell, Robin Eastwood, and Serge Gauthier. "Standardization of the Diagnosis of Dementia in the Canadian Study of Health and Aging." *Neuroepidemiology* 15, no. 5 (1996): 246–56.

Gravitz, Lauren. "Drugs: A Tangled Web of Targets." *Nature* 475 (2011): S9–S11.

Green, Robert C., V. C. Clarke, N. J. Thompson, J. L. Woodard, and R. Letz. "Early Detection of Alzheimer Disease: Methods, Markers, and Misgivings." *Alzheimer Disease & Associated Disorders* 11, suppl. 5 (1997): S1–S5, discussion S37–S39.

Green, Robert C., J. Scott Roberts, L. Adrienne Cupples, Norman R. Relkin, Peter J. Whitehouse, Tamsen Brown, Susan LaRusse Eckert, et al. "Disclosure of APOE Genotype for Risk of Alzheimer's Disease." *New England Journal of Medicine* 361, no. 3 (2009): 245–54.

Green, Robert, Scott Roberts, Jason Karlawish, Thomas Obisesan, L. Adrienne Cupples, Denise Lautenbach, Margaret Bradbury, et al. "Disclosure of APOE Genotype to Persons with Mild Cognitive Impairment (MCI)." *Alzheimer's & Dementia* 8, no. 4 (2012): 423.

Greicius, Michael D., Gaurav Srivastava, Allan L. Reiss, and Vinod Menon. "Default-Mode Network Activity Distinguishes Alzheimer's Disease from Healthy Aging: Evidence from Functional MRI." *Proceedings of the National Academy of Sciences of the United States of America* 101, no. 13 (2004): 4637–42.

Griffiths, Paul E. "Developmental Systems Theory." In *Nature Encyclopedia of the Life Sciences*, 1. London: John Wiley, 2002.

Growdon, John H. "Apolipoprotein E and Alzheimer Disease." *Archives of Neurology* 55, no. 8 (1998): 1053–54.

Guan, Ji-Song, Stephen J. Haggarty, Emanuela Giacometti, Jan-Hermen Dannenberg, Nadine Joseph, Jun Gao, Thomas J. F. Nieland, et al. "HDAC2 Negatively Regulates Memory Formation and Synaptic Plasticity." *Nature* 459 (2009): 55–60.

Guerreiro, Rita, Aleksandra Wojtas, Jose Bras, Minerva Carrasquillo, Ekaterina Rogaeva, Elisa Majounie, Carlos Cruchaga, Celeste Sassi, John S. K. Kauwe, Steven Younkin, Lilinaz Hazrati, John Collinge, Jennifer Pocock, Tammaryn Lashley, Julie Williams, Jean-Charles Lambert, Philippe Amouyel, Alison Goate, Rosa Rademakers, Kevin Morgan, John Powell, Peter St George-Hyslop, Andrew Singleton, and John Hardy. "*Trem2* Variants in Alzheimer's Disease." *New England Journal of Medicine* 368 (November 14, 2012): 117–27.

Gureje, Oye, Adesola Ogunniyi, and Lola Kola. "The Profile and Impact of Probable Dementia in a Sub-Saharan African Community: Results from the Ibadan Study of Aging." *Journal of Psychosomatic Research* 61, no. 3 (2006): 327–33.

Hacking, Ian. "Introduction." In Thomas Kuhn, *The Structure of Scientific Revolutions*, 50th Anniversary ed., vii–xxxvii. Chicago: University of Chicago Press, 2012.

———. "The Looping Effects of Human Kinds." In *Causal Cognition: A Multidisciplinary Approach*, edited by D. Sperber, D. Premack, and A. J. Premack, 351–83. Oxford: Oxford Medical Publications, 1995.

———. "Making up People." In *Reconstructing Individualism: Autonomy, Individuality, and the Self in Western Thought*, edited by T. C. Heller, M. Sosna, and D. E. Wellberry, 222–36. Stanford: Stanford University Press, 1986. http://www.icesi.edu.co/blogs/antro_conocimiento/files/2012/02/Hacking_making-up-people.pdf.

———. *The Taming of Chance*. Cambridge: Cambridge University Press, 1990.

Hallowell, Nina. "Doing the Right Thing: Genetic Risk and Responsibility." *Sociology of Health & Illness* 21, no. 5 (1999): 597–621.

Hampel, Harald, Simone Lista, and Zaven S. Khachaturian. "Development of Biomarkers to Chart All Alzheimer's Disease Stages: The Royal Road to Cutting the Therapeutic Gordian Knot." *Alzheimer's & Dementia* 8, no. 4 (2012): 312–36.

Haraway, Donna J. *Simians, Cyborgs, and Women: The Reinvention of Nature*. New York: Routledge, 1991.

Harding, J. E. "The Nutritional Basis of the Fetal Origins of Adult Disease." *International Journal of Epidemiology* 30, no. 1 (2001): 15–23.

Hardy, John A. "The Amyloid Cascade Hypothesis Has Misled the Pharmaceutical Industry." *Biochemical Society Transactions* 39, no. 4 (2011): 920–23.

———. "ApoE, Amyloid, and Alzheimer's Disease." *Science* 263 (1994): 454–55.

Hardy, John A., and Dennis J. Selkoe. "The Amyloid Hypothesis of Alzheimer's Disease: Progress and Problems on the Road to Therapeutics." *Science* 297 (2002): 353–56.

Hardy, John A., and Gerald A. Higgins. "Alzheimer's Disease: The Amyloid Cascade Hypothesis." *Science* 256 (1992): 184–85.

Hayden, Kathleen M., Peter P. Zandi, Constantine G. Lyketsos, Ara S. Khachaturian, Lori A. Bastian, Gene Charoonruk, JoAnn T. Tschanz, et al. "Vascular Risk Factors for Incident Alzheimer Disease and Vascular Dementia: The Cache County Study." *Alzheimer Disease & Associated Disorders* 20, no. 2 (2006): 93–100.

Heath, Deborah, and Karen-Sue Taussig. "Genetic Citizenship." In *A Companion to the Anthropology of Politics*, edited by D. Nguyent and J. Vincent, 152–67. London: Blackwell, 2004.

Heinik, Jeremia. "V. A. Kral and the Origins of Benign Senescent Forgetfulness and Mild Cognitive Impairment." *International Psychogeriatrics* 22, no. 3 (2010): 395–402.

Hendrie, Hugh. "Diagnosis of Dementia and Alzheimer's Disease in Indianapolis and Ibadan: Challenges in Cross-Cultural Studies of Aging and Dementia." *Alzheimer's & Dementia* 5, no. 4 (2009): P122.

Herskovits, A. Zara, and John H. Growdon. "Sharpen That Needle." *Archives of Neurology* 67, no. 8 (2010): 918–20.

Herskovits, E. "Struggling over Subjectivity: Debates about the 'Self' and Alzheimer Disease." *Medical Anthropology Quarterly* 9 (1996): 146–64.

Heyman, J. C., M. L. Barer Hertzman, and R. G. Evans, eds. *Healthier Societies: From Analysis to Action*. Oxford: Oxford University Press, 2006.

Hill, Shirley. *Managing Sickle Cell Disease in Low-Income Families*. Philadelphia: Temple University Press, 1994.

Holmes, Clive. "The Genetics of Alzheimer's Disease." *Menopause International* 8, no. 1 (2002): 20–23.

Holmes, Clive, Delphine Boche, David Wilkinson, Ghasem Yadegarfar, Vivienne Hopkins, Anthony Bayer, Roy W. Jones, et al. "Long-Term Effects of Aβ42 Immunisation in Alzheimer's Disease: Follow-Up of a Randomised, Placebo-Controlled Phase I Trial." *The Lancet* 372 (2008): 216–23.

Holstein, Martha. "Alzheimer Disease and Senile Dementia, 1885–1920: An Interpretive History of Disease Negotiation." *Journal of Aging Studies* 11 (1997): 1–13.

House of Lords. *Science and Technology Committee: 6th Report*. London: Parliament, 2006. http://www.publications.parliament.uk/pa/ld200506/ldselect/ldsctech/146/14603.htm.

Hughes, Charles. "Insanity Defined on the Basis of Disease." *The Alienist and Neurologist* 20 (1887): 170–74.

Hughes, Julian C., Stephen J. Louw, and Steven R. Sabat. *Dementia: Mind, Meaning, and the Person*. Oxford: Oxford University Press, 2006.

Huppert, Felicia A., Carol Brayne, and Daniel W. O'Conner, eds. *Dementia and Normal Aging*. Cambridge: Cambridge University Press, 1994.

Hyman, Bradley T. "Alzheimer's Disease or Alzheimer's Diseases? Clues from Molecular Epidemiology." *Annals of Neurology* 40, no. 2 (1996): 135–36.

Hyman, Bradley T., Creighton H. Phelps, Thomas G. Beach, Eileen H. Bigio, Nigel J. Cairns, Maria C. Carrillo, Dennis W. Dickson, et al. "National Institute on Aging–Alzheimer's Association Guidelines for the Neuropathologic Assessment of Alzheimer's Disease." *Alzheimer's & Dementia* 8, no. 1 (2012): 1–13.

Ignatieff, Michael. *Scar Tissue*. London: Chatto & Windus, 1993.

Ikels, Charlotte. "The Experience of Dementia in China." *Culture, Medicine, and Psychiatry* 22 (1998): 257–83.

"Iris Murdoch's Last Novel Reveals First Signs of Alzheimer's Disease." *UCL News*, December 1, 2004. http://www.ucl.ac.uk/news/news-articles/news-releases-archive/murdoch.

Jablonka, Eva, and Marion J. Lamb. *Evolution in Four Dimensions: Genetic, Epigenetic, Behavioral, and Symbolic Variation in the History of Life*. Cambridge, Mass.: MIT Press, 2005.

Jack, Clifford R., Jr., Marilyn S. Albert, David S. Knopman, Guy M. McKhann, Reisa A. Sperling, Maria C. Carrillo, Bill Thies, and Creighton H. Phelps. "Introduction to the Recommendations from the National Institute on Aging-Alzheimer's Association Workgroups on Diagnostic Guidelines for Alzheimer's Disease." *Alzheimer's & Dementia* 7, no. 3 (2011): 257–62.

Jack, Clifford R., Val J. Lowe, Stephen D. Weigand, Heather J. Wiste, Matthew L. Senjem, David S. Knopman, Maria M. Shiung, et al. "Serial PIB and MRI in Normal, Mild Cognitive Impairment and Alzheimer's Disease: Implications for Sequence of Pathological Events in Alzheimer's Disease." *Brain* 132, no. 5 (2009): 1355–65.

Jackson, Richard, and Neil Howe. *The Graying of the Middle Kingdom: The Demographics and Economics of Retirement Policy in China*. Washington, D.C.: Center for Strategic & International Studies, 2004. http://csis.org/files/media/csis/pubs/grayingkingdom.pdf.

Jackson, Richard, Keisuke Nakashima, and Neil Howe. *China's Long March to Retirement Reform: The Graying of the Middle Kingdom Revisited*. Washington, D.C.: Center for Strategic & International Studies, 2009. http://csis.org/files/media/csis/pubs/090422_gai_chinareport_en.pdf.

Jagust, William. "Aging, Amyloid and Neural Activity." *Alzheimer's & Dementia* 8, no. 4 (2012): 427.

———. "Tracking Brain Amyloid-β in Presymptomatic Alzheimer's Disease." *The Lancet Neurology* 11 (2012): 1018–20.

Jagust, W. J., S. M. Landau, L. M. Shaw, J. Q. Trojanowski, R. A. Koeppe, E. M. Reiman, N. L. Foster, et al. "Relationships between Biomarkers in Aging and Dementia." *Neurology* 73, no. 15 (2009): 1193–99.

Jasen, Patricia. "Breast Cancer and the Language of Risk, 1750–1950." *Social History of Medicine* 15, no. 1 (2002): 17–43.

Jilek, Wolfgang. "Emil Kraepelin and Comparative Sociocultural Psychiatry." *European Archives of Psychiatry and Clinical Neuroscience* 245 (1995): 231–38.

Jonsson, Thorlakur, Jasvinder K. Atwal, Stacy Steinberg, Jon Snaedal, Palmi V. Jonsson, Sigurbjorn Bjornsson, Hreinn Stefansson, et al. "A Mutation in APP Protects Against Alzheimer's Disease and Age-Related Cognitive Decline." *Nature* 488 (2012): 96–99.

Jonsson, Thorlakur, Hreinn Stefansson, Stacy Steinberg, Ingileif Jonsdottir, Palmi V. Jonsson, Jon Snaedal, Sigurbjorn Bjornsson, Johanna Huttenlocher, Allan I. Levey, James J. Lah, Dan Rujescu, Harald Hampel, Ina Giegling, Ole A. Andreassen, Knut Engedal, Ingun Ulstein, Srdjan Djurovic, Carla Ibrahim-Verbaa, Albert Hofman, M. Arfan Ikram, Cornelia M. van Duijn, Unnur Thorsteinsdottir, Augustine Kong, and Kari Stefansson. "Variant of *TREM2* Associated with the Risk of Alzheimer's Disease." *New England Journal of Medicine* 368 (November 14, 2012): 107–16. doi:10.1056/NEJMoa1211103

Kamboh, M. I. "Apolipoprotein E Polymorphism and Susceptibility to Alzheimer's Disease." *Human Biology* 67, no. 2 (1995): 195–215.

Karlawish, Jason. "Disclosing Amyloid Imaging Results." *Alzheimer's & Dementia* 8, no. 4 (2012): 423.

Karran, Eric, Marc Mercken, and Bart De Strooper. "The Amyloid Cascade Hypothesis for Alzheimer's Disease: An Appraisal for the Development of Therapeutics." *Nature Reviews Drug Discovery* 10, no. 9 (2011): 698–712.

Katzman, Robert. "The Prevalence and Malignancy of Alzheimer Disease: A Major Killer." *Archives of Neurology* 33 (1976): 217–18.

Katzman, Robert, and Katherine L. Bick. "The Rediscovery of Alzheimer Disease in the 1960s and 1970s." In Whitehouse, Maurer, and Ballenger, *Concepts of Alzheimer Disease*, 104–14.

Kaufman, Sharon. . . . *And a Time to Die: How American Hospitals Shape the End of Life.* New York: Scribner, 2005.

Kaufman, Sharon, Julie Livingston, Hong Zhang, and Margaret Lock. "Transforming the Concepts of Aging: Three Case Studies from Anthropology." In *Oxford Textbook of Old Age Psychiatry*, edited by Tom Dening and Alan Thomas. 39–55. Oxford: Oxford University Press,

Kavanagh, Anne M., and Dorothy H. Broom. "Embodied Risk: My Body, Myself?" *Social Science & Medicine* 46, no. 3 (1998): 437–44.

Keating, Peter, and Alberto Cambrosio. *Biomedical Platforms: Realigning the Normal and the Pathological in Late-Twentieth-Century Medicine.* Cambridge, Mass.: MIT Press, 2003.

Keller, Evelyn Fox. *The Century of the Gene.* Cambridge, Mass.: Harvard University Press, 2000.

———. *The Mirage of a Space between Nature and Nurture.* Durham, N.C.: Duke University Press, 2010.

Keller, Matthew C., and Geoffrey Miller. "Resolving the Paradox of Common, Harmful, Heritable Mental Disorders: Which Evolutionary Genetic Models Work Best?" *Behavioral Brain Science* 29, no. 4 (2006): 385–452.

Kerr, Anne, Sarah Cunningham-Burley, and Amanda Amos. "The New Genetics and Health: Mobilizing Lay Expertise." *Public Understanding of Science* 7, no. 1 (1998): 41–60.

Kertzer, David L., and Peter Laslett. *Aging in the Past: Demography, Society and Old Age.* Berkeley: University of California Press, 1995.

Khachaturian, Ara S., Michelle M. Mielke, and Zaven S. Khachaturian. "Biomarker Development: A Population-Level Perspective." *Alzheimer's & Dementia* 8, no. 4 (2012): 247–49.

Khachaturian, Zaven S. "Diagnosis of Alzheimer's Disease." *Archives of Neurology* 42, no. 11 (1985): 1097–1105.

Khachaturian, Zaven S. "Perspective on the Alzheimer's Disease Neuroimaging Initiative: Progress Report and Future Plans." *Alzheimer's & Dementia* 6, no. 3 (2010): 199–201.

———. "Plundered Memories." *The Sciences* 37, no. 4 (1997): 20–23.

Khachaturian, Zaven S., and Ara S. Khachaturian. "Prevent Alzheimer's Disease by 2020: A National Strategic Goal." *Alzheimer's & Dementia* 5, no. 2 (2009): 81–84.

Kier, Frederick J., and Victor Molinari. "'Do-It-Yourself' Dementia Testing: Issues Regarding an Alzheimer's Home Screening Test." *The Gerontologist* 43, no. 3 (2003): 295–301.

Kirkwood, Tom. *Time of Our Lives: The Science of Human Aging*. Oxford: Oxford University Press, 1999.

Kitwood, Tom. *Dementia Reconsidered: The Person Comes First*. Maidenhead: Open University Press, 1997.

Klunk, William E., Henry Engler, Agneta Nordberg, Yanming Wang, Gunnar Blomqvist, Daniel P. Holt, Mats Bergström, et al. "Imaging Brain Amyloid in Alzheimer's Disease with Pittsburgh Compound-B." *Annals of Neurology* 55, no. 3 (2004): 306–19.

Kolata, Gina. "Alzheimer's Tied to Mutation Harming Immune Response." *New York Times*, November 14, 2012.

———. "Drug Trials Test Bold Plan to Slow Alzheimer's." *New York Times*, July 16, 2010.

———. "Early Tests for Alzheimer's Pose Diagnosis Dilemma." *New York Times*, December 17, 2010.

———. "F.D.A. Sees Promise in Alzheimer's Imaging Drug." *New York Times*, January 20, 2011.

———. "Finding Suggests New Target for Alzheimer's Drugs." *New York Times*, September 1, 2010.

———. "Guidelines Seek Early Detection of Alzheimer's." *New York Times*, July 14, 2010. http://query.nytimes.com/gst/fullpage.html?res=9F06E6D91F3AF937A25754C0A9669D8B6.

———. "In Preventing Alzheimer's, Mutation May Aid Drug Quest." *New York Times*, July 11, 2012.

———. "In Spinal-Fluid Test, an Early Warning on Alzheimer's." *New York Times*, August 9, 2010.

———. "Insights Give New Hope for New Attack on Alzheimer's." *New York Times*, December 13, 2010.

———. "New Scan May Spot Alzheimer's." *New York Times*, July 13, 2010.

———. "Rare Sharing of Data Led to Results on Alzheimer's." *New York Times*, August 12, 2010, Health/Research sec. http://www.nytimes.com/2010/08/13/health/research/13alzheimer.html.

———. "Vast Gene Study Yields Insights on Alzheimer's." *New York Times*, April 3, 2011.

Konrad, Monica. *Narrating the New Predictive Genetics: Ethics, Ethnography, and Science*. Cambridge: Cambridge University Press, 2005.

Kuhn, Thomas. *The Structure of Scientific Revolutions*. Chicago: University of Chicago Press, 1962.

Lambek, Michael. *Knowledge and Practice in Mayotte: Local Discourses of Islam, Sorcery and Spirit Possession*. Toronto: Toronto University Press, 1993.

Landecker, Hannah. *Culturing Life: How Cells Became Technologies*. Cambridge, Mass.: Harvard University Press, 2007.

Langreth, Robert. "Eli Lilly Alzheimer's Disease Failure Bolsters Amyloid Theory Skeptics—Forbes." *Forbes*, August 17, 2010. http://www.forbes.com/sites/robertlangreth/2010/08/17/eli-lilly-alzheimers-failure-bolsters-skeptics-on-amyloid-theory/.

Larusse, Susan, J. Scott Roberts, Theresa M. Marteau, Heather Katsen, Erin L. Linnenbringer, Melissa Barber, Peter Whitehouse, Kimberly Quaid, Tamsen Brown, Robert C. Green, and Norman R. Relkin. "Genetic Susceptibility Testing versus Family History-Based Risk Assessment: Impact on Perceived Risk of Alzheimer's Disease." *Genetic Medicine* 7 (2005): 48–53.

Latour, Bruno. *Pandora's Hope: Essays on the Reality of Science Studies*. Cambridge, Mass.: Harvard University Press, 1999.

———. *The Pasteurization of France*. Cambridge, Mass.: Harvard University Press, 1988.

"Leading Edge: How Much Is Dementia Care Worth?" *The Lancet Neurology* 9, no. 11 (2010): 1037.

Leale, M. "The Senile Degenerations, Their Symptom-Complex and Treatment." *International Clinics* 4 (1911): 37–47.

Lee, Virginia M.-Y. "Tauists and Baptists United—Well Almost!" *Science* 293 (2001): 1446–47.

Leibing, Annette, and Lawrence Cohen, eds. *Thinking about Dementia: Culture, Loss, and the Anthropology of Senility*. New Brunswick, N.J.: Rutgers University Press, 2006.

Levine, Judith. "Managing Dad." In Ballenger et al., *Treating Dementia*, 116–24.

Lewontin, Richard C. "Science and Simplicity." *New York Review of Books* 50 (2003): 39–42.

Libon, David J., Catherine C. Price, Kenneth M. Heilman, and Murray Grossman. "Alzheimer's 'Other Dementia.'" *Cognitive and Behavioral Neurology* 19, no. 2 (2006): 112–16.

Liddell, M. B., S. Lovestone, and M. J. Owen. "Genetic Risk of Alzheimer Disease: Advising Relatives." *British Journal of Psychiatry* 178 (2001): 7–11.

Lin, Haiqun, Charles E. McCulloch, Bruce W. Turnbull, Elizabeth H. Slate, and Larry C. Clark. "A Latent Class Mixed Model for Analysing Biomarker Trajectories with Irregularly Scheduled Observations." *Statistics in Medicine* 19, no. 10 (2000): 1303–18.

Lippa, Carol F., and John C. Morris. "Alzheimer Neuropathology in Nondemented Aging Keeping Mind over Matter." *Neurology* 66, no. 12 (2006): 1801–2.

Lippman, Abby. "Led (Astray) by Genetic Maps: The Cartography of the Human Genome and Human Care." *Social Science & Medicine* 35, no. 12 (1992): 1469–96.

Lishman, W. A. "The History of Research into Dementia and Its Relationship to Current Concepts." In Huppert, Brayne, and O'Conner, *Dementia and Normal Aging*, 41–56.

Lock, Margaret. "The Alienation of Body Tissue and the Biopolitics of Immortalized Cell Lines." *Body & Society* 7, nos. 2–3 (2001): 63–91.

———. "Centering the Household: The Remaking of Female Maturity in Japan." In *Re-Imagining Japanese Women*, edited by Anne Imamura, 73–103. Berkeley: University of California Press, 1996.

———. *Encounters with Aging: Mythologies of Menopause in Japan and North America*. Berkeley: University of California Press, 1993. http://www.ucpress.edu/book.php?isbn=9780520201620.

———. "Interrogating the Human Genome Diversity Project." *Social Science & Medicine* 39, no. 5 (1994): 603–6.

———. "The Lure of the Epigenome." *The Lancet* 381 (2013) 1896–1897.

Lock, Margaret, Julia Freeman, Gillian Chilibeck, Briony Beveridge, and Miriam Padolsky. "Susceptibility Genes and the Question of Embodied Identity." *Medical Anthropology Quarterly* 21, no. 3 (2007): 256–76.

Lock, Margaret, Julia Freeman, Rosemary Sharples, and Stephanie Lloyd. "When It Runs in the Family: Putting Susceptibility Genes in Perspective." *Public Understanding of Science* 15, no. 3 (2006): 277–300.

Lock, Margaret, Stephanie Lloyd, and Janalyn Prest. "Genetic Susceptibility and Alzheimer Disease: The Penetrance and Uptake of Genetic Knowledge." In Cohen and Leibing, *Thinking about Dementia*, 123–54.

Lock, Margaret M., and Vinh-Kim Nguyen. *An Anthropology of Biomedicine*. Oxford: Wiley-Blackwell, 2010.

Lock, Margaret, Allan Young, and Alberto Cambrosio, eds. *Living and Working with the New Medical Technologies: Intersections of Inquiry*. Cambridge: Cambridge University Press, 2000.

London, Alex John, and Jonathan Kimmelman. "Justice in Translation: From Bench to Bedside in the Developing World." *The Lancet* 372 (2008): 82–85.

Lopez, Lorna M., Sarah E. Harris, Michelle Luciano, Dave Liewald, Gail Davies, Alan J. Gow, Albert Tenesa, et al. "Evolutionary Conserved Longevity Genes and Human Cognitive Abilities in Elderly Cohorts." *European Journal of Human Genetics* 20, no. 3 (2012): 341–47.

Lumey, L. H. "Decreased Birthweights in Infants after Maternal in Utero Exposure to the Dutch Famine of 1944–1945." *Paediatric and Perinatal Epidemiology* 6, no. 2 (1992): 240–53.

Lunn, Stephen. "End of Alzheimer's Curse 'a Decade Away.'" *The Australian*, September 19, 2001.

Mahley, R. W. "Apolipoprotein E: Cholesterol Transport Protein with Expanding Role in Cell Biology." *Science* 240 (1988): 622–30.

Mahley, R. W., K. H. Weisgraber, and Y. Huang. "Apolipoprotein E4: A Causative Factor and Therapeutic Target in Neuropathology, Including Alzheimer Disease." *Proceedings of the National Academy of Sciences* 103, no. 15 (2006): 5644–51.

Mandavilli, Apoorva. "The Amyloid Code." *Nature Medicine* 12, no. 7 (2006): 747–51.

Manolio, Teri A., Francis S. Collins, Nancy J. Cox, David B. Goldstein, Lucia A. Hindorff, David J. Hunter, Mark I. McCarthy, et al. "Finding the Missing Heritability of Complex Diseases." *Nature* 461 (2009): 747–53.

Marks, Harry M. *The Progress of Experiment: Science and Therapeutic Reform in the United States, 1900–1990*. Cambridge: Cambridge University Press, 1997.

Márquez, Gabriel García. *One Hundred Years of Solitude*. New York: HarperPerennial, 1991.

Martin, Paul M. V., and Estelle Martin-Granel. "2,500-Year Evolution of the Term Epidemic." *Emerging Infectious Diseases* 12, no. 6 (2006): 976–80.

Massoud, Fadi, Gayatri Devi, Yaakov Stern, Arlene Lawton, James E. Goldman, Yan Liu, Steven S. Chin, and Richard Mayeux, "A Clinicopathological Comparison of Community-Based and Clinic-Based Cohorts of Patients with Dementia," *Archives of Neurology* 56, no. 11 (1999): 1368–73.

Mastroeni, Diego, Andrew Grover, Elaine Delvaux, Charisse Whiteside, Paul D. Coleman, and Joseph Rogers. "Epigenetic Mechanisms in Alzheimer's Disease." *Neurobiology of Aging* 32, no. 7 (2011): 1161–80.

Mastroeni, Diego, Ann McKee, Andrew Grover, Joseph Rogers, and Paul D. Coleman. "Epigenetic Differences in Cortical Neurons from a Pair of Monozygotic Twins Discordant for Alzheimer's Disease." *PLoS ONE* 4, no. 8 (2009). http://www.ncbi.nlm.nih.gov/pmc/articles/PMC2719870/.

Mattick, John S. "Challenging the Dogma: The Hidden Layer of Non-Protein-Coding RNAs in Complex Organisms." *BioEssays* 25, no. 10 (2003): 930–39.

———. "The Hidden Genetic Program of Complex Organisms." *Scientific American* 291, no. 4 (2004): 60–67.

Maurer, Konrad, and Ulrike Maurer. *Alzheimer: The Life of a Physician and the Career of a Disease*. New York: Columbia University Press, 1986.

Maurer, Konrad, Stephan Volk, and Hector Gerbaldo. "Auguste D: The History of Alois Alzheimer's First Case." In Whitehouse, Maurer, and Ballenger, *Concepts of Alzheimer Disease*, 20–29.

Maury, C.P.J. "The Emerging Concept of Functional Amyloid." *Journal of Internal Medicine* 265, no. 3 (2009): 329–34.

Mayr, Ernst. "Cause and Effect in Biology." *Science* 134 (1961): 1501–6.

McGeer, Edith G., and Pat L. McGeer. "Neuroinflammation: Alzheimer Disease, and Other Aging Disorders." In *Pharmacological Mechanisms in Alzheimer's Therapeutics*, edited by A. C. Cuello, 149–66. New York: Springer, 2007. http://dx.doi.org/10.1007/978-0-387-71522-3.

McGowan, Patrick O., Aya Sasaki, Ana C. D'Alessio, Sergiy Dymov, Benoit Labonté, Moshe Szyf, Gustavo Turecki, and Michael J. Meaney. "Epigenetic Regulation of the Glucocorticoid Receptor in Human Brain Associates with Childhood Abuse." *Nature Neuroscience* 12, no. 3 (2009): 342–48.

McGowan, Patrick O., and Moshe Szyf. "The Epigenetics of Social Adversity in Early Life: Implications for Mental Health Outcomes." *Neurobiology of Disease* 39, no. 1 (2010): 66–72.

McKeown, Thomas. *The Modern Rise of Population*. New York: Academic Press, 1976.

McKie, Robin. "Discovery of 'Methuselah Gene' Unlocks Secret of Long Life." *The Observer*, February 3, 2002.

McMenemy, W. H. "Alzheimer's Disease: Problems Concerning Its Concept and Nature." *Acta Neurologica Scandinavica* 39 (1963): 369–80.

Medical Research Council. *Senile and Presenile Dementias: A Report of the MRC Subcommittee, Compiled by W. A. Lishman*. London: Medical Research Council, 1977.

Mercier, C. *Sanity and Insanity*. London: Walter Scott, 1895.

Mesulam, M. Marsel. "Neuroplasticity Failure in Alzheimer's Disease: Bridging the Gap between Plaques and Tangles." *Neuron* 24, no. 3 (1999): 521–29.

Michie, Susan, Harriet Drake, Theresa Marteau, and Martin Bobrow. "A Comparison of Public and Professionals' Attitudes towards Genetic Developments." *Public Understanding of Science* 4, no. 3 (1995): 243–53.

Midgley, Mary. *Science and Poetry*. London: Routledge, 2001.

Mintun, M. A., G. N. Larossa, Y. I. Sheline, C. S. Dence, S. Y. Lee, R. H. Mach, W. E. Klunk, C. A. Mathis, S. T. DeKosky, and J. C. Morris. "[11C]PIB in a Nondemented Population: Potential Antecedent Marker of Alzheimer Disease." *Neurology* 67, no. 3 (2006): 446–52.

Mirowski, Philip, and Robert Van Horn. "The Contract Research Organization and the Commercialization of Scientific Research." *Social Studies of Science* 35, no. 4 (2005): 503–48.

Mitchell, John J., Annie Capua, Carol Clow, and Charles R. Scriver. "Twenty-Year Outcome Analysis of Genetic Screening Programs for Tay-Sachs and Beta-Thalassemia Disease Carriers in High Schools." *American Journal of Human Genetics* 59, no. 4 (1996): 793–98.

Mitchell, Sandra D. *Unsimple Truths: Science, Complexity and Policy*. Chicago: University of Chicago Press, 2009.

Miyagi, S., N. Iwama, T. Kawabata, and K. Hasegawa. "Longevity and Diet in Okinawa, Japan: The Past, Present and Future." *Asia-Pacific Journal of Public Health* 15, suppl. (2003): S3–S9.

Möller, Hans-Jurgen, and Manuel B. Graeber. "Johann F.: The Historical Relevance of the Case for the Concept of Alzheimer Disease." In Whitehouse, Maurer, and Ballenger, *Concepts of Alzheimer Disease*, 30–46.

Moreira, Tiago. "Truth and Hope in Drug Development and Evaluation in Alzheimer disease." In Ballenger et al., *Treating Dementia*, 210–30.

Moreira, Tiago, Carl May, and John Bond. "Regulatory Objectivity in Action Mild Cognitive Impairment and the Collective Production of Uncertainty." *Social Studies of Science* 39, no. 5 (2009): 665–90.

Moreira, Tiago, and Paolo Palladino. "Ageing between Gerontology and Biomedicine." *BioSocieties* 4, no. 4 (2009): 349–65.

Morris, John C. "The Relationship of Plaques and Tangles to Alzheimer Disease Phenotype." In *Pathobiology of Alzheimer's Disease*, edited by Alison M. Goate and Frank Ashall, 193–218. London: Academic Press, 1995.

———. "Revised Criteria for Mild Cognitive Impairment May Compromise the Diagnosis of Alzheimer Disease Dementia." *Archives of Neurology* 69, no. 6 (2012): 700–708.

Morris, John C., Catherine M. Roe, Elizabeth A. Grant, Denise Head, Martha Storandt, Alison M. Goate, Anne M. Fagan, David M. Holtzman, and Mark A. Mintun. "Pittsburgh Compound B Imaging and Prediction of Progression from Cognitive Normality to Symptomatic Alzheimer Disease." *Archives of Neurology* 66, no. 12 (2009): 1469–75.

Morris, John C., Catherine M. Roe, Chengjie Xiong, Anne M. Fagan, Alison M. Goate, David M. Holtzman, and Mark A. Mintun. "APOE Predicts Aβ but Not Tau Alzheimer's Pathology in Cognitively Normal Aging." *Annals of Neurology* 67, no. 1 (2010): 122–31.

Mosammaparast, Nima, and Yang Shi. "Reversal of Histone Methylation: Biochemical and Molecular Mechanisms of Histone Demethylases." *Annual Review of Biochemistry* 79 (2010): 155–79.

Mudher, Amritpal, and Simon Lovestone. "Alzheimer's Disease—Do Tauists and Baptists Finally Shake Hands?" *Trends in Neurosciences* 25, no. 1 (2002): 22–26.

Mukherjee, Siddhartha. *The Emperor of All Maladies: A Biography of Cancer*. New York: Scribner, 2010.

Munir, Kerim, Suzanne Coulter, John H. Growdon, Ann MacDonald, Patrick J. Skerrett, and Jane A. Leopold. "How to Solve Three Puzzles." *Newsweek* 151, no. 3 (January 21, 2008): 64–66.

Murphy, Tom. "Alzheimer's Drug Fails Study but Flashes Potential." *Associated Press*, August 24, 2012.

Nagy, Z., M. M. Esiri, K. A. Jobst, et al. "The Effects of Additional Pathology on the Cognitive Deficit in Alzheimer Disease." *Journal of Neuropathology & Experimental Neurology* 56 (1997): 163–70.

Naj, Adam C., Gyungah Jun, Gary W. Beecham, Li-San Wang, Badri Narayan Vardarajan, Jacqueline Buros, Paul J. Gallins, et al. "Common Variants at MS4A4/MS4A6E, CD2AP, CD33 and EPHA1 Are Associated with Late-Onset Alzheimer's Disease." *Nature Genetics* 43, no. 5 (2011): 436–41.

Nascher, Ignatz. "Senile Mentality." *International Clinics* 4 (1911): 48–59.

Nasreddine, Ziad S., Natalie A. Phillips, Valérie Bédirian, Simon Charbonneau, Victor Whitehead, Isabelle Collin, Jeffrey L. Cummings, and Howard Chertkow. "The Montreal Cognitive Assessment, MoCA: A Brief Screening Tool for Mild Cognitive Impairment." *Journal of the American Geriatric Society* 53, no. 4 (2005): 695–99.

National Institutes of Health. *Alzheimer Disease Research Summit 2012: Path to Treatment and Prevention.* Bethesda, Md.: National Institutes of Health, 2012. http://www.nia.nih.gov/about /events/2012/alzheimers-disease-research-summit-2012-path-treatment-and-prevention.

Nee, L. E., R. Eldridge, T. Sunderland, C. B. Thomas, D. Katz, K. E. Thompson, H. Weingartner, H. Weiss, C. Julian, and R. Cohen. "Dementia of the Alzheimer Type: Clinical and Family Study of 22 Twin Pairs." *Neurology* 37, no. 3 (1987): 359–63.

Neumann-Held, Eva M., and Christopher Rehmann-Sutter. *Genes in Development: Rereading the Molecular Paradigm.* Durham, N.C.: Duke University Press, 2006.

Neuropathology Group of the Medical Research Council Cognitive Function and Ageing Study. "Pathological Correlates of Late-Onset Dementia in a Multicentre, Community-Based Population in England and Wales." *The Lancet* 357 (2001): 169–75.

"New Alzheimer's Genes Identified." *CBC News—Health,* April 3, 2011. http://www.cbc.ca /news/health/story/2011/04/01/alzheimer-genes-identified.html.

Nichter, Mark. *Global Health: Why Cultural Perceptions, Social Representations, and Biopolitics Matter.* Tucson: University of Arizona Press, 2008.

Niewöhner, Jörg. "Epigenetics: Embedded Bodies and the Molecularization of Biography and Milieu." *BioSocieties* 6 (2011): 279–98.

Niewöhner, Jörg, Martin Döring, Michalis Kontopodis, Jeannette Madarász, and Christoph Heintze, "Cardiovascular Disease and Obesity Prevention in Germany: An Investigation into a Heterogeneous Engineering Project," *Science, Technology & Human Values* 36, no. 5 (2011): 723–51.

Noë, Alva. *Out of Our Heads: Why You Are Not Your Brain and Other Lessons from the Biology of Consciousness.* New York: Farrar, Straus and Giroux, 2009.

"Northern Ireland Dementia Total More Than Estimated." *BBC News,* February 3, 2010. http:// news.bbc.co.uk/2/hi/uk_news/northern_ireland/8494975.stm.

Novas, Carlos, and Nikolas Rose. "Genetic Risk and the Birth of the Somatic Individual." *Economy and Society* 29, no. 4 (2000): 485–513.

Office of Public Affairs, University of California, Santa Barbara. "Clinical Trials for Alzheimer's Disease Preventative Drug to Begin Early 2013." News release, May 21, 2012. http://www .ia.ucsb.edu/pa/display.aspx?pkey=2734.

Ortega, Francisco, and Fernando Vidal, eds. *Neurocultures: Glimpses into an Expanding Universe.* Frankfurt: Peter Lang, 2011.

"Over the Mountain to Alzheimer's." *Maclean's,* January 4, 2010. http://www2.macleans.ca /2010/01/04/over-the-mountain-to-alzheimers/.

Oyama, Susan, Paul E. Griffiths, and Russel D. Gray. *Cycles of Contingency: Developmental Systems and Evolution, Life and Mind.* Cambridge, Mass.: MIT Press, 2001.

Paradies, Yin C., Michael J. Montoya, and Stephanie M. Fullerton. "Racialized Genetics and the Study of Complex Diseases: The Thrifty Genotype Revisited." *Perspectives in Biology and Medicine* 50, no. 2 (2007): 203–27.

Parra, Mario A., Sharon Abrahams, Robert H. Logie, Luis G. Méndez, Francisco Lopera, and Sergio Della Sala. "Visual Short-Term Memory Binding Deficits in Familial Alzheimer's Disease." *Brain* 133, no. 9 (2010): 2702–13.

Parry, Sarah, and John Dupré. "Introducing Nature After the Genome." In *Nature After the Genome,* edited by Sarah Parry and John Dupré, 3–16. Oxford: Blackwell, 2010.

Patra, Prasanna Kumar, and Margaret Sleeboom-Faulkner. "Bionetworking: Experimental Stem Cell Therapy and Patient Recruitment in India." *Anthropology & Medicine* 16, no. 2 (2009): 147–63.

Peine, Alexander. "Challenging Incommensurability: What We Can Learn from Ludwik Fleck and the Analysis of Configurational Innovation." *Minerva* 49 (2011): 489–508.

Pepin, Jacques. *The Origin of AIDS*. Cambridge: Cambridge University Press, 2011.

Perry, Elaine K., Peter H. Gibson, Garry Blessed, Robert H. Perry, and Bernard E. Tomlinson. "Neurotransmitter Enzyme Abnormalities in Senile Dementia. Choline Acetyltransferase and Glutamic Acid Decarboxylase Activities in Necropsy Brain Tissue." *Journal of the Neurological Sciences* 34, no. 2 (1977): 247–65.

Perry, Elaine K., Robert H. Perry, Peter H. Gibson, Garry Blessed, and Bernard E. Tomlinson. "A Cholinergic Connection between Normal Aging and Senile Dementia in the Human Hippocampus." *Neuroscience Letters* 6, no. 1 (1977): 85–89.

Perry, George, Akihiko Nunomura, Arun K. Raina, and Mark A. Smith. "Amyloid-β Junkies." *The Lancet* 355 (2000): 757.

Petersen, Ronald C. "The Current Status of Mild Cognitive Impairment—What Do We Tell Our Patients?" *Nature Clinical Practice Neurology* 3, no. 2 (2007): 60–61.

———, ed. *Mild Cognitive Impairment: Aging to Alzheimer's Disease*. Oxford: Oxford University Press, 2003.

———. "Mild Cognitive Impairment Is Relevant." *Philosophy, Psychiatry, & Psychology* 13, no. 1 (2006): 45–49.

Petersen, Ronald C., Rachelle Doody, Alexander Kurz, Richard C. Mohs, John C. Morris, Peter V. Rabins, Karen Ritchie, Martin Rossor, Leon Thal, and Bengt Winblad. "Current Concepts in Mild Cognitive Impairment." *Archives of Neurology* 58, no. 12 (2001): 1985–92.

Petersen, Ronald C., Stephen C. Waring, Glenn E. Smith, Eric G. Tangalos, and Stephen N. Thibodeau. "Predictive Value of APOE Genotyping in Incipient Alzheimer's Disease." *Annals of the New York Academy of Sciences* 802, no. 1 (1996): 58–69.

Petronis, A. "Human Morbid Genetics Revisited: Relevance of Epigenetics." *Trends in Genetics* 17, no. 3 (2001): 142–46.

Petryna, Adriana. "Clinical Trials Offshored: On Private Sector Science and Public Health." *BioSocieties* 2, no. 1 (March 2007): 21–40.

Pickersgill, Martyn, Sarah Cunningham-Burley, and Paul Martin. "Constituting Neurologic Subjects: Neuroscience, Subjectivity and the Mundane Significance of the Brain." *Subjectivity* 4 (2011): 346–65.

Pike, Kerryn E., Greg Savage, Victor L. Villemagne, Steven Ng, Simon A. Moss, Paul Maruff, Chester A. Mathis, William E. Klunk, Colin L. Masters, and Christopher C. Rowe. "B-amyloid Imaging and Memory in Non-demented Individuals: Evidence for Preclinical Alzheimer's Disease." *Brain* 130, no. 11 (2007): 2837–44.

Pimplikar, Sanjay W. "Alzheimer's Isn't Up to the Tests." *New York Times*, July 19, 2010, Opinion sec. http://www.nytimes.com/2010/07/20/opinion/20pimplikar.html.

Pollack, Andrew. "Alzheimer's Drug Fails Its First Big Clinical Trial." *New York Times*, July 23, 2012.

Post, S. G., P. J. Whitehouse, R. H. Binstock, T. D. Bird, S. K. Eckert, L. A. Farrer, L. M. Fleck, et al. "The Clinical Introduction of Genetic Testing for Alzheimer Disease: An Ethical Perspective." *Journal of the American Medical Association* 277, no. 10 (1997): 832–36.

Powers, Evan T., Richard I. Morimoto, Andrew Dillin, Jeffery W. Kelly, and William E. Balch. "Biological and Chemical Approaches to Diseases of Proteostasis Deficiency." *Annual Review of Biochemistry* 78 (2009): 959–91.

Prainsack, Barbara, and Gil Siegal. "The Rise of Genetic Couplehood? A Comparative View of Premarital Genetic Testing." *BioSocieties* 1, no. 1 (2006): 17–36.

Price, Joseph L., Daniel W. McKeel Jr., Virginia D. Buckles, Catherine M. Roe, Chengjie Xiong, Michael Grundman, Lawrence A. Hansen, et al. "Neuropathology of Nondemented Aging:

Presumptive Evidence for Preclinical Alzheimer Disease." *Neurobiology of Aging* 30, no. 7 (2009): 1026–36.

Prince, Martin, Daisy Acosta, Helen Chiu, Marcia Scazufca, and Mathew Varghese. "Dementia Diagnosis in Developing Countries: A Cross-Cultural Validation Study." *Lancet* 361 (2003): 909–17.

Prince, Martin, Renata Bryce, Emiliano Albanese, Anders Wimo, Wagner Ribeiro, Cleusa P. Ferri. "The Global Prevalence of Dementia: A Systematic Review and Metaanalysis." *Alzheimer's and Dementia* 9 (2013): 63–75.

Purdy, Michael C. "Investigational Drugs Chosen for Major Alzheimer's Prevention Method." Washington University in St. Louis, 2012. http://www.wustel.edu.

Quaid, Kimberley A., and Michael Morris. "Reluctance to Undergo Predictive Testing: The Case of Huntington Disease." *American Journal of Medical Genetics* 45, no. 1 (1993): 41–45.

Quiroz, Yakeel T., Andrew E. Budson, Kim Celone, Adriana Ruiz, Randall Newmark, Gabriel Castrillón, Francisco Lopera, and Chantal E. Stern. "Hippocampal Hyperactivation in Presymptomatic Familial Alzheimer's Disease." *Annals of Neurology* 68, no. 6 (2010): 865–75.

Rabinow, Paul. "Afterword: Concept Work." In Gibbon and Novas, *Biosocialities, Genetics and the Social Sciences*, 188–92.

———. "Artificiality and Enlightenment: From Sociobiology to Biosociality." In *Essays on the Anthropology of Reason*, 91–111. Princeton: Princeton University Press, 1996.

Raeburn, Paul. "NY Times Strangely Quiet on Alzheimer's Test That 'Can Be 100 Percent Accurate.'" *Knight Science Journalism at MIT*, January 21, 2011. http://ksj.mit.edu /tracker/2011/01/ny-times-strangely-quiet-alzheimers-test.

Raiha, I., J. Kaprio, M. Koskenvuo, T. Rajala, and L. Sourander. "Alzheimer's Disease in Finnish Twins." *Lancet* 347 (1996): 573–78.

Rapp, Rayna. "Cell Life and Death, Child Life and Death: Genomic Horizons, Genetic Diseases, Family Stories." In *Remaking Life and Death: Toward an Anthropology of the Biosciences*, edited by Sarah Franklin and Margaret Lock, 23–60. Santa Fe, N.Mex.: School of American Research Press, 2004.

———. *Testing Women, Testing the Fetus: The Social Impact of Amniocentesis*. New York: Routledge, 1999.

Rapp, Rayna, Deborah Heath, and Karen-Sue Taussig. "Genealogical Disease: Where Hereditary Abnormality, Biomedical Explanation, and Family Responsibility Meet." In *Relative Matters: New Directions in the Study of Kinship*, edited by Sarah Franklin and Susan MacKinnon, 384–412. Durham, N.C.: Duke University Press, 2001.

Redlich, F. C. "The Concept of Health in Psychiatry." In *Explorations in Social Psychiatry*, edited by Alexander H. Leighton, J. N. Clausen, and R. N. Wilson, 138–64. New York: Basic Books, 1957.

Reiman, Eric, Francisco Lopera, Jessica Langbaum, Adam Fleisher, Naparkamon Ayutyanont, Yakeel Quiros, Laura Jakimovitch, Carolyn Langlois, and Pierre Tariot. "The Alzheimer's Prevention Initiative." *Alzheimer's & Dementia* 8, no. 4 (2012): 427.

Reiman, Eric, Yakeel T. Quiroz, Adam S. Fleisher, Kewei Chen, Carlos Velez-Pardo, Marlene Jimenez-Del-Rio, Anne M. Fagan, Aarti R. Shah, Sergio Alvarez, Andrés Arbelaez, Margarita Giraldo, Natalia Acosta-Baena, Reisa A. Sperling, Brad Dickerson, Chantal E. Stern, Victoria Tirado, Claudia Munoz, Rebecca A. Reiman, Matthew J. Huentelman, Gene E. Alexander, Jessica B. S. Langbaum, Kenneth S. Kosik, Pierre N. Tariot, and Francisco Lopera. "Brain Imaging and Fluid Biomarker Analysis in Young Adults at Genetic Risk for Autosomal Dominant Alzheimer's Disease in the Presenilin 1 E280A Kindred: A Case-Control Study." *The Lancet Neurology* 11 (2012): 1048–56.

Rheinberger, Hans-Jorg "Beyond Nature and Culture: modes of reasoning in the age of molecular biology and medicine." In Lock, Young, and Cambrosio, *Living and Working with the New Medical Technologies*, 19–30.

Relkin, Norman R. "Apolipoprotein E Genotyping in Alzheimer's Disease." *The Lancet* 347 (1996): 1091–95.

Relkin, Norman R., Younga J. Kwon, Julia Tsai, and Samuel Gandy. "The National Institute on Aging/Alzheimer's Association Recommendations on the Application of Apolipoprotein E Genotyping to Alzheimer's Disease." *Annals of the New York Academy of Sciences* 802, no. 1 (1996): 149–76.

Reynolds, Susan. *Questioning Misfortune: The Pragmatics of Uncertainty in Eastern Uganda.* Cambridge: Cambridge University Press, 1997.

Rheinberger, Hans-Jörg. "Beyond Nature and Culture: Modes of Reasoning in the Age of Molecular Biology and Medicine." In Lock, Young, and Cambrosio, *Living and Working with the New Medical Technologies*, 19–30.

Richards, Marcus, and Carol Brayne. "What Do We Mean by Alzheimer Disease?" *British Medical Journal* 341 (2010): 865–67.

Richards, Martin. "Lay and Professional Knowledge of Genetics and Inheritance." *Public Understanding of Science* 5, no. 3 (1996): 217–30.

Ritchie, K., and A. M. Dupuy. "The Current Status of APOε4 as a Risk Factor for Alzheimer's Disease: An Epidemiological Perspective." *International Journal of Geriatric Psychiatry* 14, no. 9 (1999): 695–700.

Ritchie, K., D. Leibovici, B. Lessert, and J. Touchon. "A Typology of Sub-clinical Senescent Cognitive Disorder." *British Journal of Psychiatry* 168 (1966): 470–76.

Roberts, J. S., K. D. Christensen, and R. C. Green. "Using Alzheimer's Disease as a Model for Genetic Risk Disclosure: Implications for Personal Genomics." *Clinical Genetics* 80, no. 5 (2011): 407–14.

Robertson, Jennifer. "Robo Sapiens Japanicus: Humanoid Robots and the Posthuman Family." *Critical Asian Studies* 39, no. 3 (2007): 369–98.

Román, Gustavo C., and Donald R. Royall. "Executive Control Function: A Rational Basis for the Diagnosis of Vascular Dementia." *Alzheimer Disease & Associated Disorders* 13 (1999): S4–S8.

Rose, Nikolas. *The Politics of Life Itself: Biomedicine, Power, and Subjectivity in the Twenty-First Century.* Princeton: Princeton University Press, 2006.

Rosenberg, Charles E. "The Tyranny of Diagnosis: Specific Entities and Individual Experience." *Milbank Quarterly* 80, no. 2 (2002): 237–60.

———. "What Is Disease? In Memory of Owsei Temkin." *Bulletin of the History of Medicine* 77, no. 3 (2003): 491–505.

Roses, Allen D. "Apolipoprotein E and Alzheimer's Disease: A Rapidly Expanding Field with Medical and Epidemiological Consequences." *Annals of the New York Academy of Sciences* 802, no. 1 (1996): 50–57.

———. "Apolipoprotein E and Alzheimer's Disease: The Tip of the Susceptibility Iceberg." *Annals of the New York Academy of Sciences* 855, no. 1 (1998): 738–43.

Roth, Martin. "Dementia and Normal Aging of the Brain." In Huppert, Brayne, and O'Conner, *Dementia and Normal Aging*, 57–76.

Roth, Martin, Bernard E. Thomlinson, and Gary Blessed. "Correlation between Scores for Dementia and Counts of Senile Plaques in Cerebral Grey Matter of Elderly Subjects." *Nature* 209 (1966): 109–10.

Rothschild, David. "The Practical Value of Research in the Psychoses of Later Life." *Diseases of the Nervous System* 8, no. 4 (1947): 123–28.

Runciman, David. "Will We Be All Right in the End?" *London Review of Books* 34, no. 1 (2012): 3–5.

Sato, Ashida, Laura M. Koehly, J. Scott Roberts, Clara A. Chen, Susan Hiraki, and Robert C. Green. "Disclosing the Disclosure: Factors Associated with Communicating the Results of Genetic Susceptibility Testing for Alzheimer's Disease." *Journal of Health Communication* 14, no. 8 (2009): 768–84.

Saunders, A. M., W. J. Strittmatter, D. Schmechel, P. H. St George-Hyslop, M. A. Pericak-Vance, S. H. Joo, B. L. Rosi, J. F. Gusella, D. R. Crapper-MacLachlan, and M. J. Alberts. "Association of Apolipoprotein E Allele Epsilon 4 with Late-Onset Familial and Sporadic Alzheimer's Disease." *Neurology* 43, no. 8 (1993): 1467–72.

Savva, George M., Stephen B. Wharton, Paul G. Ince, Gillian Forster, Fiona E. Matthews, and Carol Brayne. "Age, Neuropathology, and Dementia." *New England Journal of Medicine* 360, no. 22 (2009): 2302–9.

Schiffman, Mark, Nicolas Wentzensen, Sholom Wacholder, Walter Kinney, Julia C. Gage, and Philip E. Castle. "Human Papillomavirus Testing in the Prevention of Cervical Cancer." *Journal of the National Cancer Institute* 103, no. 5 (2011): 368–83.

Schipper, Hyman M. "Presymptomatic Apolipoprotein E Genotyping for Alzheimer's Disease Risk Assessment and Prevention." *Alzheimer's & Dementia* 7, no. 4 (2011): e118–e123.

Schneider, Julie A., Zoe Arvanitakis, Woojeong Bang, and David A. Bennett. "Mixed Brain Pathologies Account for Most Dementia Cases in Community-Dwelling Older Persons." *Neurology* 69, no. 24 (2007): 2197–2204.

Schneider, Julie A., Zoe Arvanitakis, Sue E. Leurgans, and David A. Bennett. "The Neuropathology of Probable Alzheimer Disease and Mild Cognitive Impairment." *Annals of Neurology* 66, no. 2 (2009): 200–208.

Scinto, Leonard F. M., and Kirk R. Daffner, eds. *The Early Diagnosis of Alzheimer's Disease*. Totowa, N.J.: Humana Press, 2000.

Scriver, C. R., and P. J. Waters. "Monogenic Traits Are Not Simple: Lessons from Phenylketonuria." *Trends in Genetics* 15, no. 7 (1999): 267–72.

Segelken, Roger H. "Alzheimer Activist." *New York Times*, September 24, 2008.

Selkoe, Dennis J. "The Pathophysiology of Alzheimer's Disease." In Scinto and Daffner, *The Early Diagnosis of Alzheimer's Disease*, 83–104.

Serematakis, Nadia C. *The Last Word: Women, Death, and Divination in Inner Mani*. Chicago: University of Chicago Press, 1991.

Shah, Yogesh. "Gray Tsunami: Challenges and Solutions of Global Aging." *DMU Magazine*, Summer 2011. http://www.dmu.edu/magazine/summer-2011/my-turn-summer-2011/gray-tsunami-challenges-and-solutions-of-global-aging/.

Sheriff, Natasija. "Dental Health Linked to Risk of Developing Dementia." *Globe and Mail*, August 22, 2012.

Shi, Hui, Christopher Medway, James Bullock, Kristelle Brown, Noor Kalsheker, and Kevin Morgan. "Analysis of Genome-Wide Association Study (GWAS) Data Looking for Replicating Signals in Alzheimer's Disease (AD)." *International Journal of Molecular Epidemiology and Genetics* 1, no. 1 (2009): 53–66.

Siegmund, Kimberly D., Caroline M. Connor, Mihaela Campan, Tiffany I. Long, Daniel J. Weisenberger, Detlev Biniszkiewicz, Rudolf Jaenisch, Peter W. Laird, and Schahram Akbarian. "DNA Methylation in the Human Cerebral Cortex Is Dynamically Regulated throughout the Life Span and Involves Differentiated Neurons." *PLoS ONE* 2, no. 9 (2007). http://www.ncbi.nlm.nih.gov/pmc/articles/PMC1964879/.

Simchowicz, T. "Histologische Studien über die senile Demenz." In *Histologische und histopathologische Arbeiten*, vol. 4, edited by F. Nissl and A. Alzheimer, 267–444. Jena: Fischer, 1913.

Slaby, Jan. "Steps towards a Critical Neuroscience." *Phenomenology and the Cognitive Sciences* 9, no. 3 (2010): 397–416.

Small, G. W., S. Komo, A. La Rue, S. Saxena, M. E. Phelps, J. C. Mazziotta, A. M. Saunders, J. L. Haines, M. A. Pericak-Vance, and A. D. Roses. "Early Detection of Alzheimer's Disease by Combining Apolipoprotein E and Neuroimaging." *Annals of the New York Academy of Sciences* 802 (1996): 70–78.

Smith, Eric E., and Steven M. Greenberg. "Beta-Amyloid, Blood Vessels and Brain Function." *Stroke* 40, no. 7 (2009): 2601–6.

Smith, Glenn E., Ronald C. Petersen, Joseph E. Parisi, Robert J. Ivnik, Emre Kokmen, Eric G. Tangalos, and Stephen Waring. "Definition, Course, and Outcome of Mild Cognitive Impairment." *Aging, Neuropsychiatry and Cognition* 3 (1996): 141–47.

Snowdon, David. "Aging and Alzheimer Disease: Lessons from the Nun Study." *The Gerontologist* 35 (1997): 150–56.

———. *Aging with Grace: What the Nun Study Teaches Us about Leading Longer, Healthier and More Meaningful Lives*. New York: Bantam Books, 2001.

Sontag, Estelle, Christa Hladik, Lisa Montgomery, Ampa Luangpirom, Ingrid Mudrak, Egon Ogris, and Charles L. White III. "Downregulation of Protein Phosphatase 2A Carboxyl Methylation and Methyltransferase May Contribute to Alzheimer Disease Pathogenesis." *Journal of Neuropathology & Experimental Neurology* 63, no. 10 (2004): 1080–91.

Soscia, Stephanie J., James E. Kirby, Kevin J. Washicosky, Stephanie M. Tucker, Martin Ingelsson, Bradley Hyman, Mark A. Burton, et al. "The Alzheimer's Disease-Associated Amyloid β-Protein Is an Antimicrobial Peptide." *PLoS ONE* 5, no. 3 (2010): e9505.

Sperling, Reisa A., Paul S. Aisen, Laurel A. Beckett, David A. Bennett, Suzanne Craft, Anne M. Fagan, Takeshi Iwatsubo, et al. "Toward Defining the Preclinical Stages of Alzheimer's Disease: Recommendations from the National Institute on Aging-Alzheimer's Association Workgroups on Diagnostic Guidelines for Alzheimer's Disease." *Alzheimer's & Dementia* 7, no. 3 (2011): 280–92.

Sperling, Reisa A., and Scott Roberts. "Disclosure of Amyloid Status in Secondary Prevention Trials for Alzheimer's Disease." *Alzheimer's & Dementia* 8, no. 4 (2012): 423.

Stanford, P. K. *Exceeding Our Grasp: Science, History and the Problem of Unconceived Alternatives*. New York: Oxford University Press, 2006.

Star, Susan Leigh. "Cooperation without Consensus in Scientific Problem Solving: Dynamics of Closure in Open Systems." In *CSCW: Cooperation or Conflict?*, edited by Steve Easterbrook, 93–106. London: Springer, 1993.

Steenhuysen, Julie. "Roche Alzheimer's Drug Picked for Major Test." *Reuters*, May 15, 2012. http://www.reuters.com/article/2012/05/15/us-alzheimers-genentech -idUSBRE84E0UJ20120515.

Stix, Gary. "Obama's War on Alzheimer's: Will We Be Able to Treat the Disease by 2025?" *Scientific American*, January 31, 2012. http://blogs.scientificamerican.com/observations/2012 /01/31/obamas-war-on-alzheimers-will-we-be-able-to-treat-the-disease-by-2025/.

Storandt, Martha, Mark A. Mintun, Denise Head, and John C. Morris. "Cognitive Decline and Brain Volume Loss Are Signatures of Cerebral Aβ Deposition Identified with PIB." *Archives of Neurology* 66, no. 12 (2009): 1476–81.

Stotz, Karola, Adam Bostanci, and Paul Griffiths. "Tracking the Shift to 'Postgenomics.'" *Community Genetics* 9, no. 3 (2006): 190–96.

Strittmatter, W. J., D. Y. Huang, R. Bhasin, A. D. Roses, and D. Goldgaber. "Avid Binding of Beta A Amyloid Peptide to Its Own Precursor." *Experimental Neurology* 122, no. 2 (1993): 327–34.

Strobel, Gabrielle. "Paper Alert: GWAS Hits Clusterin, CR1, PICALM Formally Published." *Alzheimer Research Forum*, September 7, 2009. http://www.alzforum.org/new/detail .asp?id=2233.

Strohman, R. "A New Paradigm for Life: Beyond Genetic Determinism." *California Monthly* 111 (2001): 4–27.

Sullivan, Michele G. "Studies Take Aim at Groups at High Risk for Alzheimer's." *WorldCare Clinical*, March 21, 2011.

Swartz, R. H., S. E. Black, and P. St George-Hyslop. "Apolipoprotein E and Alzheimer's Disease: A Genetic, Molecular and Neuroimaging Review." *Canadian Journal of Neurological Sciences* 26, no. 2 (1999): 77–88.

Szyf, Moshe. "The Early Life Social Environment and DNA Methylation: DNA Methylation Mediating the Long-Term Impact of Social Environments Early in Life." *Epigenetics* 8 (2011): 971–78.

Szyf, Moshe, Patrick McGowan, and Michael J. Meaney. "The Social Environment and the Epigenome." *Environmental and Molecular Mutagenesis* 49, no. 1 (2008): 46–60.

Taussig, Karen-Sue, Rayna Rapp, and Deborah Heath. "Flexible Eugenics: Technologies of the Self in the Age of Genetics." In Goodman, Heath, and Lindee, *Genetic Nature/Culture*, 58–76.

Tedde, Andrea, Benedetta Nacmias, Monica Ciantelli, Paolo Forleo, Elena Cellini, Silvia Bagnoli, Carolina Piccini, Paolo Caffarra, Enrico Ghidoni, Marco Paganini, Laura Bracco, and Sandro Sorbi. "Identification of New Presenilin Gene Mutations in Early-Onset Familial Alzheimer Disease." *Archives of Neurology* 60, no. 11 (2003): 1541–44.

Templeton, Alan R. "The Complexity of the Genotype-Phenotype Relationship and the Limitations of Using Genetic 'Markers' at the Individual Level." *Science in Context* 11, nos. 3–4 (1998): 373–89.

Teng, E. L., and H. C. Chui. "The Modified Mini-Mental State (3MS) Examination." *Journal of Clinical Psychiatry* 48, no. 8 (1987): 314–18.

Tilley, L., K. Morgan, and N. Kalsheker. "Genetic Risk Factors in Alzheimer's Disease." *Molecular Pathology* 51, no. 6 (1998): 293–304.

Torack, Richard M. *The Pathologic Physiology of Dementia, with Indications for Diagnosis and Treatment.* Berlin: Springer-Verlag, 1978.

Traphagan, John W. *Taming Oblivion: Aging Bodies and the Fear of Senility in Japan.* Albany: State University of New York Press, 2000.

Turney, J. "The Public Understanding of Science—Where Next?" *European Journal of Genetics in Society* 1, no. 2 (1995): 5–22.

"2008 Alzheimer's Disease Facts and Figures." *Alzheimer's & Dementia* 4, no. 2 (2008): 110–33.

United Nations Department of Economic and Social Affairs, Population Division. *World Population Ageing: 1950–2050.* 2002. http://www.un.org/esa/population/publications /worldageing19502050/.

Victoroff, J., W. J. Mack, S. A. Lyness, and H. C. Chui. "Multicenter Clinicopathological Correlation in Dementia." *American Journal of Psychiatry* 152, no. 10 (1995): 1476–84.

Von Dras, D. D., and H. T. Blumenthal. "Dementia of the Aged: Disease or Atypical-Accelerated Aging? Biopathological and Psychological Perspectives." *Journal of the American Geriatrics Society* 40, no. 3 (1992): 285–94.

Waalen, Jill, and Ernest Beutler. "Genetic Screening for Low-Penetrance Variants in Protein-Coding Genes." *Annual Review of Genomics and Human Genetics* 10 (2009): 431–50.

Wadman, Meredith. "Fleshing Out the US Alzheimer's Strategy." *Nature News*, January 19, 2012. http://www.nature.com/news/fleshing-out-the-us-alzheimer-s-strategy-1.9856.

Wadman, Meredith, and Nature Magazine. "U.S. Government Sets Out Alzheimer's Plan." *Scientific American*, May 23, 2012. http://www.scientificamerican.com/article.cfm?id =us-government-sets-alzheimers-plan.

Wailoo, Keith, and Stephen Pemberton. *The Troubled Dream of Genetic Medicine: Ethnicity and Innovation in Tay-Sachs, Cystic Fibrosis, and Sickle Cell Disease.* Baltimore: Johns Hopkins University Press, 2006.

Wang, Sun-Chong, Beatrice Oelze, and Axel Schumacher. "Age-Specific Epigenetic Drift in Late-Onset Alzheimer's Disease." *PLoS ONE* 3, no. 7 (2008): e2698.

Waters, C. K. "Causes That Make a Difference." *Journal of Philosophy* 104 (2007): 551–79.

Weiner, Jonathan. *The Beak of the Finch.* New York: Vintage Books, 1994.

Weiner, Michael W., Paul S. Aisen, Clifford R. Jack Jr., William J. Jagust, John Q. Trojanowski, Leslie Shaw, Andrew J. Saykin, et al. "The Alzheimer's Disease Neuroimaging Initiative: Progress Report and Future Plans." *Alzheimer's & Dementia* 6, no. 3 (2010): 202–11.

Weiss, Kenneth M., and Anne V. Buchanan. "Is Life Law-Like?" *Genetics* 188, no. 4 (2011): 761–71.

———. *The Mermaid's Tale: Four Billion Years of Cooperation in the Making of Living Things.* Cambridge, Mass.: Harvard University Press, 2009.

Wenk, G. L. "Neuropathologic Changes in Alzheimer's Disease." *Journal of Clinical Psychiatry* 64 (2003): 7–10.

Wexler, Alice. *Mapping Fate: A Memoir of Family Risk and Genetic Research.* Berkeley: University of California Press, 1995.

Whitehouse, Peter J. "Can We Fix This with a Pill? Qualities of Life and the Aging Brain." In Ballenger et al., *Treating Dementia*, 168–82.

———. "Mild Cognitive Impairment—A Confused Concept?" *Nature Clinical Practice Neurology* 3, no. 2 (2007): 62–63.

Whitehouse, Peter J., and Daniel George. *The Myth of Alzheimer's: What You Aren't Being Told about Today's Most Dreaded Diagnosis*. New York: St. Martin's, 2008.

Whitehouse, Peter J., Konrad Maurer, and Jesse F. Ballenger, eds. *Concepts of Alzheimer Disease: Biological, Clinical, and Cultural Perspectives*. Baltimore: Johns Hopkins University Press, 2000.

Whitehouse, Peter J., Donald L. Price, Arthur W. Clark, Joseph T. Coyle, and Mahlon R. DeLong. "Alzheimer Disease: Evidence for Selective Loss of Cholinergic Neurons in the Nucleus Basalis." *Annals of Neurology* 10, no. 2 (1981): 122–26.

Whyte, Susan Reynolds. *Questioning Misfortune: The Pragmatics of Uncertainty in Eastern Uganda*. Cambridge: Cambridge University Press, 1997.

Wilkinson, Alec. "The Lobsterman." *The New Yorker*, July 31, 2006, 56–65.

Winblad, B., K. Palmer, M. Kivipelto, V. Jelic, L. Fratiglioni, L.-O. Wahlund, A. Nordberg, et al. "Mild Cognitive Impairment—Beyond Controversies, towards a Consensus: Report of the International Working Group on Mild Cognitive Impairment." *Journal of Internal Medicine* 256, no. 3 (2004): 240–46.

Winnefeld, Marc, and Frank Lyko. "The Aging Epigenome: DNA Methylation from the Cradle to the Grave." *Genome Biology* 13, no. 7 (2012): 165–68.

Wisniewski, H. M., A. Rabe, W. Silverman, and W. Zigman. "Neuropathological Diagnosis of Alzheimer Disease: A Survey of Current Practices." *Alzheimer Disease & Associated Disorders* 2, no. 4 (1988): 396–414.

Yang, Benjamin. "A United Disease Theory Brings Two Groups of Alzheimer's Disease Researchers Together." *Discovery Medicine*, May 10, 2009. http://www.discoverymedicine.com/Benjamin-Yang/2009/05/10/a-united-disease-theory-brings-two-groups-of-alzheimer-researchers-together/.

Yasuda, Kazuki, Kazuaki Miyake, Yukio Horikawa, Kazuo Hara, Haruhiko Osawa, Hiroto Furuta, Yushi Hirota, et al. "Variants in KCNQ1 Are Associated with Susceptibility to Type 2 Diabetes Mellitus." *Nature Genetics* 40, no. 9 (2008): 1092–97.

Yip, Agustin G., Carol Brayne, and Fiona E. Matthews. "Risk Factors for Incident Dementia in England and Wales: The Medical Research Council Cognitive Function and Ageing Study. A Population-Based Nested Case-Control Study." *Age and Ageing* 35, no. 2 (2006): 154–60.

Yong, Ed. "Life Begins at 100: Secrets of the Centenarians." *Mind Power News*, 2009. http://www.mindpowernews.com/LifeBeginsAt100.htm.

Yoxen, Edward J. "Constructing Genetic Diseases." In *The Problem of Medical Knowledge: Examining the Social Construction of Medicine*, edited by P. Wright and A. Treacher, 144–61. Edinburgh: University of Edinburgh, 1982.

Zhang, Hong. "Who Will Care for Our Parents? Changing Boundaries of Family and Public Roles in Providing Care for the Aged in China." *Journal of Long-Term Home Health Care* 25 (2007): 39–46.

Zheng, Wei, Jirong Long, Yu-Tang Gao, Chun Li, Ying Zheng, Yong-Bin Xiang, Wanqing Wen, et al. "Genome-wide Association Study Identifies a New Breast Cancer Susceptibility Locus at 6q25.1." *Nature Genetics* 41, no. 3 (2009): 324–28.

Zuk, Or, Eliana Hechter, Shamil R. Sunyaev, and Eric S. Lander. "The Mystery of Missing Heritability: Genetic Interactions Create Phantom Heritability." *Proceedings of the National Academy of Sciences*, December 5, 2011. http://www.pnas.org/content/early/2012/01/04/1119675109.

Index

MORRIS AUTOMATED INFORMATION NETWORK

0 1004 0276436 7

DATE DUE

WITHDRAWN

OCT - - 2013